William Kitchen Parker

On the Structure and Development of the Skull in the Batrachia

William Kitchen Parker

On the Structure and Development of the Skull in the Batrachia

ISBN/EAN: 9783337308230

Printed in Europe, USA, Canada, Australia, Japan

Cover: Foto ©ninafisch / pixelio.de

More available books at **www.hansebooks.com**

ON

THE STRUCTURE AND DEVELOPMENT

OF THE

SKULL IN THE BATRACHIA.

PART III.

BY

WILLIAM KITCHEN PARKER, F.R.S.

From the PHILOSOPHICAL TRANSACTIONS OF THE ROYAL SOCIETY.—Part I, 1881.

PHILOSOPHICAL TRANSACTIONS.

I. *On the Structure and Development of the Skull in the Batrachia.*—Part III.
By WILLIAM KITCHEN PARKER, *F.R.S.*

Received April 29,—Read May 27, 1880.

[PLATES 1-44.]

INTRODUCTION.

MY first attempt at working out the morphology of the Batrachian skull (Phil. Trans., 1871), instead of satisfying my mind, only served to increase tenfold the desire to know the meaning of the mysterious changes undergone by that part—the main part —of the organisation of the Frog.

Since then no opportunity has been lost of laying up in store fresh and fresh materials for further work in this field of research. Moreover, an additional strip of ground has since then been cleared and cultivated (Phil. Trans., 1876); in that second essay I was greatly helped by Professor HUXLEY, who showed me what was wrong in the first attempt, and also cut through some of the thickest and thorniest parts of this tangled subject.

I am also indebted to him for *materials*, and also to Professor A. AGASSIZ, and Mr. GARMAN (of Harvard University, U.S.); also to Professor RUPERT JONES, Dr. GÜNTHER, Professor W. H. FLOWER, Dr. MURIE, Dr. DOBSON, Mr. T. J. MOORE (of Liverpool), W. FERGUSON, Esq. (Ceylon), JAMES WOOD-MASON, Esq. (Calcutta), ALFRED C. HADDON, Esq., and GEORGE DINES, Esq.

Primarily, the aim of this extended research into the meaning of the skull in one "Order," or main group, is to get light upon the great cranial problem. A second use will be to rectify the classification of the group itself.

To make a systematic classification of these metamorphous animals upon such characters as are *easily seen* by the Zoologist is to build upon a sandy foundation; none will be more ready to acknowledge this than those who are most familiar with the group.

Even the cranial characters, which lie deeper down, are extremely variable, so that in formulating them for any division or sub-division of the Order it is always necessary to give some qualification of the scheme, and to say that "as a rule" such and such modifications of structure exist in the group under notice.

Our "genera" suffer from this weakness; and in some cases, notably in the great genus *Rana*, there are morphological modifications and variations such as are not to be seen in *whole groups of Families* in the Osseous Fishes.

On the whole, my own views correspond very accurately with those of my friends Messrs. GÜNTHER, MIVART, and WALLACE (whose works are referred to in the bibliographical list); they, I am satisfied, will be struck with the evidence here shown of the common origin of groups of the Batrachia that now are very widely dispersed, and marked by every variety of external character.

For there can be no doubt that these curious fishy air-breathers are, as Mr. WALLACE has suggested to me, a very ancient kind of Vertebrates. They have not struggled for life through one but through many epochs; they have been put to every kind of shift to live, and with infinite readiness and adaptability they have become all things to all conditions.

Here, undoubtedly, we get light upon the mystery of the great perfection of the various organs seen in the members of so lowly a group; for they are mere *anamniotics* at the best, and their upspring has been from some of the lowest of the Vertebrate stocks.

In every kind of facility for motion, in organs of sense wonderfully perfect, in power of speech and of song, and in instincts and habits innumerable, the Frogs and Toads teach the order, and anticipate the life of the peopled kingdoms of the nobler tribes that have risen above them in the scale.

The Salamandrian tribes ("Urodela") have branched up and beyond the "Dipnoi" (*Lepidosiren*, *Ceratodus*), and more or less, as a rule, lose their gills after a time, acquiring, in each type, a fenestral passage and a stapedial plug to their ear-capsule.

They also, in harmony with their more and more terrestrial habits, acquire a rudimentary larynx, so that the beginnings of the better kinds of organs of hearing and of voice are found in them.

But the Batrachians, springing from another part of the "stock," and indeed from a far lower "node," rise high above the Salamandrians in the metamorphosis of their organs, especially those of voice and hearing; their general intelligence and their gymnastic powers are also of a much higher kind.

Supposing the Urodeles to have arisen from some archaic forms of the "Dipnoi," the height in the scale of such double-breathing Fishes must have corresponded very closely with the Crossopterygii, and these again are manifestly a mere subdivision of the great "Ganoid" order.

But the ideal *Protobatrachians* came up by the way of the Sucking Fishes (" Marsipobranchii "), and grazing the edge of the Chimæroids and ordinary Selachians, developed into the larval form (or Tadpole) *first*. I am under the impression that in many cases the tailed gill-bearing condition has only *lately* been completely departed from, and that the *anurous* form or stage of such a type as *Pseudis* has only become universal in the newer geological times.

Whilst these tailed forms—hypothetically the primary condition of the Batrachia—have been yielding to terrestrial influences, and undergoing more and more curtailment and general metamorphosis, they have also, from time to time, dropped out of their organisation much of their old bony armour; and that which has been retained has become less and less superficial (or " dermosteal ").

They have to a remarkable degree lost their *mandibular*, and in many cases their *maxillary* teeth also; this has been evidently a correlate of the development over the ventral end of the hyoid arch of a peculiar cushioned fold of the floor of the mouth—the tongue. As this fold has become a more and more effective prehensile organ, the teeth have become less and less useful to the creature, whose succulent "articulate" or "molluscous" food is caught suddenly, and swallowed whole.

The *size* of these types has evidently undergone a steady *secular* diminution; this, and in many cases an exquisite specialisation of the fingers and toes, has all been in their favour; up high in the trees of the forest, and variously painted to resemble their surroundings of bark, leaves, and flowers, these marvels are wrought in them by the "Archchemic Sun ;"—their small size, I say, their curiously mimicking coat, and their high nestling, give them a chance of life and of life's enjoyments equal to that given to any tribe of animals whatever.

As in the common living forms now, the dilemma for the larval Batrachian is to "transform or perish" in short annual periods ; so, I opine, in long-past secular periods, again and again, those tribes—the forefathers of our existing kinds—have been put to the same extremity of shifting for their lives.

As the earth was made to be inhabited, and as the evolution of its tribes takes place through the harmonious inter-action of the forces within the organisms and the influences surrounding them, it has come to pass again and again that the *extremity* of some archaic form has been Nature's *opportunity;* its threatened extinction has been the occasion of its transformation into a higher kind of being, and the drying up of the old waters has been followed by the peopling of the new land with fresh and fresh forms, enjoying in many ways " newness of life."

There are two main groups of dwarf forms, namely, the highly developed and typical Tree-frogs, and the arrested, low kinds, such as the " Engystomidæ " and "Phryniscidæ ;" these latter appear to be waifs and strays from old and extensive tribes that have gone down in the world and are becoming gradually extinct; the "Hylidæ " (like the smallest " Carinate " birds), being Frogs of high degree, are in no such danger of extinction.

The range of size is greatest in the genus *Rana*, the highest or main typical form; here we see that a Frog, as a Frog, should not be too large, nor too small; the largest have in them, as the American Bull-frog, much of an old generalised nature; whilst the smallest, as *R. cyanophlictis* and *R. pygmæa* of the Oriental region, and some very small kinds of true Frogs in both the Neotropical and Ethiopian regions, are manifestly kinds that have been arrested at a stage similar to that seen in similar-sized young of the typical or medium species.

BIBLIOGRAPHY.

I here append a list of papers and works that have been of service to me in the present work; there is here no pretence of completeness, and several of the works treat of other types of Ichthyopsida:—

AGASSIZ, Professor A., and GARMAN, S. W., Esq. 'On Fishes and Reptiles of the Lake Titacaca.' Cambridge, Mass., U.S.; Nov. 26, 1875.
ANDERSON, JOHN, M.D., F.L.S., F.Z.S. "On some Indian Reptiles." Proc. Zool. Soc., Feb. 21, 1871, pp. 149-211.
BALFOUR, F. M., M.A., F.R.S. 'On the Development of Elasmobranch Fishes.' London: MACMILLAN and Co. 1878.
BELL, THOMAS, F.R.S. 'A History of British Reptiles.' London: JOHN VAN VOORST. 1839.
BORN, Dr. GUSTAF. 'Ueber die Nasenholen und der Thranen Nasengang der Amphibien.' Leipzig: 1877.
BRIDGE, Professor T. W., B.A.
   1. "On the Cranial Osteology of *Amia calva*." Jour. of Anat. and Phys., vol. xi., pp. 605-622; pl. 23.
   2. "On the Osteology of *Polyodon folium*." Phil. Trans., Part II., 1878, pp. 683-733; Plates 55-57.
COPE, Professor E.
   1. "On the Classification of the Anura." Jour. Nat. Sc. Phil., U.S., vol. vi.
   2. "Geology of Montana." 'American Naturalist,' July, 1879, pp. 432-441.
   3. "Erpetology of Tropical America." Proc. Amer. Phil. Soc., 1879, pp. 261-277.
CUVIER. 'Ossemens Fossiles.' Atlas, plate 252. Paris: 1836.
DUGÈS. 'Recherches sur l'Ostéologie et la Myologie des Batraciens à leurs différens âges.' Paris: 1834.
GARMAN, S. W., Esq. (Cambridge, Mass., U.S.).
   1. "Peruvian Fishes and Reptiles." Bulletin of the Museum, 1875.
   2. "Fishes and Reptiles of the West Coast of South America." Proc. Bost. Soc. Nat. Hist., vol. xviii., pp. 202-205; Dec., 1875.
   3. "Reptiles and Batrachians." Ibid., vol. xviii., pp. 402-413; June 21, 1876.
   4. "On Pseudis." A note on the so-called *Batrachichthys*, described in the Ann. Mus. Nac. Rio de Janerio, vol. i., plate 6, by Dr. PIZARRO.

GEGENBAUR, Dr. CARL. 'Untersuchungen zur vergleichenden der Wirbelthiere' (Selachians). Leipzig: 1872.
GRAY, Dr. JOHN EDWARD, F.R.S. "On *Dactylethra.*" Proc. Zool. Soc., 1864; p. 458.
GÜNTHER, Dr. A., F.R.S.
   1. "Batrachia Salientia." Brit. Mus. Cat., 1858.
   2. "Reptiles of British India." Ray Soc., 1864.
   3. A series of papers descriptive of the Batrachia in the Proc. Zool. Soc., as follows:—1862, p. 190; 1863, p. 249 (and Ann. and Mag. of N. H., 1863, p. 26); 1864, pp. 46 and 303; 1868, p. 478; 1869, p. 500; 1870, p. 401; 1872, p. 586; 1873, p. 165; 1875, p. 567.
   4. "Description of *Ceratodus.*" Phil. Trans., Part II., 1871, pp. 511-571; Plates 30-42.
HUBRECHT, Dr. A. A. W. 'Kopfskelet der Holocephalen' (*Callorhynchus* and *Chimæra*). Leyden and Leipzig: 1877.
HUXLEY, T. H., LL.D., F.R.S.
   1. "The Croonian Lecture." Proc. Roy. Soc., Nov. 18, 1858.
   2. 'Elements of Comparative Anatomy.' London: 1864.
   3. 'Anatomy of the Vertebrated Animals.' London: 1871.
   4. "On *Menobranchus lateralis.*" Proc. Zool. Soc., March 17, 1874, pp. 186-204; plates 29-32.
   5. Article "Amphibia." Encycl. Brit., 9th edit., pp. 750-771.
   6. "On *Petromyzon.*" Jour. of Anat. and Phys., vol. x., pp. 412-419; plates 17, 18.
   7. "On *Ceratodus Forsteri.*" Proc. Zool. Soc., Jan. 14, 1876, pp. 23-59.
MIALL, Professor L. C.
   1. "On the Labyrinthodonts of the Coal Measures." Rep. Brit. Assoc., 1873, pp. 225-249; plates 1-3.
   2. Ibid., 1874, pp. 149-192; plates 4-7.
MIVART, Professor ST. GEORGE, F.R.S.
   1. "On the Classification of the Anourous Batrachians." Proc. Zool. Soc., 1869, pp. 280-295.
   2. "On the Common Frog." 'Nature' Series. MACMILLAN and Co. 1874.
   3. "On *Pachybatrachus robustus.*" Proc. Zool. Soc., 1868, p. 557.
MÜLLER, JOH. "Vergleichende Anatomie der Myxinoiden, der Cyclostomen, mit durchbohrtem Gaumen." Abh. Berl. Akad., 1835; Wiegm. Arch., 1836, ii., p. 245.
PARKER, W. K., F.R.S.
   1. "On the Structure and Development of the Skull of the Common Frog." Phil. Trans., 1871, pp. 137-211; Plates 3-10.
   2. "On the Skull of the Eel." 'Nature,' June 22, 1871, pp. 146-148.
   3. "On the Skull of the Salmon." Phil. Trans., 1873, pp. 95-145; Plates 1-8.

4. "On the Sturgeon's Skull." M. Micr. Jour., June 1, 1873, pp. 254-257; plate 20.
5. "On the Skull of the Batrachia." Phil. Trans., 1876, Part II., pp. 602-669; Plates 54-62.
6. "On the Skull of the Urodeles." Phil. Trans., 1877, Part I., pp. 529-597; Plates 21-29.
7. "On the Skull of Sharks and Skates." Trans. Zool. Soc., vol. x., part iv., 1878, pp. 189-234; plates 34-42.
8. "On the Skull of the Urodeles." Trans. Linn. Soc., ser. 2, Zool., vol. ii., 1879, pp. 165-212; plates 14-21.

SWINHOE, ROBERT, H.M. Consul, Formosa. "On Reptiles and Batrachians from China." Proc. Zool. Soc., 1870, pp. 409-412.

TRAQUAIR, RAMSAY, M.D.
1. "On the Osteology of *Polypterus*." Jour. of Anat. and Phys., vol. v., pp. 166-182; plate 6.
2. 'On Ganoid Fishes' (Palæoniscidæ). Palæont. Soc., 1877, pp. 1-60; plates 1-7.

WALLACE, ALFRED RUSSEL, Esq., F.L.S. 'The Geographical Distribution of Animals.' London: MACMILLAN and Co., 1876. Pp. 414-423.

WIEDERSHEIM, Dr. R.
1. 'On *Salamandrina perspicillata* and *Geotriton fuscus*.' Genoa: 1875.
2. 'Das Kopfskelet der Urodelen.' Leipzig: 1877.
3. "On *Labyrinthodon Rütimeyeri*." Abh. der Schwerz Palæont. Gesell., Zurich, 1878.
4. "Zur Anatomie des *Amblystoma Weismanni*. Zeitschr. f. Wiss. Zool., bd. xxxii., pp. 216-236; plates 11, 12.
5. 'Die Anatomie der Gymnophionen.' Jena: 1879.

WILDER, Professor BURT. G. "The Branchiæ of the Embryo *Pipa*." 'The American Naturalist,' July (or August), 1877.

Zoological list of the species whose skulls are described or referred to[*] in the present paper.

In drawing up this list I shall make free use of the labours of my friends, Dr. A. GÜNTHER, F.R.S., and Professor ST. GEORGE MIVART, F.R.S. The "Batrachia

* The species, the descriptions and illustrations of whose skulls are referred to here, and only partially described, are :—
1st, *Rana temporaria* (see Phil. Trans., 1871, pp. 137-211, Plates 3-10; and Phil. Trans., 1876, pp. 603-605, Plate 54, figs. 1, 2).
2nd, *Rana esculenta* (see HUXLEY, article " Amphibia," Encyc. Brit., 9th Edit., vol. iii., pp. 750-771).
3rd, *Bufo vulgaris* (see Phil. Trans., 1876, pp. 605-625, Plates 54, 55).
4th, *Dactylethra capensis* vel *lævis* (see Phil. Trans., 1864, pp. 625-648, Plates 56-59).
5th, *Pipa Americana* vel *monstrosa* (see Phil. Trans., 1864, pp. 648-665, Plates 60-62).

Salientia" and various papers in the 'Zoological Proceedings' give me the views of the former, and the article on the "Classification of the Batrachia" (Proc. Zool. Soc., 1869, pp. 280-295) those of the latter, author. I shall use as few of the zoological characters as possible, and refer my reader to the works just mentioned for further information.

I.—ANURA PHANEROGLOSSA.*
(INCLUDING "OPISTHOGLOSSA" AND "PROTEROGLOSSA.")
II.—ANURA AGLOSSA.

I.—PHANEROGLOSSA.

A. With teeth in the maxillaries, premaxillaries, and generally in the vomers.
   a. With sharp fingers and toes.
   b. With dilated digital disks.
B. With no teeth in either upper jaws or vomers.
   a. With sharp fingers and toes.
   b. With dilated digital disks.

I. A. a. 1.—*Frogs with sharp toes, teeth in jaws and vomers, toes more or less webbed, a bony shaft to manubrium ("omosternum"); cylindrical processes to sacral vertebra, and without parotoid glands.*

First Family. "RANIDÆ."

Genera—*Rana, Tomopterna, Pyxicephalus.*

First genus. *Rana.*
1. *R. clamata,* DAUD.—Larvæ, three stages (A, B, C). Cambridge, Mass., U.S.
2. *R. pipiens,* HARL.—Three stages of larvæ (A, B, C) and Adult. N. America.
3. *R.* ——? sp.--Larva. India.
4. *R. palustris,* LECONTE.—Newly metamorphosed young. Cambridge, Mass., U.S.
5. *R. halecina,* KALM.—Immature male. N. America.
6. *R. temporaria,* LINN.—Great Britain. Various stages described (see Phil. Trans., 1871, Plates 3-10, pp. 137-211.)
7. *R. esculenta,* LINN.—Adult. Europe (see HUXLEY, Art. "Amphibia," Encyc. Brit., 9th edit., vol. iii., pp. 750-771.)

---

* A few types enjoy some freedom of the tongue in front, namely, *Rhinophrynus* and *Xenorhina;* I have also found this character developed, in some degree, in *Rhinoderma Darwini,* but Dr. GÜNTHER tells me that he does not think it a modification of any great importance; yet the rule is for the tongue to be only free behind. Another character, namely, the possession of *mandibular teeth,* as in *Hemiphractus* and *Grypiscus* (MIVART, pp. 204, 205) is more important, but unfortunately I have not at present been able to procure either of these kinds.

8. *R. gracilis*, WIEZ.—Adult male. Ceylon.
9. *R. cyanophlyctis*, SCHNEID. (see GÜNTHER, Rept. of Brit. Ind., p. 406).— Ceylon.
10. *Rana pygmæa*, GTHR. (see P. Z. S., 1875, pp. 567, 568).—Adult male, ⅜ inch long. Anamallays Mountains, Malabar, S.W. India.
11. *R. tigrina*, DAUD.—Adult female. Ceylon.
12. *R. hexadactyla*, LESS.—Adult female. Ceylon.
13. *R. Kuhli*, SCHLEG.—Male, two-thirds grown. Ceylon.
14. *R.* ——— ? sp.—Adult female. A small species, 1 inch long. Lagos, W. Africa.

Second genus. *Tomopterna*.
15. *Tomopterna breviceps*, SCHNEID., sp. (*Pyxicephalus breviceps* in GÜNTHER's Rept. of Brit. Ind., p. 411).—(A) Half grown female. S. India. (B) Adult female. Ceylon.

Third genus. *Pyxicephalus*.
16. *Pyxicephalus rufescens*, JERDON.—Adult male. India.

I. A. *a.* 2.—*Like the last in most things, but have their toes free, or but little webbed, and the "omosternum" not ossified, or even absent.*

Second Family. "CYSTIGNATHIDÆ."

First genus. *Pseudis*.
17. *Pseudis paradoxa*, LINN., sp.—Larvæ in four stages, and adult. S. America.

Second genus. *Gomphobates*.
18. *Gomphobates* (*Leiuperus*) ——— ? sp. (see P. Z. S., 1868, p. 478, same as *Leiuperus*, " Batr. Sal.," pp. 22 and 135).—Adult (?), 10 lines long. River Plate.

Third genus. *Cystignathus*.
19. *Cystignathus ocellatus*, var. LINN., sp.—Adult male, 5¼ inches long, Dominica; with larvæ 2⅜ and 3¼ inches long. Brazils.
20. *C.* ——— ? sp.—Tadpole, 1 inch long; tail, ¾ inch; hind legs, 1 line. Lake Jaunarg, Manaoo, Brazils.
21. *C. typhonius*, DAUD., sp. (*C. typhonius*, DUM. and BIB.).—Adult female, 1½ inch long. Porto Rico.

Fourth genus. *Pleurodema*.
22. *Pleurodema Bibronii*, TSCHUDI.—Adult female, 1½ inch long. Chili.

Fifth genus. *Lymnodynastes*.
23. *Lymnodynastes tasmaniensis*, GTHR.—Adult female, 1¾ inch long. Tasmania.

DEVELOPMENT OF THE SKULL IN THE BATRACHIA.

Sixth genus. *Camariolius.*
24. *Camariolius tasmaniensis* (?) (see GÜNTHER on "Batrachians from Australia," P. Z. S., 1864, pp. 46-49, plate 7, fig. 3; there called *Pterophrynus tasmaniensis*. Dr. GÜNTHER is nearly satisfied that my specimens belong to that species).—(A) Adult female, $\frac{7}{8}$ inch long; and (B) Larva, $\frac{3}{4}$ inch long; legs, $\frac{3}{16}$ inch. Australia.

Seventh genus. *Cyclorhamphus.*
25. *Cyclorhamphus marmoratus*, DUM. and BIB.—Adult female, $1\frac{3}{8}$ inch long. Vinco Caya, Peruvian Andes. Height 16,000 feet.
26. *Cyclorhamphus culeus*, GARMAN.—Larva, $3\frac{1}{4}$ inches long; tail, 2 inches; hind legs, 7 lines. Puno, Lake Titacaca, Peru.

The next group differs from the Ranidæ in having the sacral apophyses dilated.

I. A. a. 3.—Third Family. "DISCOGLOSSIDÆ."

First genus. *Discoglossus.*
27. *Discoglossus pictus*, OTTH.—Adult male, $2\frac{1}{8}$ inches long. S. Europe.
Second genus. *Pelodytes.*
28. *Pelodytes punctatus*, DAUD., sp., FITZ., gn.—Adult male, $1\frac{1}{2}$ inch long. Europe.
Third genus. *Xenophrys.*
29. *Xenophrys monticola*, GTHR. (see Rept. Brit. Ind., plate 26, figs. II, II', p. 414).—Adult male, 3 inches long. Darjeeling.
Fourth genus. *Calyptocephalus.*
30. *Calyptocephalus Gayi*, DUM. and BIB.—(A) Adult female, $5\frac{1}{2}$ inches long, and (B) Tadpole, $4\frac{3}{4}$ inches long; tail, $2\frac{3}{4}$ inches; hind legs, $\frac{3}{8}$ inch. Chili.

I. A. a. 4.—Fourth Family. "ALYTIDÆ."

In this family the Ranian type is modified by some new characters. The sacral apophyses are dilated, and there are "parotoids" as in the Toad; ribs present, and vertebræ *opisthocœlian*.

Genus *Alytes.*
31. *Alytes obstetricans*, LAUR., sp.—Adult female, 1 inch 10 lines long. Europe.

I. A. a. 5.—Fifth Family. "HYPEROLIIDÆ."

This type differs from the last in many things, notably in possessing, besides the neck-glands and wide sacrum, which is "not much extended" (GÜNTHER, "Batr. Sal.," p. 39), procœlian vertebræ, and no ribs (MIVART, p. 291).

Genus *Hyperolius* (formerly *Uperoleia*; the species of so-called *Hyperolius* described in the "Batrachia Salientia" are now put into the genus *Ruppia*); they are narrow-backed Tree-frogs.

32. *Hyperolius marmoratus*, GRAY, sp.—Adult female, 1¼ inch long. Paramatta, Australia.

I. A. a. 6.—Sixth Family. "BOMBINATORIDÆ."

This type has dilated sacral apophyses but no parotoids, it has ribs, and abnormal (*opisthocœlian*) vertebræ, but no columella, or tympanic cavity.

Genus *Bombinator*.
33. *Bombinator igneus*, RÖSEL, sp.—Adult female, 1¾ inch long. Europe.

I. A. a. 7.—Seventh Family. "PELODATIDÆ."

The last has one of the softest, this has one of the hardest, skulls. There is no cavum tympani, but there is a small bony columella. There are no ribs, and the vertebræ are procœlian; it has a broad sacrum, but no parotoids.

Genus *Pelobates*.
34. *Pelobates fuscus*, LAUR., sp.—Adult male, 2 inches 5 lines long. Europe.

I. A. b.—*With dilated digital disks.*

First sub-division of Tree-frogs.

First Family. "POLYPEDATIDÆ."

Ear perfect. Sacral apophyses cylindrical. "Omosternum" with a bony shaft (MIVART, p. 292). "No parieto-frontal fontanelle" (except in such small types as the Australian *Ruppia bicolor*). No neck-glands.

First genus. *Polypedates*.
35. *Polypedates chloronotus*, GTHR. (see P. Z. S., pp. 567–576, plate 65, fig. A). Adult male, 2 inches long (GÜNTHER gives the length of the body in the male as 51 millims., and of the female 93 millims; p. 570). India.
36. *Polypedates maculatus*, GRAY, sp.—Adult male, 2 inches 1 line long. India.
Second genus. *Rhacophorus*.
37. *Rhacophorus maximus*, GTHR.—Adult male, 3¼ inches long. N. India.
Third genus. *Ixalus*.
38. *Ixalus variabilis*, GTHR.—Adult female, 1 inch 1 line long. Ceylon.

Fourth genus. *Hylarana*.
    39. *Hylarana malabarica*, Dum. and Bib. sp.—Young, ¾ inch long. India.
    40. *Hylarana temporalis*, Gthr.—Adult male, 2¼ inches long. Ceylon.

Fifth genus. *Rappia*.
    41. *Rappia* ——? sp.—Adult female, ⅜ inch long. Lagos. (Probably *R. lagoensis*, Gthr., see P. Z. S., 1868, pp. 478-490, plate 40, fig. 2.)
    42. *Rappia* (*Hyperolius*) *bicolor*, Gray, sp.—Adult female, ⅜ inch long. Dog-trap Road, Paramatta, Australia. (See "Batr. Sal.," p. 89; Dr. Günther there remarks that, "this species is very probably the type of a separate genus, but the condition of this single specimen does not enable me to give the characters with certainty." The fact of the case is, that the skull of this specimen is extremely membranous, as much so as in that of *Acris Pickeringii* (see Plate 19, figs. 6-10, and Plate 30, figs. 1-5). The *Rappia* from Lagos had, on the contrary, a highly ossified skull, with a well-covered roof (see Plate 28, figs. 6-10).

Second Family. "Hylodidæ."

(I make up this family of Professor Mivart's sub-families Acridina and Hylodina.) No bone in the manubrium ("omosternum"). Vertebræ procœlian; fontanelle very open (*Acris*), or covered (*Hylodes*).

First genus. *Hylodes*.
    43. *Hylodes martinicensis*, Bib., sp.—Adult female, 1½ inch long. Martinique.
Second genus. *Acris*.
    44. *Acris Pickeringii*, Holb.—(A) Adult female, 10 lines long (⅚ inch); and (B) larva, 1 inch 2 lines long; tail, ¾ inch; hind legs, ¼ inch. Cambridge, Mass., U.S.

Second sub-division of Tree-frogs.

Family "Hylidæ."

These have the sacral apophyses dilated; and have no parotoids. The ear is perfect, and, as a rule, the fontanelle is open; I find it covered in *Hyla rubra*, and almost covered (by sculptured solid bones) in *Nototrema*.

First genus. *Hyla*.
    45. *Hyla Ewingii*. Dum. and Bib.—Adult female, 1½ inch long. Van Diemen's Land.
    46. *H. phyllochroa*, Gthr. (see P. Z. S., 1868, p. 481).—Adult female, 1 inch 5 lines long. Cape York, Australia.

47. *Hyla arborea*, LINN., sp.—Adult male, 1¾ inch long. S. Europe.
48. *H. albomarginata*, SPIX.—Adult female, 2¼ inches long. Brazils.
49. *H. rubra*, DAUD.—Adult male, 1 inch 11 lines long. S. America.
50. *H.* ——? sp.—Tadpole, 1 inch long; hind legs, 5 lines. Rio Janeiro.

Second genus. *Litoria*.

51. *Litoria marmorata*, DUMÉRIL.—Adult male, 1½ inch long. Australia. In this type the digital disks are very small and the head is very large in proportion to the body. I suspect that it comes very close to an "Oxydactyle" Frog from the same region, viz.: *Myxophyes fasciolatus*, GTHR. (see P. Z. S., 1864, p. 46, plate 7, fig. 1).

Third genus. *Nototrema*.

52. *Nototrema marsupiatum*, DUM. and BIB., sp.—(A) Adult male, 1¾ inch long. S. America; and (B) Tadpole, 2⅓ inches long; body, 1 inch; tail, 1⅓ inch; hind legs 13 lines; *right* fore leg free, ½ inch long; *left* fore leg still under the operculum.

Third sub-division of Tree-frogs.

Family "PELODRYADIDÆ."

Parotoids present; other characters *Hyline*.

First genus. *Pelodryas*.

53. *Pelodryas cœruleus*,[a] WHITE, sp. (see "Batr. Sal.," p. 199, and plate 9, fig. B.)—Adult male, 3 inches long. N. S. Wales.

Second genus. *Phyllomedusa*.

54. *Phyllomedusa bicolor*, BODDAERT.—Adult female, 3½ inches long. Santarem, River Amazon, lat. 2° 20′ S., S. America.

I. B. a.—*Toothless* "Anura." *Without digital disks. Oxydactyle Toads.*

First Family. "BUFONIDÆ."

Typical Toads, with parotoids, and processes of sacral vertebræ dilated. Ears perfect. Toes webbed. Skull generally strongly roofed; an open fontanelle in *Bufo calamita*.

First genus. *Bufo*.

55. *Bufo pantherinus*, BOIE. Adult female, 4¼ inches long. Africa.
56. *B. melanostictus*, SCHNEID.—Half-grown male, 2½ inches long. India.
57. *B. agua*, LATR.—Old female, 6½ inches long; and younger female, 5 inches long. S. America.

[a] Evidently named from a bleached *spirit specimen*; the living Frog is bright green.

58. *B. chilensis*, Tschudi.—(A) Adult male, 3 inches long; and (B) Tadpole ⅝ inch long; tail, ½ inch; hind legs, 1/20th of an inch. Arequipa, Peru.
59. *Bufo lentiginosus*, Shaw.—(A) Tadpole, ⅝ inch long; hind legs, 1½ inch long; and (B) newly metamorphosed young male, ⅝ inch long. Penekese Island Mass., U.S.
60. *B. vulgaris*, Laur.—Adult and Tadpoles many stages. England.
61. *B. calamita*, Laur.—Adult female, 2¾ inches long. England.
62. *B. ornatus*, Spix.—Adult female, 2⅛ inches long. S. America.

Second genus. *Otilophus*.
63. *Otilophus margaritifer*, Lath.—(A) Half-grown female, 1½ inch long; Venezuela; and (B) adult female, 2⅜ inches long (Hyrtl's prepn. Mus. Coll. Surg., Eng.), Brazils.

Second Family. "Rhinodermatidæ."

This group agrees with the "Bufonidæ" in many things, but there are no parotoids, and the ear is less perfectly developed.

First genus. *Rhinoderma*.
64. *Rhinoderma Darwinii*, Dum. and Bib.—Adult male, 1 inch long. Chili.

Second genus. *Diplopelma*.
65. *Diplopelma ornatum vel rubrum*, Dum. and Bib., sp.—Adult male, 11 lines long. India.
66. *Diplopelma Berdmorei* (?), Blyth (Günther, Rep. of Brit. Ind., p. 417).—Adult female, 1 inch 1 line long. Moulmin, Tenasserim.

Third Family. "Brachycephalidæ."

These Toads have "no tympanum nor cavum tympani";—"free toes, with the processes of sacral vertebræ dilated, and without parotoids." ("Batr. Sal.," p. 45.)

Genus *Pseudophryne*.
67. *Pseudophryne Bibronii*, Gthr.—Adult female, 1 inch long; adult male, ⅝ inch long. N. S. Wales.

Fourth Family. "Phryniscidæ."

Ear imperfect (sometimes with a columella, as in *Phryniscus cruciger*); webbed toes; dilated sacral apophyses; and no parotoids.

Genus *Phryniscus*.
68. *Phryniscus cruciger*, Martins and Günther.—Adult male, 1¼ inch long. Interior of Brazils.
69. *P. varius*, Stan., sp.—Adult female, 1⅔ inch long. Costa Rica.
70. *P. lævis*, Gthr.—Adult female, 1¾ inch long. Ecuador.

Fifth Family. "ENGYSTOMIDÆ."

Ear rather imperfect; dilated sacral apophyses; without parotoids; and with free toes.

Genus *Engystoma*.

71. *Engystoma carolinense*, HOLBR.—Adult male, 11 lines long. Florida.

I. B. b. *Toads with digital disks*.

First Family. "HYLÆDACTYLIDÆ."

Ear perfect; webbed toes; dilated sacral apophyses; and no parotoids.

Genus *Callula*.

72. *Callula pulchra*, GRAY.—Adult female, $2\frac{5}{8}$ inches long. Pegu.

Second Family. "HYLAPLESIDÆ."

Ear perfect; toes free; processes of sacral vertebræ cylindrical; and without parotoids.

Genus *Hylaplesia*.

73. *Hylaplesia tinctoria*, SCHNEID., sp.—Adult female, $1\frac{1}{8}$ inch long. South America.

II.—ANURA AGLOSSA.

Without tongue; ear perfectly developed; Eustachian tubes united; cavum tympani entirely bony.

a. *With maxillary teeth*.

Family "DACTYLETHRIDÆ."

With webbed toes; dilated processes of sacral vertebræ; and without parotoids.

Genus *Dactylethra*.

74. *Dactylethra lævis*, DAUD., sp. (*D. capensis*, CUV., see Phil. Trans., 1876, pp. 625–648, Plates 56–59.)—Adult and four larval stages. S. and W. Africa.

b. *Without maxillary teeth*.

Family "PIPIDÆ."

With webbed toes; and with the processes of the sacral vertebræ dilated; no parotoids.

Genus *Pipa*.

75. *Pipa Americana*, SEBA. (see Phil. Trans., 1867, pp. 648–665, Plates 60–62.)—Adult, ripe young, and larvæ 9 lines long.

DEVELOPMENT OF THE SKULL IN THE BATRACHIA. 15

ON THE TYPICAL BATRACHIAN SKULL.

In the following descriptions I shall take the skull of *Rana temporaria* as the "norma" or pattern form: it is the best on the whole; one or two exceptional characters exist in it, viz.: the mark of the originally separate metapterygoid (a rare character); the annulus tympanus is not perfect until old age; and the stylo-hyal is a long while fusing with the floor of the skull and ear-capsule, if indeed it ever becomes fused; but, with these exceptions, this may be taken as the highest kind of Batrachian skull, and the best rule to measure the others by.

ON THE EARLY STAGES OF THE BATRACHIAN SKULL.

No known kind of Vertebrate shows so many and such instructive stages in its development as the Batrachian.

The skull of the newly-hatched embryo of the common Frog or Toad is strictly comparable to that of a larval Lamprey (*Ammocœtes*), whilst that of the well-grown Tadpole comes very close to the skull of the adult *Petromyzon*.

But whilst the ordinary Batrachia have a very Petromyzine or *suctorial* larva, the "Aglossa" in their early stages have a cranium and face very similar to what is seen in the "Siluroid" Teleostean Fishes, and their skull suggests to the observer the most probable form of endocranium likely to have existed in such Ganoids as *Pterichthys* and *Coccosteus*.

Both these kinds of larval endocranium (*chondrocranium*) are figured in my "Batrachian Skull," Part II. (Phil. Trans., 1876, Plates 54 and 55, *Bufo vulgaris*; Plates 56-58, *Dactylethra*; and Plate 60, fig. 3, *Pipa*).

My present business is with the suctorial type of larval Batrachian skull; for full details and figures of the wide-mouthed *Siluriform* type of skull I must refer the reader to the paper just referred to.

My most successful dissection of the earliest *cartilaginous* skull in these types was that of *Bufo vulgaris* (op. cit., Plate 55, figs. 1, 2); the embryo was 4 *lines* (⅓rd of an inch) in total length; the next to this (Plate 55, fig. 3) was of a Tadpole of the same species, 5 *lines* long.

I shall make the first of these my *First Stage*, referring the reader to the plates in the published paper.

Of the skull in more advanced stages the present work will give many instances and illustrations; after describing the simple foundation, as seen in *Bufo*, I shall describe a series of stages in various species of the genus *Rana*.

The larval skulls of other types will be described in their proper order, with the adult condition of the skull in the same and other species.

I must here remark that the early cranium of the Aglossal types, especially that of *Dactylethra*, is in some respects more Petromyzine than that of an ordinary Tadpole.

In some things the chondrocranium of the Tadpole of *Bufo vulgaris* comes nearest to that of the Lamprey, as in the fusion together of the various plates of cartilage; but the skull of the larval *Dactylethra* comes very near to that of the adult *Petromyzon* in several respects.

This is seen in the complete fusion of the elements of the branchial skeleton, and in the histological condition of those bars and pouches: these are composed, as in the extra-branchial basket-work of the Lamprey, of a very light kind of cartilage, with large cells and scarcely any intercellular substance: the large amount of *superficial* cartilage shows also the same relationship.

The Common Toad differs from most of the Opisthoglossa in retaining the primordial "pedicle" to the pier of the mandible, which is, from the first, continuous with the trabecular bar: this is a remarkable survival of a suctorial character—even after metamorphosis.

SKULL OF LARVAL BATRACHIANS.

First Stage.— *Craniofacial cartilages of embryo of* Bufo vulgaris *soon after hatching;* $\frac{1}{3}rd$ *of an inch in total length.*

Nearly all the truly cartilaginous part of this cranium ("Skull of Batrachia," Plate 55, figs. 1, 2) lies at the base of the fore half of the head (*tr.*); two globes of soft cartilage unfinished in their upper third, are to be seen right and left in the hind half; these are the very distinct auditory capsules (*au.*).

At present the whole body of the embryo exists as a sort of tail-like appendage to the huge and precocious head: the development of cartilage appears, therefore, almost entirely in front of the axis of the organism (*notochord*), which stops at the post-pituitary space, *before* the "pituitary body" is formed.

Thus the cranium, as yet, is nearly all *pro*-chordal; and the *para*-chordal part, like the side-walls and roof of the cranium, is still membranous.

Yet these prochordal tracts, or "trabeculæ cranii" are manifestly true *paraxial* elements or parts; they are homologous with the paired cartilages that appear in any region along the sides of the sheath of the notochord.

All growths *above* these (dorsad) are of the nature of neural arches; all growths *below* these (ventrad) are " visceral " or " pleural " arches.

But these rods are *continuous*, and they are *not parallel* the one with the other; how is this to be explained on the theory of their *paraxial* nature ?

A consideration of the development of this, or of any, Vertebrate embryo, will help us to understand the meaning of these first foundations of the cranium.

The neural axis rapidly enlarges at its cephalic end, forms three vesicles there, the hind, mid, and fore brain, and this beaded structure is bent suddenly upon itself so that the fore brain is turned downwards and somewhat backwards, and the mid brain is tilted upwards and forwards, and forms the *actual* end of the embryo, lying against what is called the "frontal wall."*

The fore and hind vesicles are thus brought into contiguity, and the organic apex, or real fore end of the embryo, is a little in front of the hind vesicle; it looks downwards and a little backwards.

At that point a remarkable process takes place; the notochord turns downwards, taking, as the skeletal axis, the same direction as the neural axis; the fore brain after a time sends a budding process *out*, and the oral mucous membrane sends a budding process *in*, and the two, gradually becoming mutually engrafted, the one upon the other, open freely into each other, and then the lower process becoming closed below, we have the pituitary body formed.

In this *quasi-archaic* condition of the head and its parts, the *body* is a mere appendage, solid hyaline cartilage forms first where most pressure is, and the paraxial bands grow first of all from near the apex of the notochord to near the frontal wall.

But the fore brain swells and hangs down; the skeletal parts respond to this condition and wind their way round its base, embracing its sides, and then meet, or nearly so, in front, to diverge again in the nasal region of the face.

After a few days the paired bands have developed backwards under the hind brain (ibid., figs. 3 and 4), and behind the head they appear in patches that alternate with the somatomes (muscle-segments, &c.) along the spinal region.

Then, in Tadpoles *five-twelfths of an inch long* (op. cit., Plate 55, fig. 3), the chondrocranium is simply a pair of planks on which the brain lies; it has become much straighter, but the mid brain lies high, still.

Strength, however, has been gained by fusion at two points (in the antorbital and postorbital regions), and between these two points the bars are becoming crested; these ridges are the beginning of the ethmoidal and sphenoidal side walls.

Returning to the earliest stage we find that there are, indeed, two pairs of cartilage on each side nearly equal, with a third pair of much shorter bars.

In these *temporary sucking fishes* the chief parts of the organism are all crowded into the head and throat; all else is their paddling fin, and the creature is called a *bullhead*, on account of this cephalic preponderance.

The second, more external, cartilaginous band (*pd., q.*), carrying at its end a short inturned segment (*Mk.*), is the "pier" of the mandible, with its free swinging joint, the rudiment of the lower jaw.

Under that pier, one-fourth of the way from its distal end, the third cartilage is seen (*e.hy.*); this is the free joint of the hyoid (*lingual*) arch whose pier or *epi-hyal* element does not appear for three or four months to come, and when developed is devoted to new purposes, does not become the *practical* "suspensorium" of the lingual arch at all, but forms part of the outworks of the auditory labyrinth.

* See Balfour's 'Elasmobranchs,' plate 7.

Behind these parts, in relation to the hinder visceral clefts, four other arches are beginning to form. These, the branchials, are still more feebly developed, never acquire an *upper* element ("opi-branchial"), but in their arrest and feebleness are supplemented by four pairs of *extra-branchial* cartilages or pouches; these I shall describe in the developed embryos of the larger species of the genus *Rana*.

Until then, also, I shall not describe the other superficial cartilages that are found in the Tadpole's head, viz: that over the "1st cleft," and those so largely forming the suctorial mouth of the Tadpole.

In this early stage the cranial nerves pass over the trabeculæ through the *membranocranium*.

Here, if anywhere, we ought to find rudiments of *pre-oral* arches, if there be any; there are certain things in other kinds of embryos, and even adult forms, that suggest such a possibility, but none here.[*]

At present the mandibular pier has two points of fusion with the trabecula; the hinder of them is the pedicle (*pd*.), and the front one the palato-pterygoid bar (*p.pg*.); the pedicle will be largely absorbed, and the palato-pterygoid enormously developed during the metamorphosis of the Tadpole.

Before passing to the description of the well-developed Tadpole, I must refer again to the intermediate stages (ibid., Plate 55, figs. 3, 4, 5).

On its way to the *coronoid region* of the mandible, the temporal muscle passes over the palato-pterygoid bar, and under a leafy growth of the "suspensorium"—the "orbitar process" (*or.p*.). In *Bufo vulgaris* the apex of this leaf coalesces with the ethmoidal cartilage, a rare character. This fact, and its nonconformity with what is seen in the Tadpole of the Frog, were pointed out to me by Professor HUXLEY five or six years ago.

In other species (even of *Bufo*) I do not find this very peculiar condition of things, but the process itself is well developed, even in the wide-mouthed larvæ of the "Aglossa" (ibid., Plates 56-61).

In the half-grown larvæ figured in my paper (ibid., Plate 55, figs. 4, 5, 5*a*), we see the formation of the "fenestra ovalis" as an oblique lateral cleft in the auditory capsule; and of the "stapes" as the solidification of the soft tissue left in this space. In larvæ three-fourths grown ossification begins in the tissue over and under the chondrocranium.

---

[*] Professor MILNES MARSHALL and the writer, from a consideration of the "segmental nerves" of the head, from the development of the mouth itself and of the lacrymal and nasal passages, and from what is seen in various cartilages around and in front of the mouth, hold to the *opinion* that there are rudiments of at least *two*, perchance *three*, pre-oral visceral arches.

But Professor HUXLEY and Mr. F. M. BALFOUR will not see with our eyes, and are not in the least satisfied with the evidence which seems so conclusive to us. The Lamprey and the Tadpole are indeed great stumbling-blocks; if they represent primordial or archaic forms, and if what we see in other kinds bear to be interpreted as specialisations having relation to modification of the oral aperture and its framework, the *onus probandi* will still rest with my talented young friend MARSHALL and me.

SKULLS OF LARVAL "RANIDÆ."

Chondrocranium of perfect larva of Bull Frogs.

First Stage.—1. *Tadpole of* Rana clamata (A), *whose total length was* $3\frac{1}{4}$ *inches; of the tail*, 2 *inches; and of the hind legs*, 5 *lines* ($\frac{5}{12}$ *inch*).

The skull (Plate 2, figs. 5, 6) has now attained its full cartilaginous (*Petromyzine*) condition, and has gone beyond it; there is a pair of cartilage-bones, and there are three membrane bones. I shall deal with these signs of metamorphic progress after describing the chondrocranium.

The cranium, proper, is now a small flat-bottomed *boat*, with a semi-cartilaginous *deck*; this has been constructed upon the original frame in the following manner.

The large open space in the orbital region has acquired a thin floor of cartilage, and this median ("intertrabecular") tract has grown forwards to the region of the inner nostrils (*i.n.*), where the trabeculæ have grown together, and are diverging again as free, flat, broad-ended "cornua" (*c.tr.*).

Behind, the occipital arch is as perfect as may be, for the auditory capsules (*cb.*) fit into large fenestræ in its sides; its top (*s.o.*) is flat, and extends to the foremost third of the capsules; its floor is flat, and contains the diminishing notochord (*nc.*) in its middle: in this kind this rod scarcely acquires any true cartilage in its outer sheath.

In the large orbital region the side walls are cartilaginous, except where the optic nerves (II.) pass out of their "fenestræ"—spaces three or four times as large as the actual *foramina*.

As in the occipital region, so also in this, there is a cartilaginous "tegmen;" its halves have not united in the middle, and a bilobate "fontanelle" (*fo'.*) exists between these growths and the supraoccipital cartilage. Then there is a large oval fontanelle (*fo.*) or membranous space up to a short distance of the end of the cranial cavity, where a tract of cartilaginous roof margins this egg-shaped space.

Here there is a perfect cartilaginous cincture, which becomes in the adult the "os en ceinture;" its front end is extended as a cartilaginous beak formed from below upwards as a crested tract of the "intertrabecula;" this is the mesethmoid (*p.e.*) in rudiment; the lateral regions are the right and left "ecto-ethmoids" (*al.e.*).

These lateral parts grow outward into wings whose scooped front margins form a back wall to the nasal sacs; the olfactory nerves pass to these sacs through "fenestræ" that lie between the middle and outer parts of the cincture.

The nasal roofs are still soft, and in the dissection the olfactory sacs were removed to display the cartilages.

But the internal nostrils (*i.n.*) are shown in these figures; they are large and neatly circular; much of their margin is seen to be formed by the narrow, clinging part of the trabeculæ, and the curved *spatulæ* that grow out from the ethmoidal cincture. On the outside the quadrate cartilage (*q.*) sends a process inwards which is tied by a ligament to the corresponding trabecular horn (*c.tr.*), and thus the circle is completed. From this part the cornua turn downwards in a gentle manner, as they diverge, and end in a broad, sub-emarginate flap.

The sclerotics and nasal roofs, which were beginning to chondrify, having been removed, we have only the auditory sacs to describe in this stage.

They are now thoroughly cartilaginous, and quite confluent with the cranial walls, roof, and floor. Moreover, they are swollen with the growth within of the three wide canals, and the sacculated base of the membranous labyrinth (*a.s.c.*, *h.s.c.*, *p.s.c.*, *vb.*).

A crescentic flap of cartilage grows outwards from the horizontal canal; this is the "tegmen tympani," under that roof we see the large fenestra ovalis, and the well-fitting oval stapes (*vb.*, *st.*). The fore angle of the *tegmen* has a flap of cartilage growing from it, the result of coalescence; this, the rudimentary "annulus," will be described soon.

Leaving out some of the smaller passages, we may refer to the double passage for the 9th and 10th nerves (fig. 6, IX., X.), the single passage for the 7th and 5th (VII., V.) and the fenestra for the 2nd nerve (II.): the 8th and 1st are out of sight in this view.

Each of the huge "mandibular piers" is two-thirds the length and two-thirds the breadth of the cranium proper. Besides the two earliest conjugations with the basal cartilage, each bar articulates by its "otic process" with the cartilage that projects from the tegmen tympani, and is tied, as above stated, near its distal end to the cornu trabeculae.

It is thus swung at four points from the basis-cranii, but of these, the "pedicle" is the true apex or dorsal end (fig. 6, *pd.*), it lies beneath the parting of the facial and trigeminal nerves (VII., V.), and runs parallel for some distance with the front face of the auditory capsule.

Between the pedicle and the palato-pterygoid bar there lies a large subocular fenestra (*s.o.f.*), or membranous space, parallel with the skull; and behind the pedicle a smaller oblique space is formed, bounded behind by the ear-sac.

The large "suspensorium" is thick at its outer edge, and thin at its inner, which rises somewhat from its smooth, sinuous, upper face; on this the long temporal muscle rests, as it passes from its *origin* in the post-orbital region, to its *insertion* on the "coronoid" crest of MECKEL's cartilage (*mk.*).

The suspensorium is terminated by a reniform condyle for the mandible (*q.*, *mk.*), and opposite the ethmoid it has on its under surface, at the outer edge, a pyriform flattish facet for the hyoid plate (*hy.f.*).

Over this *secondary* condyle, which projects outwards, the suspensorium is developed into a large rounded leaf, with a broad adherent base, a ribbed edge, and a hollow upper surface—this is the "orbitar process" (*or.p.*), which bends over the temporal muscle and is tied by its apex to the skull wall in front.

Its fore margin sends out a snag; another follows this near the condyle for the lower jaw, and opposite this is a third, which has been described as tied by a ligament to the cornu trabeculæ.

The apex of the orbitar process rests upon a longitudinal crest of the palato-pterygoid band; this is the rudiment of the post-palatine (*pt.pa.*); it projects, now, into the subocular space.

The whole of the structure just described is not merely an hypertrophied *epi-visceral* element, it has given rise to a large *extra-visceral* outgrowth.

The "cerato-visceral" element is one-third the length and one-third the breadth of its own upper piece—but it is thicker; this is the free mandible or MECKEL.'s cartilage (*mk.*). This bar is strongly curved, having a long, terete, angular process behind its condylar face, which is a deep notch, like that of the human ulna. The shaft of the cartilage is both dilated and thick, and its subconcave end articulates with one side of the large sucking disk, the corresponding inferior labial (*l.l.*).

This huge, almost horizontal mandibular arch, with its free bars turned inwards at little more than a right angle, is flush, in front, with the trabecular end of the chondrocranium.

The *four* points of attachment to the basis-cranii of this long suspensorium shows a modified condition of that complete fusion of the two regions seen in the Chimæroids.

In *Dactylethra* (ibid., Plate 56) this state of things is intensified, and in *Pipa* (ibid., Plate 60, fig. 3) the membranous space between the basal and lateral cartilages is reduced only to a small crescentic slit.

The larval *Ranine* skull may therefore be said to be, in this respect, a specialisation and *dissection*, so to speak, of the Chimæroid type of skull; on the whole, it is much more like that of the Lamprey.

This immense development of the first epi-visceral element is correlated with the suppression of all those that should succeed; only afterwards, when this part has been relatively lessened, and greatly modified, does even that of the next succeeding arch appear, not then as a mere facial bar, but specialised to auditory purposes, as the "columella."

But the "cerato-hyal" (Plate 2, fig. 7, *c.hy.*) is at present very large, several times larger than its counterpart of the mandibular arch (*mk.*).

It has a dilated, fan-shaped distal expansion (fig. 7), a gently convex pyriform condyle (*hy.c.*), and a curved upper spike, the rudiment of the stylo-hyal (*st.h.*) or upper end of the bar. The basi-hyal (*b.hy.*) is still membranous, but it is followed by a solid basi-branchial (*b.br.*); these parts will be described in a corresponding larva of *R. pipiens*, and in a riper larva of this same species.

The ossifications in this stage are, first, a roundish patch of bone on each side, formed in the substance of the cartilage, close inside the aperture for the 9th and 10th nerves; these are the ex-occipitals (*e.o.*, IX., X.).

The next are formed in membrane, outside the cartilage; they are the fronto-parietals (*f.p.*), thin wedge-shaped shells of bone on the roof; and the parasphenoid (*pa.s.*), a large dagger-shaped bone lying under the floor; its point is rounded and the median hind process is a low triangle; the "basi-temporal wings" are broad and oblique at their outer end.

These three are subcutaneous ossifications, but there is also a series of subcutaneous cartilages.

The suctorial mouth is mainly formed by a coalesced upper pair and a distinct lower pair of these, here called "labials" (*u.l., l.l.*).

The upper piece shows where it was once in two distinct parts; it is a broad crescentic arched flap, with outer angles, an emarginate hind, and a round front, margin. Towards the sides there is a small *fenestra*, which is the beginning of a new subdivision; the angles answer to the upper angular cartilages of the Lamprey's mouth, and the main part to the "anterior dorsal cartilage" of that Fish. The "lower labials" form together a horseshoe, and are fixed in between the mandibles; they have a thick lower edge and a concave inner face (fig. 2, *l.l.*).

The wedge-shaped ray of cartilage, which is confluent with the "tegmen tympani," and articulates with the *otic angle* of the suspensorium, answers to the mandibular ray (the "spiracular cartilage") of the Shark: it becomes the "annulus." The extrabranchials will be described in the next instance; part of the first (*e.e.br*[1].) is shown in fig. 7.

Second Stage.—2. *Tadpoles of* Rana pipiens (A *and* B), *from* $3\frac{3}{4}$ *inches to* 5 *inches long, with hind legs appearing.*

In the largest of these I shall show the details of the structure of the palate, mouth, and throat (Plate 1, figs. 3–5); and in the lesser specimens the skull and face (Plate 3, figs. 1–3).

On the whole, the structure of the chondrocranium is very similar to that just described; there are, however, some very instructive differences.

The skull (Plate 3, figs. 1–3) is altogether rounder, less angular, and free from the projecting snags; the occipital arch is wider and flatter, the cranial cavity not so oblong, but is pinched in in the middle. The internal nostrils (*i.n.*) are oblique and subreniform, converging towards each other behind; thus the palato-pterygoid bar (*p.py.*) is longer and more oblique, and its post-palatine rudiment (*pt.pa.*) projects less into the larger and more oval subocular space. The cornua trabeculæ (*c.tr.*) have wider ends and with a more pronounced outer angle, the outer wings of the ethmoid (*al.e.*) are not so well developed, and the vertical septum (*p.e.*) is shorter.

In the side walls (fig. 3) the cartilage is low, and the hinder roof-cartilage (" tegmen cranii"), with its two fontanelles, is seen to grow forwards from behind, and not to be continuous with the low side wall.

In this stage, even, besides the centres of bone that grow inside the 9th and 10th nerves (*e.o.*, IX., X.), there is on each side a bony patch in front of the ear-capsules;

they are crescentic and enclose the outside of each "foramen ovale;" these are the prootics (V. *pr.o.*).

The fronto-parietals (*f.p.*) are less developed; each bone is a sharp style, growing broader behind, and shell-like.

Also the parasphenoid, below (fig. 2, *pa.s.*), is altogether a smaller bone, with more slender processes; in front it stops short obliquely: here there will appear an additional and rare centre—the "pro-parasphenoid" (see fig. 5, *pa.s'.*).

The auditory capsules are relatively less than in *R. clamata;* the "tegmen tympani" is a rounder, less pronounced growth, and the position of the stapes and its fenestra (*st.*) is more oblique and inferior in position.

The narrow-waisted cranium, and the more out-turned suspensorium, make the subocular fenestra of a very regular oval shape instead of oblong.

The orbitar processes (*or.p.*) are smaller, and the quadrate region in front of them shorter; and the spur (*pr.pa.*) which by ligament is tied to the cornu trabeculæ is a mere blunt projection. The palatine ridge ("post-palatine rudiment," *pt.pa.*) projects backwards less, but is larger than the outer wing of the ethmoid (*d.e.*).

The arcuate form of these great suspensoria throws the quadrate condyles further (obliquely) inwards, and thus shorter free mandibles (*mk.*) are needed. Here the condyle for the hyoid (*hy.f.*) is only two-thirds the distance from the hinge for the lower jaw, as compared with the former instance.

Except in *Pseudis*, to be afterwards described, there is no better kind of Tadpole than this of the American Bull-frog for showing the structure of the post-oral arches.

The figure (Plate 2, fig. 8) was taken from a dissection of a larva $3\frac{3}{4}$ inches long; it is drawn as it appeared from above, after the gill-tufts were cleared away, and the whole basket-work somewhat flattened out for display.

The "cerato-hyals" (*c.hy.*) are roughly hourglass shaped; very solid in the middle, and twisted; they flatten out into massive slabs at each end.

The natural position of those bars is shown in Plate 3, fig. 3, articulating by a somewhat convex condyle with the flattish facet under the fore part of the suspensorium.

Outwards and backwards from the joint the "stylo-hyal" process (*st.h.*) forms a large irregular triangle; below, the "hypo-hyal" region is greatly outspread or pedate. Each lobe projects inwards at it foremost third, and the right and left lobes are united by an isthmus of simple (*embryonic*) cartilage—the "basi-hyal" tract (*b.hy.*).

Wedged in behind this band of arrested tissue we see a large oval azygous cartilage with a rounded lobe growing from its hinder broad end: this is the 1st basi-branchial with a rudiment of the 2nd (*b.br.*).

Overlying this rudiment, and articulating with the main piece, there is a pair of winged cartilages, closely applied each to each at the mid-line: these are the common "hypo-branchial" plates (*h.br.*).

Broad in front, and rapidly narrowing backwards, they each have *four* small finger-shaped cartilages attached to them, the hindmost of these is confluent with the great plate and the rest are merely articulated with it: these are the rudimentary "cerato-branchials" ($c.br^{1-4}$.).

These rays or rudimentary *intra-branchial* arches lie in the floor of the throat, *above* and *within* the branchial pouches; but outside and below these there are four pairs of large *extra-branchial* cartilages ($ex.br^{1-4}$.) that form most of the framework of the pouches and lie close within the outer skin—are, indeed, *subcutaneous* cartilages, like the "labials," and do not belong to the category of true "visceral arches."

The foremost of these remains free distally, articulating with the widest part of the hypo-branchial plate; but the rest have become confluent with it: they all four coalesce with each other, above, like their counterparts in the Lamprey's branchial basket-work.

The *first* and *fourth* of these cartilages are bags or pouches folded into many hills and hollows; the two in the middle are widish bars.

The first and second from their hind margin, and the other two from their fore margin, send inward tooth-like rudiments of branchial rays: these agree with the binding parts that make the bars into a basket-work in the Lamprey.

In the Sharks (see "On the Selachian Skulls," T. Z. S., vol. x., plate 38, figs. 1 and 2; and GEGENBAUR's 'Selachians,' plates 11-20) we see that the extra-branchials are feeble, the intra-branchials *typical*, and the branchial rays distinct; in the Skates or Rays these fringing cartilages are much dilated externally, and these dilatations are articulated together and suppress the extra-branchials.

The two upper labials (Plate 3, figs. 1-3, *u.l.*) are but little united at the middle, and each has a fenestra in it; on the right side there is a lesser hole.

The labials of the sucking disk (*l.l.*) are larger than in the last instance, and this corresponds with the fact that this species is very generalised and archaic.

The spiracular cartilage (Plate 3, figs. 1, 2, *sp.c.*) is confluent with both the "tegmen tympani" and the "otic process" of the suspensorium.

When these parts are studied, not only after dissection but also with all their enclothing structures, we find that the ordinary Batrachian larva is indeed a kind of "Cyclostomous" fish, and the Lamprey and this larva mutually explain each other.

Palatal and lingual aspects and a longitudinally vertical section of the head of the largest of these Tadpoles (5 inches long) show many instructive things.[*]

In the first view (Plate 1, fig. 3), the inferior arches have been removed bodily, and the palate turned upwards; we thus get a view of the horny jaw-plates, the oral papillæ, the quadrate hinge, the condyle for the hyoid, the inner nares, and the whole of the palate up to the faucial region.

The upper part of this *Petromyzine* mouth shows not only the principal crescentic

[*] Professor HUXLEY (on "Petromyzon," Jour. of Anat. and Phys., vol. x., plates 17, 18, pp. 412-429) has served as "pioneer" in this part of the thicket; and a cleared space, such as he has made, makes my work comparatively easy.

horny plate, convex outside and concave within, but also lesser denticulate horny productions (*o.r.*) on the outer folds of the lips, and numerous thick papillæ (*o.p.*), laterally.

The main horny plate (Plate 1. fig. 3) forms a sheath to the upper labials (*u.l.*); behind this the mucous membrane is seen as a bilobate cushion, marked off behind by a deep sulcus which runs from one quadrate hinge (*q.c.*) to the other.

Behind this sulcus the mucous membrane is raised into a second cushion, which is concave in front, and which passes backwards into the large palatal roof. At the midline its foremost third is marked off by an "anterior palatal velum" (*a.v.*) ("median triangular papilla" of HUXLEY, *op. cit.*, p. 429); this is covered with small papillæ, is semi-oval in shape, and is curved forwards; its attached edge is crescentic, the convexity looking backwards.

In front of this flap there are two long clefts (*i.n.*) which approach each other at the mid-line, and then greatly diverge, forwards, nearly reaching the exposed quadrate hinge (*q.*). Each cleft is lipped in front, and this lip runs back again along the cleft dividing it into two chinks; this apparently double slit is guarded by two large papillæ: it is the internal nostril. In the chondrocranium this passage appears as a large round hole between the trabecular cornua and the quadrate (Plate 3, figs. 1, 2, *i.n.*, *c.tr.*, *q.*) and only partially enclosed by cartilage in front. I have just described these parts.

At the sides, opposite the "anterior velum," the condyle on the suspensorium for the hyoid cornu is exposed (*hy.f.*).

The interorbital region of the palate is lower than either the internasal or the interauditory; it is one-fourth longer than broad, is covered with papillæ, and has the largest of these productions postero-laterally.

The hind part of this palatal tract is, properly speaking, *faucial*; it is beneath the ear-sacs, and reaches the basioccipital region. This is separated from the proper palate by two "posterior vela" (*p.v.*); these are crescentic folds of mucous membrane, with their convexities forward, or opposite to that of the "anterior median velum" (*a.v.*).

The floor of the mouth is shown from above (fig. 4); the free mandibles (*mk.*) and the hyoid cornua (*c.hy.*, *hy.c.*) have their upper or articular ends exposed.

The inferior horny plate, with the outer denticles of the folds of the lower lip (Plate 1, fig. 4) are shown, as well as the right and left crops of labial papillæ; the main horny plate is modelled on the horse-shoe shaped lower labials (*l.l.*).

In this view all the post-oral clefts are seen, for at present the pouch-shaped (blind) first cleft (*cl¹.*) is seen in position. I have found it open externally in no stage, although the skin over it is very thin at the time of hatching. (See my first paper on the "Frog's Skull," Phil. Trans., 1871, Plate 3, fig. 10, *cl¹.*)

This low position of the first cleft is due to the extreme length and horizontal position of the "pier" of the mandibular arch (see Plate 3, figs. 1, 2), the hinge of which (fig. 1, *q.*) nearly reaches the upper lip; afterwards, in the adult, it is beneath the ear.

This cleft (Plate 1, fig. 4, *cl¹.*) is small, crescentic, and with its convexity looking outwards and forwards; all the other clefts ($cl^{.\,3-5}$) have the same form and direction, but they are larger, and open externally, although covered afterwards with the opercular skin, which only opens on the *left side*. The second cleft (cl^{2}.) is between the hyoid cornu and the first branchial pouch; it runs into the floor some distance from the hyoid facet.

Between the articular surfaces of the free mandibles (*mk.*) the mucous membrane is raised into two crescentic folds; these are the lateral rudiments of the tongue (*tg.*); if these lobes simply grew forwards the Frog would have a bilobate tongue with its *free ends* growing forwards—it would be *protero-glossal*.

Between these lobes there are several large papillae; behind them the mucous membrane thinly veils the hyo-branchial arches (Plate 1, fig. 4, and Plate 2, fig. 8).

At the mid-line there are, behind the papillae, first the basi-hyal (*b.hy.*) and then the double basi-branchial (*b.br.*); whilst on each side are the broad hyoid cornua (*c.hy.*) and the sub-median and lateral elements of the branchial arches (*h.br.*) and its rays —all that exists of the *endoskeletal* framework of these parts. Thus in the larval condition the Frog has the dorsal portion of its hyoid and branchial arches suppressed: four large "extra-branchials" (Plate 1, fig. 4, *br.p.*; and Plate 2, fig. 8, *ex.br*$^{1-4}$.) supplement the deficiency, and those bars, the first and last of which are cochleate, are homologous with half of the branchial basket-work of the Lamprey, namely, with the four bars of its framework from the *first* to the *third* pouch inclusive.

In his valuable paper on the Lamprey (Journ. of Anat. and Phys., vol. 10, pp. 412-429), Professor HUXLEY does not notice the existence of arches homologous with the proper branchials of Selachians, Ganoids, and Teleostei; but his description of the correspondence of these parts with those of the Lamprey must be given here:—

"In the present stage the branchiae of the Tadpole are, as is well known, pouches, which present no superficial likeness to the branchial sacs of the Lamprey. A septum extends inwards from the concave face of each branchial arch, and the septa of the two middle arches terminate in free edges in the branchial dilatation of the pharynx. Vascular branchial tufts beset the whole convex outer edge of the branchial arch, and are continued inwards in parallel transverse series of elevations, which become smaller and smaller towards the free edge of each septum, near which they cease.

"In the young Ammocoete the septa of the branchial chambers similarly bear vascular processes, which are first developed close to the external branchial aperture, and thence extend inwards transversely."[*]

[*] In a note here, Professor HUXLEY remarks: "If the first-formed long branchial filaments of the Ammocoete projected through the small gill-clefts outwards instead of inwards, they would resemble the first-formed 'external gills' of Elasmobranchs. And this difference of direction seems to indicate the solution of the difficulty, that external gills, which are so generally developed at first in *Elasmobranchii*, *Ganoidei*, and *Dipnoi*, are apparently wanting in *Marsipobranchii*." I may add to this note an observation of my own illustrative of the writer's remarks, namely, that in the larva of *Dactylethra* there are no branchial filaments growing from the outer face of the branchial pouches (see "Batrachia," Part II., Plate 58, fig. 1.)

"The recesses at the sides of the floor of the pharynx into which the interseptal clefts or internal branchial clefts open, answer, taken together, to the branchial canal of the Lamprey, which is not shut off from the œsophagus in the Ammocœte. The anterior boundary of each of these recesses is marked by a fold of the mucous membrane, the free edge of which projects backwards, and is produced into papilliform angulations so as to appear scalloped. The anterior face is concave. The inner angle of each fold passes into its fellow by a ridge, produced into one or two papillæ, which is closely adherent to the median part of the floor of the mouth. The outer angle is continued into a more delicate fold of the mucous membrane lining the roof of the mouth, the free edge of which also projects backwards. It is plain that these structures answer to the pharyngeal velum of the Lamprey."

The "inferior velum," or membranous fold (*l.v.*) will be best understood by reference to both the figures of its natural condition (Plate 1, fig. 4) and also to those of its dissected framework (Plate 2, fig. 8).

We see at once that this scalloped fold is formed upon the rudimentary arches that grow from the oblique sides of the hypo-branchial plates, and that there is the common pharyngeal covering of these rudimentary *intra*-branchial arches ("cerato-branchials").

They are able to be the skeleton of a free, scalloped, papillated fold in virtue of their arrested condition; if they grew upwards round the whole circle of the throat this structure could not exist.

If each of these rudimentary arches was continued upwards, surmounted on each side by an "epi-branchial," and this in turn by a "pharyngo-branchial," we should have such arches as are seen in Selachians, Ganoids, and Teleosteans.

If the lower, or *outer*, surface of these arches was beset with a double row of pectinated branchial folds, then the Ganoid and Teleostean type of gills would exist; in the Selachians ("branchiis fixis") the gill-folds are formed in a double series of pouches, the common framework of which is formed of typical *intra*-branchial arches, each composed of *nine* pieces, on the inner side; the septa are strengthened by *five* "branchial rays," and on the outside (in the Shark) there are distinct "extrabranchials," one to each pouch, which are pointed above and pedate below.

The "branchial canal" of the Lamprey is correlated with suppression of the "cerato-branchial" rudiments, such as are seen in the Tadpole; but there is a hyoid arch.

With respect to the branchial tufts that are so copiously developed within and at the edges of the gill-pouches of the "Phaneroglossa," I find that my own views are in accord with those of Mr. Balfour.[*]

There are *true* and *false* "external gills": the first, only, are present in the "Urodeles," and for a short time in the newly-hatched Tadpole (see Phil. Trans., 1871, Plate 3, figs. 2 and 10, *br*¹., *br*².); these are developed from the epidermis—are *epi-blastic*. But the tufts that break forth from the clefts in the more developed Tadpole,

[*] I have recently had my mind set at rest upon this subject by my talented friend.

and the long, *precocious*, filamentous gills of the embryo Selachian, are, like gills generally, of *hypoblastic* origin.

The vertical longitudinal section (fig. 5) may be now described.

The large upper labial (below *e.tr.*) is seen in section; it is the "anterior dorsal cartilage" of J. MULLER: below, the lower labial is seen,[*] where it joins its fellow to form the imperfect "annular cartilage" (see also fig. 4, *l.l., mk.*).

MECKEL's cartilage is seen behind this in the section; it lies in the side of the mouth, and its main function is to serve as a suspensorium to the "annular cartilage" (the corresponding half of the imperfect ring) (see also Plate 2, fig. 6, *l.l., mk.*).

The raised cushion behind these cartilages is the rudimentary tongue (fig. 4, *tg.*); in the back of this the basal cartilages are placed.

In most things I agree with Professor HUXLEY as to his harmony of the Lamprey's and the Tadpole's mouth. There is some reason to suppose that the antero-inferior median cartilage belongs to the mandibular arch. The basi-hyal of Fishes is always a "glosso-hyal;" it projects forwards beyond the arch to which it belongs, and so does the first basi-branchial.

Any *paired* plate of cartilage lying in the same plane, but projecting outwards and backwards from the basi-branchial bar (or bars) would be "hypo-visceral;" any rods attached above these and lying in the lower half of the sides of the pharynx would be "cerato-viscerals."

In the Lamprey (HUXLEY, *op. cit.*, plate 17, fig. 1) this anterior median cartilage, below, is outside the lingual rod, and is probably a basi-mandibular; and the great lingual cartilage (*k.*) is an undivided piece, with neither osseous nor fibrous segmentations, and is evidently an *intra-visceral* element.

In the Tadpole (Plate 2, fig. 8, *b.hy.*) the anterior basal piece does not chondrify until metamorphosis takes place; it is a thick, narrow, conjugating band, composed of large cells of simple cartilage.

But the basi-branchial of the Tadpole (*b.br.*) is an oval mass of cartilage, with a hinder bud that grows downwards beneath the large hypo-branchial plates (*h.br.*); this bud does not become a separate second basi-branchial. These two semi-distinct rudiments do not represent more than the end of the long lingual cartilage of the Lamprey; and if that fish had paired cartilaginous plates lying on the same plane as the lingual cartilage and attached to its sides, behind, they would be hypo-branchial plates, for they would belong to the *inner* category of cartilages, and not to those forming the branchial basket-work.

In the Tadpole the heart (fig. 5, *h.*) is roofed over by the "hypo-branchial" plates, and the edge of the "inferior velum" (fig. 4, *i.v.*) is seen completing this roof and also covering the inner branchial openings in front.

Where the right and left *vela* unite at the mid-line, there they are directly in front of the larynx (*l.x.*); afterwards, the hypo-branchials become the "thyro-hyals."

[*] The inner face of the lower labial is lettered *mk.*, by mistake, in Plate 1, fig. 5.

Third Stage.—1 (continued). *Tadpole of* Rana clamata (B) *3 inches 5 lines long; tail* 2¼ *inches long; hind legs,* 1⅜ *inch long.*

In this stage (Plate 4, figs. 1–4) we see some metamorphic progress beyond the last; the general form of this as compared with Stage A (same species) is not much altered (see Plate 2, figs. 5–7), but there are some new parts, and some changes in the old.

The *time*, indeed, has been short during which these changes have taken place, for these Tadpoles even at their height, as larvæ, were beginning rapidly to be transformed. The exoccipitals (*e.o.*) are larger, but the prootics are not visible; outside the frontal region the premaxillaries (*px.*) have appeared in addition to the three investing bones seen in the last stage. But the thing of most importance is the rapid growth of the "intertrabecula" as a vertical nasal septum (*s.n.*), and the solidification of the membranous roof of the nasal sacs into a pair of *ear-shaped* nasal cartilages (*nt.*) whose antero-external notch forms the inner boundary of the outer nostril (*na., e.n.*). These elegant shells of cartilage are becoming confluent with the top of the septum internally, and with the ethmoidal "wing" (*al.e.*), behind: they are true "paraneurals," and answer to the "sclerotics" and "periotics."

Another change is the retreating of the orbitar process away from the side of the skull, so that the "post-palatine" rudiment is exposed (*or.p., pt.pa.*); the "pre-palatine" rudiment (*pr.pa.*) in front of the internal nostril (*i.n.*) is also further from the trabecula.

The fore margin of the chondrocranium is very similar to what it will be eventually, but the rounded end of the intertrabecula is still distinct from the nasal roofs (*s.n., na.*), and the cornua trabeculæ (*c.tr.*) are still undivided at their end.

The great upper labial (*u.l.*) has lost the fenestræ that showed signs of division, right and left, and is now, more than ever, one large saddle-like plate.*

The mandibular and hyoid arches are much in the same state as in the younger larvæ (Stage A), but the branchials are altering preparatory to extensive absorption.

Compared with those of the Bull-frog (Second Stage, Plate 2, fig. 8), it will be seen

* In this and the last two stages (A and B) we have a very instructive analysis, so to speak, of the complex nasal capsule in the higher Vertebrates, *e.g.,* of that in the Mammals. It is composed primarily of the trabeculæ with their lateral (antorbital) wings, where the cranium closes in in front of the vertical "intertrabecular" middle wall; of the nasal roofs which may grow down into side walls; and of the rich outgrowths of turbinals that grow from the front of the antorbital wings of the ethmoid (*upper and middle turbinals*); and from the inside of the outer wall (*inferior turbinals*).

Other turbinals may spring from the roof (*nasal turbinals* of the Rodents), or inside the outer nostril (*ali-nasal turbinals of Pig and Bird*).

Then the narial valves are formed of segments or subdivisions of the upper labials, the whole structure being roofed by the nasals, floored by the vomer or vomers and palatine plate of the maxillaries, and walled in by the maxillaries and premaxillaries.

Lastly this complex labyrinth of cartilage may, itself, become bony, to a greater or less degree, and even may coalesce, more or less, with the investing bones.

Anyhow, the plan and pattern, and the *numbered elements* of this labyrinth, are to be seen in their first simplicity in the nasal region of the Tadpole.

that the first and second extra-branchials (Plate 4, fig. 3, $e.e.br^{1-3}$.) have lost much of their pouch-like shape, and are but little broader than the second and third; they all become very narrow bands, and thus vanish, with the exception of a common remnant that becomes fused with the permanent hyo-branchial plate.

The third intra-branchial ($c.br^3$.) is very long, as long indeed as the counterpart bars are in Selachians, Ganoids, and Teleostei; in its normal position, round the pharynx, this bar just reaches half way up the side, and in those Fishes is surmounted by the epi-branchial.

In the *Dactylethra* Tadpole (Phil. Trans. 1876, Plate 58, fig. 1) the whole skeleton of the branchial pouches is one continuous structure; here the intrabranchials are nearly segmented off from the extra-branchials, but there is an isthmus of cartilage uniting them in *Rana*, as may be seen in the section (fig. 4, $c.br^1$., $ex.br^1$.).

I must pass again to *Rana pipiens* for my next stage; the reader will, however, easily eliminate the non-essential specific variations, at any stage, from those more important step by step processes that transform a Cyclostomous skull into one with a widely gaping mouth and jaw-hinge far behind, like that we see in certain aberrant Selachians—*Notidanus and Cestracion*.

Fourth Stage.—2 (continued). *Tadpoles of* Rana pipiens (C) *with tail lessening and all the legs free.*

This stage (Plate 3, figs. 4-13, and Plate 9, figs. 1-6), of this species, shows that in the largest and most generalised of the genus the branchial apparatus remains in full development and function much longer than in the lesser Frogs. They are two or three years (according to Wyman) before they have finished their metamorphosis, and that excellent naturalist kept them (in captivity) six or seven years as Tadpoles.

In *R. clamata* (Stage B) the nasal roofs just appear as the branchial pouches are dwindling away. In *R. pipiens* the whole nasal labyrinth is perfected before any sign of diminution appears in the gill-pouches.

This indicates that the Tadpole of the Bull Frog is well fitted out with special sense organs long before it finishes its metamorphosis, and suggests the probability of a still slower metamorphosis in former epochs. *Pseudis* offers a remarkable confirmation of this view.

Comparing this with the two last stages we shall see both the general progress and the specific differences that are found to be even in larvæ of the same genus.

The most striking change is the elongation of the palato-pterygoid band (Plate 3, figs. 4-6, *p.pg*.), thrusting the suspensorium outwards. Then the "waist" of the cranial box is narrower, and its sides more chondrified (fig. 6).

The roof bones are segmented across (fig. 4, f., p.), and, contrary to wont, the frontals are only half the size of the parietals.

Below, the parasphenoid (fig. 5, *pa.s*.) has become more developed, and foliated behind; in front, the membranous layer seen in the earlier larva (fig. 2) is now partly

DEVELOPMENT OF THE SKULL IN THE BATRACHIA. 31

occupied with a new (and *rare*) bony centre—the "pro-parasphenoid" (*pa.s'.*) The endosteal tracts in front of and behind the auditory capsules (*pr.o., e.o.*) are enlarging, and the "pedicles" (*pd.*) in front of the former are becoming very attenuated. The rest of the hind part of the cranium is but little altered, but in front the nasal capsule is almost finished (fig. 4).

The nasal roofs (*al.n.*) are now completely confluent with the septum and the lateral wings of the ethmoid; nevertheless the "cornua trabeculae" (*c.tr.*) that form the floor of the nasal cavities are undivided in front. The ear-shaped nasal roofs are seen to have curved down, round the external nostril (*e.n.*), but the terminal halves of the trabecular cornua are interposed between the nasal capsules and the upper labials.

These, however (figs. 4–6, *u.l.*), are seen to be breaking up into an inner and an outer pair; these will disappear; the *new labials* and the newly formed premaxillaries have been removed. The inferior arches (Plate 3, figs. 5, 6, and Plate 9, fig. 1) are but little altered from what they were in the Second Stage. The drawing of the branchial arches was made from an outspread preparation which was figured from below; they are seen to turn obliquely backwards as they ascend to the sides of the basis cranii.

Tooth-like processes of simple cartilage are interdigated with each other in the large clefts; but the branchiae which grow on them, and which run inwards in rows, have been removed.

But truly cartilaginous rays grow from the bars under these processes; these are partly shown in Plate 9, fig. 3, *br.r.*

The internal nostril (Plate 3, fig. 5; Plate 9, fig. 2, *i.n.*) is a curious oblong slit, diverging forwards, and protected by one or more papillae.

A series of transversely vertical sections (Plate 3, figs. 7–13; and Plate 9, figs. 4–6) show the solidity of this chondrocranium.

The *1st section* (Plate 3, fig. 7) is through the fore part of the nasal capsule, and catches the nasal roofs (*al.sp.*) both above, and where they turn over to form the narial rim, or "alinasal" fold. They have coalesced with the intertrabecular wall, *now* the septum nasi (*s.n.*).

A branch of the orbito-nasal nerve (5¹) is seen emerging between the folds; the cavity between these folds externally is the outer nostril; another space is seen below this and above the thick cornua trabeculae (*c.tr.*), which runs inwards to the inner nostril. At present the trabecular horns are not modified into the flat nasal floor, but are attached to the structure above by fibrous bands.

The *2nd section* (fig. 8) is close behind the nostril, and behind the second fold of cartilage; on one side the wide part of the roof has been (obliquely) cut through; the septum (*s.n.*) is still some distance from the cornua (*c.tr.*; the cross band is coloured, by mistake).

The *3rd section* (fig. 9) is similar; it is also oblique, and the hinder part of the capsule is partly severed.

The *4th section* (fig. 10) is through the widest part of the nasal roof, where these cartilages (fig. 9, *al.sp.*) have coalesced with the ethmoidal tract (*al.e.*); here the septum (*s.n.*) as it becomes "perpendicular ethmoid" (*p.e.*) is confluent below with the coalesced trabeculæ (*tr.*). On one side the section is seen to pass into the palato-pterygoid band (*p.pg.*).

The *5th section* (Plate 9, fig. 4), rather more than half of which is shown, is through the ethmoidal part of the skull proper, where the nasal cavities (*n.p.*) are closing in. The perpendicular ethmoid is seen to be very thick, and the trabecula (*tr.*) to pass into the solid root of the palato-pterygoid bar (*p.pg.*) from which ascends, suddenly, on the outside, the strong arcuate "orbitar process" of the suspensorium (*or.p.*).

The *6th section* (Plate 3, fig. 11) is through the solid ethmoid where it closes in the skull, leaving only the olfactory nerve-passages (1); here the skull is as solid as that of a *Chimæra*. The palato-pterygoid bars (*p.pg.*) are cut through, outside the "postpalatine rudiment" (see Plate 3, fig. 1, *pt.pa.*), whose form, as a gentle elevation, is there shown.

The *7th section* (Plate 3, fig. 12) is behind the palato-pterygoid, and through the fore part of the *skull-barge*, where the olfactory crura (1) are given off. The cavity is elliptical, the wide wall thick, the roof (anterior "tegmen," *t.cr.*) and the floor are gently concave.

On the roof we see the fore end of the frontals (*f.*), and under the floor the parasphenoid (*pa.s.*).

The *8th section* (Plate 3, fig. 13) is behind the "tegmen" and through the fore part of the "fontanelle," covered by the widest part of the frontals (*f.*); the bone below is widening, it is the parasphenoid (*pa.s.*). The elliptical cavity is deeper, the walls thinner, and the floor is now convex.

The concavity, below, in the two last sections, is due to the drawing in of the trabeculæ, and the high, vertical form of the "intertrabecular" tract in the fore skull, is now a bulging floor.

The *9th section* (Plate 9, fig. 5) is through the *middle tegmen*, between the great median and the lesser paired "fontanelles," and also through the widening parietals (*f.*, by mistake for *p.*). Here the roof passes into the wall, downwards; then there is a large *fenestra optica* on each side, with the optic nerves (II.) emerging; and then the rounded trabeculæ, with the bulging "intertrabecular tract," floored by the parasphenoid (*tr.*, *pa.s.*).

The *10th section* (Plate 9, fig. 6) is through a lateral fontanelle, the hind part of the parietal (*p.*), and the auditory capsule, confluent with the basal plate (*iv.*). The anterior and horizontal canals (*a.s.c.*, *h.s.c.*) are cut through, and the general cavity of the vestibule (*au.*) laid open.

The "tegmen tympani" (*t.ty.*) is seen projecting from the outside of the capsule, beyond the horizontal canal; and the basal plate, or "investing mass," is seen to grow outwards as a floor to the tympanic cavity (*ty.f.*). This plate is deficient, more or less, behind, the capsule resting on an imperfect floor below, and projecting into a *fenestra* in the wall of the skull.

Fifth Stage.—1 (continued). *Advanced Tadpoles of* Rana clamata (C), *all limbs free:* 3¼ *inches long; tail,* 1½ *inch long.*

If this (Plate 4, figs. 5–7) be compared with the Third Stage, the second larva of *R. clamata,* (figs. 1, 2) we shall see what approach is made towards the adult condition. The angular and almost oblong form of the whole structure is changed for a short oval; and the facial outworks are only half the real and relative width they possessed then. The auditory capsules are of greater extent, antero-posteriorly, the ex-occipitals (*e.o.*) are larger, and the prootics (*pr.o.*) have begun on the outside of the *foramen ovale* (V.). The form of the cranial cavity is narrower in front compared with the hind part, and the whole ethmo-nasal structure in front of the cavity is shorter, and much modified.

Besides the septum nasi (*s.n.*) the whole intertrabecular space is filled in in front; the "cornua" (*c.tr.*) are only half as long, and are bifurcated, the outer fork has become the large arcuate angle of the "subnasal lamina," and the lesser (inner) fork is the hooked "pro-rhinal" (*p.rh.*) which lies inside the premaxillary (*px.*), and curves backwards.

The internal nares (*i.n.*) are now relatively much forwarder, are more elongated, and are placed quite transversely.

The front wall of the nose is completed by fusion of the roof with the septum, and a growth of cartilage over it to the lower face of the labyrinth. The external nostrils (*e.n.*) are now also much nearer the front of the head, for the large labial plate has been absorbed and two pairs of small new plates that form the outworks of the nose have appeared. The inner pair (*u.l*[2].) are applied to the inner face of the apex of each nasal process of the premaxillaries (*px.*); whilst the outer forms a shell-like valve to the external margin of the nostril (*u.l*[2]*., e.n.*).

The nasal bones (a film over *na.*) have appeared like small shells. The condyle of the quadrate (*q.*) is now opposite the hinder margin of the nasal roof, instead of the front margin; it is twice as far from the roof, and its condyle looks obliquely outwards, and not inwards as before. The elongated and oblique palato-pterygoid bar now has the "post-palatine rudiment" as a ridge looking outwards and backwards, and the "pre-palatine spur" (*pr.pa.*) nearly touches the tip of the horn of the trabecula (angle of "sub-nasal lamina," *c.tr.*).

Tracing the narrowing and retreating suspensorium backwards, we see that the orbitar process, and condyle for the hyoid (*or.p., hy.f.*) are now opposite, not the antorbital region, but the middle of the subocular fenestra.

Further back, the true "otic process" is small and rounded (fig. 5, *ot.p.*), and the spiracular cartilage, now detached from the tegmen tympani, applies itself to the side of the otic process (*sp.c., ot.p., t.ty.*); it is now a trifoliate rudiment of the "annulus." Below (fig. 6, *pd.*), the pedicle is forming its condyle, to glide upon a pre-auditory process of the basal plate, whilst the dorsal part of the pier is now an attenuated thread of cartilage running under the prootic to the front of the foramen ovale (V.).

A remnant of the notochord (*nc.*) still persists; it lies on a thin junctional tract of the investing mass through which it is very visible.

The fronto-parietals (*f.p.*) are but little changed, and are not yet segmented. The parasphenoid (*pa.s.*) has enlarged in its basi-temporal wings, in correspondence with the enlarging ear-capsules.

There is only a stapes in the middle ear; no epi-hyal element has appeared.

In accordance with the out-turning of the quadrate condyles, the lower jaws (fig. 7, *mk.*) are longer; but the whole arch is largely extended by the strangely altered form of the original suctorial "horseshoe" formed by the "lower labials" (*l.l.*). Now, they are nearly straight, and stand across between the ventral ends of the mandibles. These latter bars are invested on their inner and lower face by the rudimentary "articulare" (*ar.*). The maxillaries (*mx.*) also have appeared above.

Sixth Stage.—3. *Half metamorphosed larva of* Rana ——? *sp.* (*India*); $1\frac{3}{4}$ *inch long*; *tail*, $\frac{1}{4}$ *inch; all the legs free.*

The less important *specific* modification here seen (Plate 4, figs. 8–10) is the flat, wide cranium; the more important morphological changes are self-evident. The transversely extended auditory masses, and the wide flat occipital arch, are now largely becoming ossified by the two pairs of perineural centres (*pr.o., e.o.*) that gradually more or less occupy the occipital, periotic, and post-sphenoidal regions, protecting those parts of the head where the 5th to the 10th nerves emerge. The rest of the cranial "barge" is well chondrified, but has no bone as yet; the roof bones (*f.p.*) are still large; the floor bone (*pa.s.*) is a short, blunt, dagger, with a very wide "guard," and a very short handle.

Over the nasal roof the nasal bones (*n.*) are now crescentic shells of bone; the premaxillaries and maxillaries (*p.x., mx.*) are now well developed, but the vomers are not apparent. A thin squamosal (*sq.*) like a "pre-operculum" runs down the suspensorium, and the dentary (fig. 10, *d.*) has appeared on the outside of the mandible, distally; the *coalesced* "lower labial" is becoming ossified, and the dentary is grafting itself upon the endosteal patch, so as to form the "mento-Meckelian bone;" the "articulare" (*ar.*) is lengthening with the jaw.

The fore parts of the trabeculæ, now completely confluent with the nasal roofs, are broken up into a remarkable cervicorn structure. The inner angle of each is now the finger shaped, curved "pro-rhinal" (*p.rh.*), and the outer angle, which finishes the nasal floor in front, is divided into three lobes, one upper and two lower, that turn outwards, and are imbedded in the fore end of the maxillary. The two pairs of upper labials ($u.l^1.u.l^2.$) are quite normal now, perfecting the antero-external edge of the nostril (fig. 8). Behind the pro-rhinals a pair of "fenestræ" (fig. 9, *n.n.*) are seen.

The "pre-palatine" (*pr.pa.*) has now escaped far away from the trabecular angle in front; it is a sharp falcate process at the fore end of a narrow band of cartilage which runs backwards until it is now opposite the prootic bone; there it ends in the suspensorium, which has become a stunted, triradiate cartilage, one-third of its former length.

The antorbital pedicle ("ethmo-palatine") (*e.pa.*) is now a flat tape, narrowest in the middle; the "post-palatine" (*pt.pa.*) is a similar tape passing insensibly into the still narrower "pterygoid" (*pg.*), which latter passes into the suspensorium where it subdivides into its *three forks*.

The lowest part ends in the reniform quadrate condyle (*q.*); and the position of the hinge is opposite the prootic and the foramen ovale (*pr.o.*, V.), instead of being in its old place opposite the *front of the nasal capsules.*

This is, truly, only part of the way back of the condyle (see in the adult *Rana pipiens*, Plate 8, where it gets some distance behind the occipital condyle), but this is a good distance to be travelled whilst the legs have been growing and emerging, and the lungs gradually rendering the gills unnecessary.

The spiracular cartilage (*sp.c.*) has begun to take on its crescentic form, and is now an evident *annulus tympanicus*. The epi-hyal has not yet appeared, ready to become the columella; the cerato-hyals (fig. 10, *c.hy.*) are dislocated from the suspensorium, the *hinge* having become absorbed; they lie behind the first pair of clefts, which, of course, are between them (on each side) and the corresponding suspensorium.

This latter part, the mandibular "pier," has lost all its dorsal end, the narrow upper tract of the "pedicle" having been completely absorbed. The stunted, *amputated* part has now a flat condyle on its end, which glides on the facet formed for it over the front of the ear-capsule below.

Above, the "otic process" (or *third* part) of the suspensorium (*ot.p.*) has crept close to the fore edge of the tegmen tympani, ready for fusion; it is already invested with the squamosal (*sq.*), which lies in front of a gentle ridge—all that remains (now) of the "orbitar process." The arches, pouches, sub-basal and basal plates of the branchial cartilages are reduced to a lozenge-shaped plate, ending in a pair of diverging rods, and ready to unite, in front, with the attenuating cerato-hyals (fig. 10, *c.hy., b.h.br., t.hy.*).

I shall next describe the condition of the skull in young Frogs, when that which was left unfinished in the cranium, on their assumption of terrestrial life, will be seen to have gone on unto perfection; and the "headstone" brought on to this graceful piece of vegetative architecture.

Skulls of Young and Adult "Ranidæ."

Seventh Stage.[*]—4. *Skull of* Rana palustris (*Cambridge, Mass., U.S.*). *Young recently metamorphosed;* 11 *lines long.*

The figures of this (Plate 5, figs. 6–10) and the next (figs. 1–5) stages show the endocranium with the outer bony laminæ removed from one side, and retained on the other; and these, whether they are bones that are permanently, or only for a time, distinct.

---

[*] If the three early stages of the larval skull (in *Bufo vulgaris*) were added to these, this would be the *Tenth Stage*.

All the structures are now rapidly passing into their permanent form, although the individuals are, at this time, very small.

The endocranium is here of the typical form; a flat-bottomed "barge," with one large anterior, and two small posterior membranous spaces in the "deck."

The larger lateral expansions behind are connected with the lesser lateral expansions in front, by an elegant bow of cartilage, the pterygo-palatine (*pg., e.pa.*); the hinder expansions are caused by the impaction of the *ear-masses* into the sides of the skull, and the expansions in front are due to the *nasal* growths; from the open spaces between, on each side, the eye-balls have been removed.

The two pairs of ossifications in the endocranium, the prootics and the ex-occipitals (*pr.o., e.o.*), are considerably larger than in the last stages; the hinder pair have quite encircled the twin passage for the 9th and 10th nerves; a trace of the notochord (*nc.*) still remains.

There is no girdle-bone yet; thus, with the exception of the "centres" just mentioned, the endocranium is a "chondrocranium." The form of the contents of the ear-capsule is well seen through the semi-transparent cartilage, and the form and extent of the originally separate nasal roofs, as distinguished from the pre-cerebral region of the cranium, are well seen. Also below (fig. 7), the manner in which the "intertrabecula" has filled in the space between the "cornua trabeculae," and the sub-division of the root of the "horns" into a pro-rhinal hook, and a subnasal outer angle (*p.rh., s.n.l.*), are clearly seen; also the divisions of the upper labials (fig. 6, *u.l¹.u.l².*), and their relation to the outer nostril (*e.n.*).

The "ethmo-palatine" bar (*e.pa.*) is lobate; the pre-palatine is spiked, and the post-palatine is a gently lessening bar, which with the pre-palatine *region* in front and pterygoid *region* (*pg.*) behind, forms a most elegant subocular "bow."

Behind, this bowed bar ends in three lobes; the quadrate with its reniform condyle (*q.c.*) running outwards, and downwards, and backwards; the "pedicle" (*pd.*) pedate, with a flat inturned facet below; and the "otic process" above, clamping the outer and front end of the "tegmen tympani."

For articulation with the quadrate we see the cylindroidal condyle of the mandible (fig. 8, *ar.c.*); the under and inner face of the long arched rod is invested by the "articulare" (*ar.*) nearly to the fore end; the "dentary" (*d.*) runs along the distal half on the outside, and it is grafting itself, near its lower end on the lower labial, which is not quite confluent with the mandibular rod (*mk.*), and which is itself ossifying, to become the "mento-Meckelian" bone (*m.mk.*).

The obliquely semi-oval stapes (figs. 7 and 10, *st.*) has now wedged in, between its fore margin and the ear-capsule, the proximal end of the "columella" (*epi-hyal element*).

This structure is not yet finished, and has a subdivision in it very rarely seen in adults, but which is normal in certain fishes (*e.g.*, the "Acipenseridae").

These two parts are, morphologically, the proximal or "hyomandibular," and the distal, or "symplectic." As a rule a short segment is cut off close to the stapes, the morphological "pharyngo-hyal;" this has not taken place, at present, in this young Frog.

The proximal piece is the medio-stapedial (fig. 10, *m.st.*, *i.st.*), and it is largely ossified; the distal piece is not; it is the "extra-stapedial" (*e.st.*)—a pedunculated sub-circular disk; as yet, there is no supra-stapedial band growing up from the inner side.*

The "cerato-hyal" (figs. 7, 9, *st.h., c.hy.*) is not yet as narrow as it will be, but from hanging by its primary joint under the *antorbital region*, it now is articulated to the tympanic floor under the columella and behind the Eustachian opening (*eu.*), to which it now forms the normal boundary as part of the hyoid arch, that is to say, it is behind the "1st visceral cleft;" over that cleft, outside, the "spiracular cartilage" is closing round the membrana tympani so as to form the "annulus" (fig. 6, *a.ty.*).

The cerato-hyal band still retains an auriform hypo-hyal lobe (fig. 9, *c.hy., h.hy.*), but the basi-hyal conjugation (*h.hy.*) is still a tract of simple cartilage running into the dilated basi-branchial plate (*b.br.*); attached to this we see on each side the remains of the branchial pouches, and another remnant is seen also on each side of the paired hypo-branchial plates (*h.br.*), which now end in long sigmoid rods (*t.hy.*), the still unossified "thyro-hyals."

This is the most instructive intermediate stage of these parts I have as yet succeeded in dissecting out; if the figures of these parts be compared with what is seen in the larva on one hand, and in the adult on the other, the value of this stage will be self-evident.

The investing bones, the true "parosteal" plates, are nearly all present; those of the lower arches have been already described.

The frontals (fig. 6, *f.p.*, by mistake) are now, *for a while*, distinct from the parietals; the latter lie over the paired fontanelles, and the frontals cover the larger space (*fo.*). The nasals, premaxillaries, and maxillaries (*n., p.x., mx.*) are now quite normal, the quadrato-jugal (*q.j.*) is appearing, and the squamosal (*sq.*) is acquiring a *supra-temporal* plate. Below, the parasphenoid (fig. 7, *pa.s.*) is quite normal, having all the hinder processes, and there is no *pro*-parasphenoid. The "vomers" (fig. 7, *v.*) are thin crescentic plates notched in front. The palatine and pterygoid bones (*pa., pg.*) are *now* quite separable from the subjacent cartilage, but that is only a temporary state of things in most cases.

Eighth Stage.—5. *Skull of* Rana halecina (*N. America*); 1¾ *inch long.*

This young Frog was about twice the size of the last, but the *species* are quite distinct; this is the most welcome intermediate form I have found between the typical *R. temporaria* and the gigantic *R. pipiens*.

The cranium is much more advanced, but it is still unfinished; in this respect it is a good link between the young *R. palustris* and the adult *R. temporaria.* In some

* In *Pseudis* ("Cystignathidæ") I shall be able to show *two* earlier stages of the columella than this (Plates 11 and 12).

respects it corresponds with the Oriental Frogs, and differs from most of those of the Nearctic and Palæarctic regions; this is seen in the presence of an "anterior superorbital" expansion of the chondrocranium (Plate 5, fig. 1).

This is an unusually narrow and elongated skull, and thus differs from that of *R. pipiens* which is short and wide; the last was elongated but did not narrow in so much in front. This sub-triangular form, the relatively elongated nasal region, and the gradual, but great, widening from before backwards of the cranial "barge," are very characteristic things in the skull of this species. Add to these the rudimentary superorbital plate in front at the narrow end of the skull, the wide temporal region, the heart shaped great fontanelle, and the small lesser fontanelles, and we get a number of things worth notice. Yet these are of *secondary morphological* importance, their value is largely *Zoological* and *Taxonomic*. In the nasal region we see that the outer angle of the "subnasal" cartilage (*s.n.l.*) has formed a retral lobe—a part very distinct in the adult *R. temporaria*. The pro-rhinals (*p.rh.*) are rather retral than out-turned; the upper and lower cartilaginous laminæ are curiously alternated as to their *wide* and *narrow* ends; the upper is narrow in front, and the lower wide, and *vice versâ*.

The bending of the palato-pterygoid "bow" is greater here than in the last, and the proper suspensorium is modified by the more backward position of the quadrate condyle (*q.c.*), by the fusion of the otic process with the "tegmen tympani" (fig. 1, between the letters *a.s.c.* and *h.s.c.*), and the greater perfectness of the articular facet of the pedicle (*pd.*).

The ex-occipitals (*e.o.*) are now large, and leave only a narrow oblong basioccipital space, and a wider triangular superoccipital tract of cartilage; they have risen over their own roof, and up the inside of the "epiotic" region, partly walling in the posterior canal (*p.s.c.*).

The prootics (*pr.o.*) are now typical, they enclose the great foramen ovale (V.) for its outer half below, and run round the front from thence, above; further out they have climbed up on to the auditory capsule, flanking the anterior canal on its outside, and covering the ampulla of the horizontal canal (*a.s.c.*, *h.s.c.*).

A pair of "sphenethmoidal" centres have appeared in the chondrocranium under the superorbital cave, and behind the ethmo-palatine bar (*e.pa.*); these are the symmetrical rudiments of the "girdle-bone;" they are "ecto-ethmoids" now, and are the proper side-wall bones protecting their own nerves (the 1st), just like the prootics *of the Frog*, the alisphenoids of Teleostei and higher types, the orbito-sphenoids of many types, and the ex-occipitals here and everywhere. They are separated by a tract equal to their own width below.*

---

* There is much variety in the formation of the *girdle-bone*, but these are its most essential parts; it may have, however, a median element below (as in *Pseudophryne Bibronii*), or an azygous plate above (*Rana temporaria*, *Rappia* (*Hyperolius*) *bicolor*). In *Dactylethra* ("Batrachian Skull," Part 2, Plate 59, fig. 1, *s.eth.*) this is a large **T**-*shaped* bone, it is free from the cartilage below it, which is not ossified so far forwards.

The outworks of this chondrocranium are being converted into bone at various points; the quadrato-jugal (*q.j.*) has grafted itself on to the quadrate (distal part of suspensorium), the long falcate palatine (*pa.*) is grafting itself on to the ethmo-palatine bar, and the pterygoid (*pg.*)—a long sigmoid bone with an internal snag—is doing the same to the "pterygo-quadrate" region.

In the Batrachia generally, the pedicle (*pd.*) is also more or less ossified by the proper *osteo-pterygoid;* but here, as in *Rana temporaria* and *R. pipiens*—that *old fish-bone*—the "metapterygoid" (*mt.pg.*) breaks out again; the possession of this centre is, perhaps, the only blot in the escutcheon of our native species,—in all other things the Batrachia may be said to "set their clocks" by that of *Rana temporaria*.

The mandible (fig. 3) has all its parts perfect or typical.

The columella is much more developed than, but is extremely similar to, the last; it has the distinct *distal* segmentation (Plate 5, fig. 5), but the distal piece is now contracted, like a "spatula," and has sent from its inner face a supra-stapedial band of cartilage upwards to coalesce with the under surface of the tegmen tympani (*e.st.*, *s.st.*).

Also the large solid mass of cartilage at the stapedial end of the bony medio-stapedial (*m.st.*) has become a separate "inter-stapedial" (*i.st.*), articulating with the stapes (*st.*).

The "annulus" (fig. 1, *a.ty.*) is a perfect ring of cartilage as in other North American kinds, and as in many of those from the Oriental region.

The "hyo-branchial" apparatus (fig. 4) is now quite normal; all the parts are fused together, are quite chondrified; and the "thyro-hyals" are now (typically) ossified; the hyoid bar (*st.h.*) is distinct from the ear-capsule.

The nasals and premaxillaries, and their contiguous labials (*n.*, *px.*, *u.l^a.u.l^b.*) are quite normal, so also are the maxillaries (*mx.*); but I can find no "septo-maxillaries"—small, variable, and inconstant bones; the quadrato-jugals (*q.j.*), as I have just mentioned, have united with the quadrate cartilage; and the squamosals (fig. 1) are truly elegant Batrachian bones, with a twisted sub-falcate upper, and a flat sigmoid lower, part.

The "vomers" (Plate 5, fig. 2, *v.*) have their characteristic spikes, fore, middle, and hinder; the two latter enclosing the inner nostril (*e.n.*, by mistake) largely; and a postero-internal, thick, dentigerous lobe.

The parasphenoid (fig. 2, *pa.s'.*, *pa.s.*) like that of *Rana pipiens* (Plates 3 and 8) shows the very rare condition of a distinct bony centre for the point of this little dagger.

These are the conditions of the growing skulls in this *genus*; afterwards I shall show that many of the dwarfed and more or less generalised types are arrested, in some things, at certain stages that correspond with growing stages of the typical kinds.

Skulls of adult individuals of the genus Rana

a. European species.

6. *Rana temporaria.*

I have just referred to the characters of this (typical) species, described and figured in my first paper (Phil. Trans., 1871, Plates 3–10, pp. 137–211). I shall use it as a *norma*, after eliminating its *one* aberrant character, viz.: its sub-distinct "metapterygoid."*

7. *Rana esculenta.*

For a description of the skull of this species the reader is referred to Professor HUXLEY's article on the "Amphibia" (Encyc. Brit., vol. ix., pp. 750–771).†

b. Oriental species.

8. *Rana gracilis.*—Adult male; 1¾ inch long. Ceylon.

The skull in this small species (Plate 6, figs. 6–10) differs but little from that of the Common Frog (*op. cit.*, Plate 9); its endocranium is more ossified, but its mandibular hinge does not reach so far back: the general form is very similar, being half a rather long ellipse. Between the oval occipital condyles there is a large right-angled space (figs. 6, 7), and the outline of the foramen magnum (*f.m.*) above is a large crescentic emargination.

The superoccipital soft tract is twice as wide as the basioccipital, but they are both of small extent. The whole occipito-otic mass is a large transversely-placed oval, the ends of which are nearly ossified above, but with a large tract of cartilage below, which runs across from side to side between the prootics and ex-occipitals (*pr.o., e.o.*). The passages for the main nerves (II., V., IX., X.) are large and very clearly seen in the lower view (fig. 7). The optic nerve passes through a large membranous fenestra, which has a narrow tract soft behind it, and a broad tract in front; half the interorbital region is unossified, but the foramen ovale (V.) lies in the centre of the large postorbital face of the prootic.

Above, from beyond the horizontal canal to the edge of the tegmen tympani, there is a soft tract; but from thence inwards to the lesser fontanelles all is bony, and the prootics and ex-occipitals are quite blended, walling in the canals and covering most of the hind archway of the skull. The fore half of the interorbital region is occupied by the common "sphenethmoid," or girdle-bone (*eth.*), which does not take up all the ethmoidal region; it projects in front above, but is emarginate below. The fore part of the ethmoid, all the nasal region, and the transverse ethmo-palatine bars

---

* In my figures of the nasal region of this species ("Batrachia," Part 2, Plate 54, figs. 1, 2) the 2nd upper labial is figured too high up, and the pro-rhinals are lettered *e.tr.*

† This species has the pro-rhinals distinct, as Professor HUXLEY's figure (p. 755, fig. 9) correctly shows. This distinctness is seen again in *Dactylethra* (Phil. Trans., 1876, Plate 59, fig. 6, *u.l'.*); it is there lettered *u.l'.* by mistake.

DEVELOPMENT OF THE SKULL IN THE BATRACHIA. 41

are unossified; the *floor* is wide in front and narrower behind, whilst the contrary is seen in the *roof*. The subnasal angles are well developed; the pro-rhinals (*p.rh.*) are large, elegantly pedate, and turned outwards and downwards: there is a distinct prenasal rostrum more than half their length. The cake-shaped inner, and the shell-like outer, upper labials (*u.l'.u.l².*) are well developed. The large palato-suspensorial arch is but little affected by the bones investing it,—the palatal and the pterygoid (*pa., py.*); but these are normal and well-developed: there is no metapterygoid, and the part above the quadrate hinge is largely ossified by the quadrato-jugal (*q., q.j.*). The hinge reaches to opposite the stapes; its stem is strongly clamped on the inside by the pterygoid, and on the outside by a well-developed squamosal (*sq.*), whose supra-temporal and postorbital regions are rather larger than in *R. temporaria*. Over its descending part is the "annulus" (*a.ty.*), which is also larger than in the type, and it is also completed into a ring above. The Eustachian opening (*eu.*) is large, and the stylo-hyal end of the hyoid bar (*st.h.*) is confluent with the lower part of the ear-capsule behind this passage.

The stapes and columella are large and well developed, but the solid inter-stapedial mass of cartilage is not segmented off from the medio-stapedial bar (fig. 10, *i.st.,m.st.,st.*); the extra-stapedial (*e.st.*) is not spatulate, but orbicular, and the strong supra-stapedial (*s.st.*) is confluent with the auditory roof. The mandible (fig. 8) has the dentary broad and ascending behind the mento-Meckelian; the hyo-branchial plate (fig. 9, *h.hy.*) has a broader hypo-hyal lobe than in the Common Frog.

The fronto-parietals (*f.p.*) are rounder and thinner, the parasphenoid relatively larger, but the præmaxillaries, maxillaries, and nasals agree very closely with those of the type; the quadrato-jugal, however, differs: it is largely grafted on the quadrate cartilage (figs. 6, 7, *f.p., n., pr., mx., pa.s., q.j., q.*).

This kind differs from *R. temporaria* in a few points, viz.:—

1. It has a prenasal rostrum.
2. No septo-maxillaries.
3. The supra-stapedial is confluent with the "tegmen."
4. The inter-stapedial is not distinct.
5. The stylo-hyal is confluent with the ear-capsule.
6. It has a crested dentary.
7. Quadrate partly ossified.

9. *Rana cyanophlyctis.*—Male; 1¾ inch long.* Ceylon.

This is another smaller kind of Indian Frog; it is, according to Dr. GÜNTHER (ibid.,

* Dr. GÜNTHER ("Reptiles of British India") gives the following measurements (from *snout* to *vent*) of the *largest specimens* known to him of the Indian species of *Rana* here to be described:—*a. R. Kuhlii*, 4½ inches; *b. R. hexadactyla*, 5¼ inches; *c. R. cyanophlyctis*, 1⅝ to 2¼ inches; *d. R. tigrina*, 6 to 7 inches; *e. R. gracilis*, 1¾ inch. *R. pygmæa* (see GÜNTHER, P. Z. S., 1875, p. 568) measures (adult female with ripe "ova") only 25 millims. long, or one inch.

p. 406), much like the large *R. hexadactyla*, but always remains small: its skull (Plate 10, figs. 7-10) does not suggest a specific identity with that type, of which it might be mistaken to be the young. A young individual of the large kind of this size would have a very similar skull, yet there is no reason to suppose that the two are identical; nevertheless, this is an *arrested* form.

Besides the external characters by which this kind differs from the last—*R. gracilis*—the skull itself shows a difference of character, although on the whole the variation seen is very much what would be found between an adult and a young of this somewhat larger species.

The occipito-otic regions (Plate 10, figs. 7, 8) are here less solid than in *R. gracilis* and much less ossified, for much of the horizontal, and nearly all the posterior canals (fig. 7, *pr.o.*, *e.o.*) are uncovered by bone, so that the four "centres" are distinct, even above. The cranial "barge" does not widen out so much towards the postorbital region, and the girdle-bone (*eth.*) is of less extent.

In front there is a "prenasal rostrum," but it is shorter, and the pro-rhinals are shorter and broader; among the outworks of the nose, viz.: the upper labials (*u.l¹.u.l².*) we find a small sigmoid "septo-maxillary" (*s.m.x.*). The other more constant, investing bones are very similar, but the parietal region is narrower, and the parasphenoid (*pa.s*) is less elegant. The squamosal (*sq.*) covers very little of the "tegmen tympani;" the "annulus" (*a.ty.*) is even larger, and a perfect ring; the quadrato-jugal scarcely affects the quadrate at all, and the hyoid cornu (*c.hy.*) keeps free, above. The stapes (fig. 10, *st.*) is less, and the "columella" (Plate 6, fig. 12, and Plate 10, fig. 10) is very inferior in development to that of *R. gracilis*, yet its inter-stapedial segment (*i.st.*) can be seen to be distinct from the medio-stapedial (*m.st.*). But the form of the main rod is more simple and undeveloped; the extra-stapedial (*e.st.*) is a mere rounded, soft end of the rod, and there is no "supra-stapedial" band.

The mandible is quite normal, but shows the crest on the dentary near the chin (Plate 6, fig. 11, *d.*).

The hyo-branchial apparatus (Plate 10, fig. 9) is also quite typical, it shows, however, some difference in this, that the lobes which grow out from the cerato-hyal and the base, are smaller.

There are some curious points of difference between the skull of this little Oriental Frog and that of a young *Rana temporaria* of the same size; the following are noticeable:—

1. A definite prenasal rostrum.
2. Smaller septo-maxillaries.
3. A more arrested squamosal.
4. A *crested* "dentary."
5. An arrested columella, without an ascending part, and with the distal part a mere bud.

Minute, but measurable, differences are to be found between species and species, everywhere in this outwardly similar, but inwardly most variable, "Order." The determination of these will, I am satisfied, help the Zoologist in his work of determining *species* and classifying the groups.

10. *Rana pygmæa*.--Adult male : ⅜ inch long. Anamallays Mountains, Malabar, S. W. India.

I shall, in turn, describe the skull of a true Frog as small as this from the Ethiopian, and of a sub-typical form from the Neotropical region : they will appear like the *young* of larger species; this is manifestly *old* and its skull is intensely ossified (Plate 5, figs. 11–15).

This is a very short skull; in *R. gracilis* and *R. cyanophlyctis* the width of the skull across the quadrato-jugals is exactly the same as the length. In this species the breadth is to the length as 4 to 3¼.

In *R. cyanophlyctis* the condyle of the quadrate ends opposite the exit of the 9th nerve ; in *R. gracilis* it ends opposite the same point ; in *R. pygmæa* it does not reach further back than the front third of the stapes—two-fifths of the distance between the 5th and 10th nerves, at their exit. The extremest case of wide gape is in *R. pipiens*, where this joint is far behind the occipital condyles (see Plate 8).

This is almost the only sign of *arrest* in the skull of this small, stout species,* but there are some things in it which have a look of *generalisation* about them, as if this were an old offshoot from the stock.

The occipital condyles (*oc.c.*) are oval, small, and very wide apart ; the emargination of the roof, over the foramen magnum (*f.m.*) is slight ; that of the floor is a more gentle arcuation of the transverse margin of the skull-floor.

The whole cranium, proper, is flat and wide, only gently narrowing in at the temples, and a little more in front of the eyes ; thus the ear-masses are wide apart, and with their drums outside and a little in front of them, have a very elegant appearance (Plate 5, fig. 11). These four regions are nearly of the same size.

The common bony occipito-auditory mass is cleft in its fore margin below (fig. 12), and bi-emarginate in its fore margin above (fig. 11). This widely extended postcranial region is capped with cartilage at three places on each side. These unossified tracts are—*a*, the occipital condyles (*oc.c.*) ; *b*, the edge of the tegmen tympani (fig. 11, *t.ty.*) ; and *c*, the facets for the "pedicle" below (fig. 12, *pd.*). The roofcartilage has evidently been absorbed considerably, between the temporal regions, thus the three fontanelles all run into one, which is an oval. with rounded lobes behind.

The alisphenoidal region is ossified half way to the optic opening (II.) ; the girdlebone (*eth.*) just reaches that passage in front ; thus there is a half belt of cartilage in

---

* I am of opinion that there are forms which, although highly specialised, are arrested as compared to their large nearest congeners; this must be due to *dwarfing*, and is to be distinguished from *archaic*, or truly generalised conditions.

the orbito-sphenoidal region, and this grows backwards as a sharp wedge, between the right and left bony tracts below (fig. 12).

The "girdle-bone" runs well into the true fore edge of the ethmoidal region, up to the autogenous nasal roof, and outwards into the "ethmoidal wings," where they pass into the *palatal pedicle* of the face (*e.pa.*). There is no "prenasal rostrum" formed by the intertrabecula in front, but the pro-rhinals (*p.rh.*) are good-sized hooks, and the upper and lower laminæ of cartilage are broad and well-developed. The outworks of the nose, the two pairs of upper labials (*u.l¹.u.l².*), are quite normal.

The *outer* less, and the *inner* more, bent arches of the face are very beautifully formed; they are strong, but not thick and solid. The palatine (*pa.*) and the pterygoid (*py.*) have affected the cartilaginous bar very little; the former is falcate, with an unusual curve; the latter is short, both in its process for the "pedicle" and that for the quadrate region.

The bones that form so perfect a semi-oval outline are quite normal also (*px., mx., q.j.*); the hindermost or quadrato-jugal has largely infected the quadrate with bony matter. The Eustachian tube (*eu.*) in the round notch of the pterygoid (*py.*) is a short oval with its greatest diameter parallel with the axis of the skull; it is generally oblique in position (see Plates 6 and 7).

Over the semi-osseous suspensorium the squamosal (*sq.*) binds by a very sigmoid, broadening stem: above, it is small on the tegmen tympani, but has a long, out-turned, postorbital snag. It is enclothed with a large and very broad "annulus," which, however, differs from the Oriental *Ranæ* generally, but agrees with the English kind, by being imperfect above; its crura are unusually wide apart.

The stapes (Plate 5, figs. 12 and 15 *st.*) is large, oblique, sub-uncinate behind, and has a sinuous front margin. It has the intercalary inter-stapedial (*i.st.*) very distinct, large, and ovoidal; the medio stapedial bone (*m.st.*) is pistol-shaped, with a very large handle; the extra-stapedial (*e.st.*) is small and orbicular, and its supra-stapedial process (*s.st.*) is free above, as in *R. temporaria*, but it is here a short rounded bud.

Still there is in this kind a very highly developed "middle ear," and it is not in any sense a *low* kind of Frog.

The nasals (*n.*) and the fronto-parietals (*f.p.*) are well-developed, but not very thick; the parasphenoid (*pa.s.*) is, like the skull it forms a floor to, broad and strong; it is trifurcate in front, sinuous laterally, and notched twice on each side behind: has a triangular "handle" at its end, and a wide rounded "guard"; it nearly reaches the foramen magnum, behind, but in front it reaches only half-way along the girdle-bone. The vomers (*v.*) are quite normal, but very wide apart.

The mandible (fig. 13) is perfectly normal, and *does not* show any dentary apophysis.

The hyo-branchial plate (fig. 14) is like that of the two last species, and differs but little from that of the common kind; the dorsal end of the hyoid band does not coalesce with the auditory floor behind the Eustachian tube (fig. 12, *eu., st.h.*).

As compared with that of *R. temporaria*, this skull differs very little in essentials.

Leaving out of question mere *size*, *form*, and *degree of ossification*, we have only the following things really worthy of notice, viz. :—
1. There is no division of the upper "fontanelle."
2. There are no "septo-maxillaries."
3. The "quadrato-jugal" largely ossifies the lower part of the suspensorium.
4. The "supra-stapedial" is very short.
5. The horns of the "annulus" are wide apart.

11. *Rana tigrina.*—Adult female ; 5¼ inches long.   Ceylon.

This was a large specimen, but not equal to some examined by Dr. GÜNTHER.[*] In coming next to this kind I have passed from the smallest to the largest of the Frogs of India.

As in the last, I have to be careful to distinguish those cranial characters that are due to *mere size* from those that lie deeper, and are of more impor.ance.

Of this I am certain, viz. : that although there is a special density and strength in the bony elements of the larger skulls, yet that is not necessarily connected with an extensive and generalised ossification, which, in some cases, shows most in the dwarfed kinds. The last species is an instance at hand, for its *endocranial landmarks* are much more obliterated than in this gigantic species.

In the largest specimens the skull cannot come far short in size of that of *Rana pipiens*; it is a *stronger* structure, and the cranium proper is much larger in proportion to its facial outworks (compare Plate 6, figs. 1–3, with Plate 8, figs. 1–4).

The two differ, however, in a much more important point, for this is a *severely typical* kind ; the American Bull-frog, on the other hand, is very archaic or generalised.

Although I have taken the medium-sized Common Frog as the best typical form, I do, nevertheless, consider that in *Rana tigrina* and in the *helmeted* Frogs (*e.g.*, *Ceratophrys* and *Calyptocephalus*) we come across the most perfect examples of Batrachian cranial architecture ; moreover, they are not deficient in respect of morphological characters that are deeper than, and almost independent of, mere size and bulk.

The general outline of this skull (Plate 6, figs. 1–3) is more than half of an oval, rapidly narrowing towards either end, as the outline of the Hen's egg does towards one end.

The whole form is broad behind, but narrowing rather rapidly forwards ; the length of the skull itself as compared to its greatest breadth is as 7½ to 9, but if the measurement be made from one quadrate condyle to the other, as 8 to 9, for these hinges lie some way behind the occipital hinges.

The ovoidal occipital condyles (*oc.c.*) are well seen in both the upper and lower aspects, for they are large and turned over the edge of the occipital floor, so as to fit deeply into the athantal concavities—a thing answering to the great strength of this Frog.

These condyles are separated by an emarginate tract less than their own width ;

[*] He gives 6 to 7 inches. mine was therefore only three-fourths the size of the largest.

this is greatly in contrast with what is seen in *R. pygmaea*, for it has, like many other small Batrachians, very little condyles, very wide apart.

The lower face of the occipito-auditory mass runs across at exactly a right angle to the axis of the skull, but when its whole extent is seen from above its outer angles are seen to be bent forwards.[a]

This lower face, from one "stylo-hyal" to the other, is a broad tract, one-third longer and one-third broader than the straight and almost evenly oblong interorbital region. The prootics and ex-occipitals (*pr.o.*, *e.o.*) of the same side are confluent, but there is a distinct synchondrosis both above and below (figs. 1, 2). A slight edge of cartilage remains to the floor of the tympanum, and a larger tract to its "tegmen" (fig. 3); nevertheless the ossification is intense, and reaches in front within a short distance of the optic foramen (II.). Also the girdle-bone is more than twice as extensive as in the medium type; it abuts upon the optic passage behind, and besides the whole ethmoid, with its outer wings, runs half way along the proper nasal roof, floor, and middle wall (figs. 1, 2, and 3).

A small superorbital projection can be seen over the inner canthus; these, and the wings of the ethmoid, up to the point where segmentation takes place in *Bufo*, are ossified.

As to the fore and middle parts of the endocranium, we see some curious results of overgrowth. First, the cranial "barge" appears to be very small and contracted as compared with the huge arches of the facial outworks; and, in the second place, the nasal labyrinth is greatly exposed by the retirement from it of the facial arches, in their wide sweep outwards. In the smaller kinds (see Plate 6, figs. 6 and 7) the nasal labyrinth is carefully packed between the laminae of the premaxillaries and maxillaries.

In the upper view (fig. 1) we see the nose lying naked, for the most part, between the bony boundaries; the outer angles of the floor (*e.tr.*) lying on the palatine lamina of the maxillaries, but not hidden by the upper plate. The ascending part of the outer angle of the floor (fig. 3, *a.w.*) is considerably ossified; it reaches the roof (*al.n.*) above, and coalesces with it. There is a "prenasal rostrum" (in front of *s.n.*) of moderate size, and large subretral pro-rhinals (*p.rh.*); the labials (*u.l*[1]., *u.l*[2].) also are well developed round the circular aperture (*e.n.*).

The cartilaginous palato-quadrate arch, on each side, is almost eaten away by the ectosteal palatine and pterygoid (*pa.*, *pg.*), the one thick and spatulate, the other a strong trimidate bone.[†]

---
* In the lower view (fig. 2) I have only figured the parts that come immediately under notice; the distant upper parts are left out, as they are not necessarily seen by the eye when discriminating the lower surface. The upper view must be compared with this to get a complete idea of the form.

† In this and in a large number of Batrachia these two bones are as truly ectosteal as the sheaths of the columella or of the thyro-hyal, and *become* truly homologous with the perichondral laminae of the prootic exoccipitals and sphenethmoid. I therefore colour them, throughout, as endoskeletal bones. In the "Shoulder-girdle" we see the same temporary distinctness of the perichondral, from the endosteal tracts of bone, in the "pro-coracoid" and "supra-scapula." (See my "Shoulder-girdle and Sternum." Plates 5-7.)

Above (fig. 1), and below (fig. 2), we can trace here and there some remains of the cartilage; above, in the ethmo-palatine, the pre-palatine, and the post-palatine, on the supero-external edge of the pterygoid; below, on the end of the "pedicle," its articular facet; and above and below the graft of the quadrato-jugal; the latter being the coating of the large bilobate hinge (*q.c.*).

These inner works of the skull, so largely converted into solid bone, are built over by very strong membrane-bones, which give the finish and the beauty to this style of cranial architecture.

Below, in relation to the skull proper, the parasphenoid (fig. 2, *pa.s.*) is of the usual relative size, but looks small as compared with the large bones of the face; it is typical in form, but unusually strong.

The vomers (fig. 2, *v.*) here attain a very large relative and real size, for they meet at the mid-line (compare them with those of *R. gracilis*, fig. 7, *v.*); their snags are large and strong, and the oblique dentigerous lobe of each is of considerable extent.

Above, the fronto-parietals (fig. 1, *f.p.*) form a strong roof with a notch in front, the remains of the frontal suture, but are wholly coalesced beyond this; they end, behind, in two broad wings which spread over the hinder region of the cranium, almost to the end. At first, hollow in the middle, in the postorbital region they develop a sagittal crest, which opens out into two temporal wings. The temporal part dips into the orbit (fig. 3, *f.p.*) and then rises over the ear-masses, moulding itself on to their sinuosities. The sides are notched, and the end has a concave margin.

The nasals in front (figs. 1 and 3, *n.*) form cross bars which are nearly as straight as the wings of the parasphenoid; they meet along the middle, run their pointed ends far forward, are triangular behind, where they overlap the fronto-parietals, and with them cover the girdle-bone—all but the superorbital projections and the middle part in front of the fontanelle. The fontanelles are presumably like those of the lesser kinds, but they are quite covered over.

The investing bones, outside, differ from the type in relative size and density; otherwise they are quite normal: there is no septo-maxillary, the palatal ingrowths are well developed, the nasal process of the premaxillaries *and of the maxillaries* (*pr.*, *m.x.*) is strong; the quadrato-jugal (*q.j.*) is grafted on to the quadrate, and the three regions of the squamosal (*sq.*) are extremely well developed and give the highest idea of the Batrachian form of these bones. So, also, the various condyles—of the occiput, the pedicle, and the mandible—are well and typically formed. The mandible (fig. 3) is crested both in its dentary and coronoid regions.

The stapes and its additions are of an average size (figs. 3 and 5); the stapes (*st.*) has its fore margin oblique; jammed in between it and the fenestral edge of the capsule there is the large seed-like inter-stapedial (*i.st.*); then, distinct from it, the pistol-shaped medio-stapedial (*m.st.*), whose unossified part passes into the broad spatula of the extra-stapedial (*e.st.*), which sends upwards its supra-stapedial (*s.st.*); this is ligulate, and confluent above. The stylo-hyal end of the hyoid tape is confluent, but the main part of it has been absorbed (fig. 4), otherwise, the basal plate is typical.

The "annulus-tympanicus" is large and a complete ring (fig. 1 *a.ty.*), the Eustachian openings (*eu.*) are large, oval, and oblique.

The greater part of the difference between this skull and that of the Common Frog depends upon the intensity of ossification and the large size of the investing bones—all correlated to the great size of this type. Other more important differences there are, viz.:—

1. A definite prenasal rostrum.
2. No "septo-maxillaries;" the nasal cartilage ossified in that region.
3. A crested dentary.
4. Much of the cerato-hyals absorbed.
5. Stylo-hyal end of these bars confluent, above.
6. Supra-stapedial confluent, above.
7. Quadrate partly ossified.
8. Rudiment of superorbital.

12. *Rana hexadactyla.*—Adult female; 5¼ inches long. Ceylon.

This specimen was a little less than that of *R. tigrina*, and the skull (Plate 7, figs. 1–5) is slightly longer, but its greatest breadth is only as 8 to 8¾; the quadrate condyles only project one millimetre, or about half a "line" beyond the occipital condyles; in *R. tigrina* they project twice as far backward.

In *outline*, besides its narrower form, the nasal end is broadly truncate, and these two modifications cause a third, viz.: the unusual straightness of the sides of the skull.

It is altogether a feebler skull, with less intense ossification, as well as being straighter and narrower; moreover, it is not quite symmetrical; yet its facial plates (fig. 3, *mx.*, *sq.*) are deep and well developed.

But these are mere superficial differences between the two; there are others that lie deeper down. Gentle enough are these morphological variations, and no bar to the supposition that the two species originally sprung from one common stock, yet they are not to be lightly passed over, for they are full of interest.

The occipital condyles (*oc.c.*) are rather larger and more under the skull than in the last; also the emargination between them is deeper, these are due to some differences in the working of this hinge.

The whole occipito-auditory mass is less oblong, its front faces being oblique as they pass into the interorbital region, exposing the foramina ovalia (V.), and the occipital condyles and epiotic eminences (*ep.*) over the posterior canals project more backwards: moreover, the bones of the two sides are confluent over the foramen magnum (*f.m*).

Yet there is more cartilage between the 2nd and 5th nerves (II., V.), and also a more evident "pterotic" eminence projecting behind the squamosal, at the end of the tegmen tympani (fig. 1, *sq.*).

The form of the canals within is less obscured by the bony growths and coverings than in the last, so that the general surface of this part is more uneven.

The girdle-bone (*eth.*) reaches from the optic fenestra to the middle of the septum nasi (*s.n.*); it is therefore less in anterior extent, but it is visible between the vomers (fig. 2, *v.*), in front of the descending bar of the nasal (figs. 1 and 3, *n.*), and also behind that bar, ossifying the wing of the ethmoid up to the ethmo-palatine suspensorium.

The whole interorbital part of the cranium is narrower and more pinched in the middle, and the relative narrowness of this trough approximates to what I shall show in my next instance but one, viz.: in *Rana pipiens* (Plate 8).

Here, again, a small superorbital projection (*s.ob.*) is seen outside the meeting of the nasals and frontals, as in the last.

In front, the prenasal is a mere bud, little more pronounced than in *R. temporaria*; the pro-rhinals (*p.rh.*) are much like those of the last kind. The outer angle of the subnasal lamina (fig. 1) is more extended and quite exposed, and this (*trabecular*) plate also projects outwards further back, where the girdle-bone ends; it joins the roof above by an ascending plate (*n.w.*), which is ossified for some extent at its root.

The labials are very similar in both species; the upper is hidden behind the premaxillary.

In *R. tigrina* (Plate 6, fig. 2, *pa.*) the palatines turn forwards and outwards; in this kind they run straight across; they also come nearer together; but on the whole these bones and the pterygoids (*pg.*) are very similar in both species : there is, however, more cartilage left unossified in this. The gliding joints of the pedicles (fig. 2, *pd.*) are nearer together in this narrower skull, and the Eustachian opening (*eu.*) is thrown obliquely backwards instead of forwards, outside, by the straighter hind process of the pterygoid.

The hinge of the quadrate (*q.c.*) is less deep; its substance is more ossified by the quadrato-jugal (*q.j.*).

Altogether the parasphenoid (*pa.s.*) is slenderer and more elegant; it is equally subcarinate, and has the *blade* and *handle* longer in proportion to the *guard* than in *R. tigrina*: here we see the most perfect form of this bone in the Batrachia, the basitemporal processes being slender at first and then dilating outwards.

The vomers (Plate 7, fig. 2, *v.*) are not so large as in the last, do not come so close together, and their inner edge is sinuous, not arcuate.

The fronto-parietals (fig. 1, *f.p.*) are less dense, are distinct in their fore half, and the temporal fossae are bounded, above, each by its own parietal ridge; the hinder spreading part is altogether less.

The nasals (*n.*) are larger, broader before and behind, and their facial process (fig. 3) forms a more perfect suture with the maxillary.

The nasal processes of the premaxillaries (fig. 1, *p.x.*) are less, but the palatal (fig. 2) are larger, and the *right* bone is much the larger of the two where it joins the palatal part of the maxillary; there is a small septo-maxillary also on the right side (fig. 1, *s.mx.*).

The maxillaries and quadrato-jugals (*mx.*, *q.j.*) are much alike in both species; the squamosal (*sq.*) ends abruptly over the auditory mass, and not as a triangular process (see also Plate 6, fig. 1); its postorbital process is deeper; the mandible (fig. 3) has a crest to the dentary, and the coronoid process of the articulare is high.

The "annulus" (*a.ty.*) is large and complete. The parts of the "middle ear" are similar to those of the last, except that the inter-stapedial (fig. 5, *i.st.*) is larger and is well ossified, the medio-stapedial (*m.st.*) straighter, and the extra-stapedial (*e.st.*) broad and orbicular; the supra-stapedial (*s.st.*) is coalesced above.

The stylo-hyal ends of the hyoid bands are confluent above; the whole bar is perfect and normal on the left side, but on the *right* it is absorbed from the hypo-hyal nearly to the stylo-hyal regions (Plate 7, fig. 4, *c.hy.*, *h.hy.*).

The differences to be seen between this and the last are largely due to diminished size and strength; these variations, as compared to what is seen in the lesser types, are largely due to their greater bulk. But there is a residuum of variations that cannot fairly be put down to these causes. There are in this species :—

1. A small but definite prenasal rostrum.
2. A small septo-maxillary on the *right side* only, but the nasal angle in its ascent has a solid bony mass formed in it.
3. A crested dentary.
4. The right cerato-hyal absorbed.
5. Stylo-hyal confluent, above.
6. Supra-stapedials confluent, above.
7. Inter-stapedial well ossified.
8. Quadrate partly ossified.
9. Rudiment of a superorbital plate.

13. *Rana Kuhli.*—Male; 2½ inches long; two-thirds grown. Ceylon.

This large, but not full grown, Frog belongs to the most aberrant species of its genus, and its peculiarities are the stronger expression of what is more feebly seen to particularise the other Oriental kinds of *Rana*.

We shall see in this type how near the characters of a species of this genus may come to those of types which lie at the furthest distance from the model-form ;— borderers and mixed breeds, so to speak, that help to break the isolation of the main group, and to show its affinities to the groups that encompass it round about.

As to general form, this skull (Plate 7, figs. 6–10) is half a moderately long ellipse, and has very neat outline.

Its regions, fore, middle, and hinder, are in contrast with what is seen in the two Bull-frogs just described and also in the lesser species, for the middle region is long, relatively, beyond that of any kind I know, each orbital region being large enough to ensocket a pair of eye-balls twenty times the size of those that do lodge in them. This middle region shows a cranial trough, not high (fig. 8) but narrow, and with an approach to the outline of an hour-glass.

The length is a fraction less than the greatest breadth, and the quadrate condyles (*q.c.*) just reach to a supposed line running across the double hole for the 9th and 10th nerves (IX., X.).

Thus the gape is not that of a very characteristic Frog, but rather that of an arrested or a somewhat generalised type. Laterally seen (fig. 8), the skull is arched, having strong deep curved planks of bone built round and over it; this is a correlate of its powerful mandible with its *mimetic canine tooth* (*d.*). Everything in the side view speaks of strong pterygoid and temporal muscles.

The outer bones are unusually strong, so are the *intermediate* pterygoids (*pg.*), but the ossification of the endocranium is exactly like that of an adult Common Frog, and very inferior in degree to what is seen in the two large kinds, and in the dwarf species (*R. pygmaea*, Plate 5, figs. 11, 12).

The occipital condyles (*oc.c.*) are large, low in position, near together, and with the short interspace straight. There is a considerable basi- and supraoccipital tract unossified, and the prootics and ex-occipitals (*pr.o., e.o.*) are divided below, and only slightly confluent above.

There are the three normal fontanelles above, and the side walls (fig. 8) are well built up to the roof, under the edges of which there is on each side a very definite "wall-plate." The arched form of the skull is combined with considerable overlapping of the roof and floor (figs. 7, 8, *f.p., pr.s.*), the investing bones being applied above and below very closely round the endocranium (*eth., o.s.*) so as to leave the interorbital wall uncovered to an unusually small extent. The short and not very broad nasal region lies well within the outer bones; there is no "rostrum," but the pro-rhinals (*p.rh.*) are well developed and unusually long and projecting; the labials (*a.l¹.a.l².*) are normal. The "girdle-bone" (*eth.*) reaches only to the front of its own region and half-way back to the optic nerves; it does not ossify the very definite angular supraorbital projection (*s.ob.*), which here attains to a distinctness almost equal to what it has in the "Hylidae."

The more rounded form of the great orbital space in front, enclosed there by the ethmo-palatine bar, gives rise to a sickle-shaped palatine bone (*pa.*), and this is followed at the outer edge by a very remarkable pterygoid (*pg.*). The processes of this bone that enwrap the pedicle (*pd.*), and bind upon the inside of the suspensorium to the hinge (*q.c.*), are short but normal, but the *intra-jugal* portion of the bone is of great depth (fig. 8, *pg.*), and strongly inbent.

In the "axil" of the pterygoid the Eustachian passage is small and round; the stylo-hyal end of the hyoid (*st.h.*) has coalesced with the tympanic floor, and bends strongly back to keep to the margin of this passage as the hinder half of its natural skeleton.

The quadrate region of the suspensorium (*q.*) is not ossified by the "quadrato-jugal" (*q.j.*)—shaped like a Serpent's tooth; the saddle-shaped condyle (*q.c.*) is very large, but does not reach to the extreme angle of the suspensorium (figs. 7 and 8).

The "annulus," like the Eustachian passage, is small, the band itself is wide (fig. 6, *a.ty.*), and it is a perfect ring; the whole tympano-Eustachian cavity is small, but the columella and stapes (fig. 10) are of the average size; the latter (*st.*) is almost a long triangle, but the posterior and inferior edges are rounded: the anterior margin is concave, fitting into the inter-stapedial.

This latter part (*i.st.*) has a deep saddle-shaped condyle for the stapes; it is unossified. The medio-stapedial (*m.st.*) is nearly all ossified; it is gently arched, and not much expanded where it articulates with the inter-stapedial.

The extra-stapedial (*e.st.*) is very large, elegantly heart-shaped, *peltate* in arrangement—its cartilaginous "handle" passing from the medio-stapedial bone into the inner face of the main plate obliquely; and it covers half the concave, wide "annulus."

There is no supra-stapedial band—a part which is seen even in the peltate extra-stapedial of *Bufo vulgaris* (Phil. Trans., 1876, Plate 54, figs. 7 and 8); but here we have, suddenly, as it were, the columella of *Pipa* and *Dactylethra* over again, with the difference of a short and soft inter-stapedial (Phil. Trans., 1876, Plates 59 and 62). The extra-stapedial part is mostly orbicular in the Oriental *Ranæ*, but not large.

The mandible (fig. 8) is normal, with the exception of the fore part of the dentary (*d.*); the *crest* which grows up from that bone in front in other species is here a high strong spur—an imitation, in bone, of a large *canine tooth*: in old age (according to Dr. GÜNTHER) it becomes capped with tooth-substance. It "cuts the gum" early, and soon dents the side of the palate, which becomes hollow to receive it. Moreover, the bones all round this excavated part are greatly modified; this is partially seen *on one side* in *R. hexadactyla* (fig. 2), with its much smaller dentary crest.[*]

Where the palatine plates of the premaxillaries and maxillaries meet (fig. 7, *px.*, *mx.*), their processes project far inwards; this is to leave room for the socket of the *quasi-canine*; the hole is bordered on the outside by the dentary edges of these bones at their junction.

Not only so. Where they meet on the outside (fig. 8), there they rise high at their junction, and leave an angular space or gap in the toothed margin of the fore face. There are no septo-maxillaries; both the quadrato-jugal (*q.j.*) finishing the cheek, and the squamosal (*sq.*), in the temporal region, are very strong and elegant bones, but the latter is very remarkable, even for a Frog. The bone seen from above (fig. 6, *sq.*) is a sickle, its rough notched handle lying on the tegmen tympani; running obliquely

---

[*] Dr. GÜNTHER looks upon this *quasi-canine* as a special thing—a tendency, so to speak, to produce a tooth, here; my own mind leans to the opinion that it is not a *rudiment*, but a *remnant*; nearly all the Indian kinds show it more or less, and their *Common Parent* may have had large genuine teeth in the front.

into it, we see the postorbital process which is the curved blade; together, they are *half as long* as the whole skull (see fig. 8). Also the angle at which the stem runs down over the massive quadrate and its condyle is very remarkable (compare figs. 3 and 8, *sp.*), for the axis of the postorbital process is almost coincident with it, and between the stem and the supra-temporal plate the space is only half a right angle.

The roof and floor have been wrought over the inner skull, above and below, so as to half hide it; thus the parasphenoid and the fronto-parietals are bony troughs or hollow splints. The latter (*f.p.*) are quite distinct, right and left, very long, narrow, narrowest in the middle, not very wide in the temporal region, where their edge is emarginate, curving with the canal below; whilst in front they are overlapped by the nasals (*n.*) and hide the girdle-bone: outside they do not hide the superorbital cartilages (*s.ob*). The nasals form a pair of large wings and are highly arched and strong (fig. 8); the outer part is an elegant facial hook, binding on the upper edge of the maxillary; the two bones meet all along, forming a nasal suture.

The parasphenoid (figs. 7 and 8, *pa.s*) is a very long trough of bone of the usual shape, with well-formed basi-temporal wings that are trilobate.

The hyo-branchial plate (fig. 9) is quite normal in form; there is a slight fissure in the substance of the "hypo-hyal" lobe (*h.hy.*); a division of the hyoid cornua into an outer and an inner tape not uncommon in the Batrachia: the outer is a *remnant* of the pectinate inter-branchial cartilage such as we see in the Chimæroids and their kindred.

Compared with the type, *Rana Kuhli* shows the following divergences of character:—
1. No septo-maxillaries.
2. Dentary with a very large tooth-shaped process.
3. Articulation of pre-maxillaries and maxillaries very wide and high, to admit of the process of lower jaw.
4. Stylo-hyal confluent, above.
5. Extra-stapedial a very large cordate disk, peltate, and without supra-stapedial.
6. Superorbital cartilaginous, and very distinct.
7. Eustachian openings very small.

I have already mentioned the peculiar form of the skull, and of certain individual bones, but the above are the most important morphological variations. This skull is altogether very instructive and suggestive; *R. Kuhli* is certainly a "borderer," and I suspect it has retained some very archaic characters that have been more or altogether obliterated in the other species of its genus and its territory.

c. North American ("Nearctic") species.

2 (continued).* *Rana pipiens*, Harl.—(*R. catesbiana*, Shaw; *R. mugiens*, Merr.).

This specimen was considerably larger than those of the Indian Bull-frogs just described.

* See p. 22 where the larva of this species is described.

I have already shown the characters of two kinds from this "region" (*R. palustris* and *R. halecina*), but they were young; yet some unexpected and very important points were elicited, and will be referred to in the description of this large and most characteristic kind. In this species we see those peculiarities of the Batrachian skull which set it by itself, making it to differ from other kinds of skulls, carried to an extreme degree of development.

These things bespeak a highly specialised type, and indeed this American Bull-frog is a *Frog* of the Frogs; his shape, dress, voice, and carriage, all combine to make him the representative of his group or "Order."

On the other hand, like all giants, there is much in him that bespeaks an ancientness (as if he only remained of the remnant of the giants), and was somewhat out of place among the more proud and elegant dwellers in the marshes and miry places of this, the newest, epoch.

In outline this skull (Plate 8) is half an ellipse and very regular; its greatest breadth is, as to its axial length, as $7\frac{1}{2}$ to $5\frac{3}{4}$. If the *length* were measured up to a line passing from one quadrate condyle to the other, then the skull would be *one-tenth* longer than when measured up to the convexity of the occipital condyles. In my specimen the former reach *one-fifth* of an inch further back than the latter.

In *R. tigrina* (Plate 6, figs. 1, 2) this distance is only *one-sixth* of an inch; *relatively*, however, the distance is the same, for the Indian skull was smaller; in *R. hexadactyla* (Plate 7, figs. 1, 2) the distance is only half as much (*one line*).

This extreme extension of the gape backwards, during metamorphosis, is in remarkable contrast with what I find in my oldest specimen of the adult of a small toothless Australian Batrachian, viz.: *Pseudophryne Bibronii*. In this kind, in which the *female* is one inch long, and the *male* three-quarters of an inch, I find that the quadrate condyle reaches very little more than half the relative distance attained to in *R. pipiens* and *R. tigrina*, or only *two-thirds* as far back as the occipito-atlantal hinge, measuring from the front of the snout, backwards. In that kind the whole suspensorium is arrested in its backward movement, when it forms a right angle with the axis of the skull: this is very similar to what is seen in the skulls of "Caducibranchiate Urodeles," whose gape is so much less than in the "Anura."*

The next thing that strikes the eye, after the great extent of the gape, is the very small size of the cranial "barge" as compared with the huge facial outworks; its average width is a quarter of an inch, scarcely more than that of the Common Toad, with a head little more than half the length of this Bull-frog.

The occipital condyles are large, near together; more shown below than above, and with a gentle emargination between them. The whole occipito-auditory region, right and left, is marked by great hills and hollows, and jutting snags, very unlike what is

---

* It is a general rule in this "Order" that the gape is relatively, as well as really, larger, the larger the species becomes; and in very dwarfed kinds the contracted gape is a correlate of several other arrests in the development of the parts of the skull.

seen in the skull of small kinds, and bringing into prominence parts that also project largely in osseous Fishes. The epiotic region burrowed by the posterior canal (*ep.*) stands out, divergingly, on each side of the great archway, with its transversely oval entrance (fig. 4). The anterior canal sweeps round the front; outside it there is a great hollow, and then the periotic mass rises over the horizontal canal, this it does still more at the edge (*tegmen tympani*). The outer margin grows backwards into a large torete unossified "pterotic" ridge (*pt.o.*)—the part that is ossified by the *pterotic bone* in Teleostei. The whole structure is, on the whole, bony up to the fore margin of the foramen ovale, and these right and left masses are confluent over the foramen magnum, but not below (fig. 2); there is a small tract of cartilage in the basioccipital region.

Below, the twin post-aural nerve passages (fig. 2, IX., X.) are wide apart, and the antero-external margin of the ear-capsule is bevelled away and covered with a plate of cartilage for articulation with the pedicle (*pd.*). The ear-masses and intervening hind skull together make only a third of the whole width at this part; above (fig. 1) the *pterotic crests* (*pt.o.*) stand further out, where they pass beneath the squamosal (*sq.*).

Suddenly in the temporal region the skull is compressed to two-thirds its average width, then becomes of a fuller form, and gently narrows again before it spreads out into the wings of the ethmoid, opposite the closing in of the cranial trough.

The optic fenestra (II.) is moderate; it is margined by cartilage in front, above, and below, and the girdle-bone (*eth.*) has not much more interorbital space than this tract, which lies in front obliquely over the bone. There is a small ossified superorbital lobe (*s.ob.*). The cruciform bony mass only reaches in front to the proper morphological edge of the ethmoidal territory; all the true *nasal region* is unossified. The whole nasal roof is a winged sub-pentagonal tract, not ending in more than a bud of the "prenasal," and with half its septal part and a headland of its roof uncovered by the nasals (fig. 1, *s.n., n.r., n.*). The outer nostrils (figs. 1, 3, *e.n.*) are protected by the usual inner and outer upper labials (*u.l¹.u.l².*). The *roof* runs forward, narrowing; the *floor* (fig. 2, *s.n.*) is sub-quadrate, and is finished antero-externally by the curious cervicorn angles (Plate 8, fig. 2, and Plate 9, fig. 7, *s.n.l.*). The septum (*s.n.*) below is very thick in front, and narrows backwards towards the ethmoid bone; the angles of the subnasal lamina behind pass into the large, widely-extended ethmo-palatines which are curiously covered with splints, above and below.

On each side of the bud-shaped end of the septum there is a rather large fenestra (Plate 9, fig. 7, *s.n.*), and in this enlarged figure of part of this region we see the prorhinal (*p.rh.*); it is very large, sub-flabelliform, and turned (contrary to rule) inwards. In this same figure the angle of the nasal capsule, a part derived from the trabeculæ, is seen to give off a curious retral process (*al.n., s.n.l.*): this is quite normal.

The whole palato-quadrate arch (*pa., pg.*) is an immense structure; measured from the pre-palatine spike to the end of the condyle, it equals the entire length of the skull, measured from the snout to the occipital condyles, yet in front of the pre-palatine we have the fore part of the maxillary and the whole of the pre-maxillary.

All at once in this part we come across a number of bones that have no representatives in the Batrachia, generally, yet some of them correspond to familiar bones in the Teleostei. I shall name these latter in accordance with my description of the "Salmon's Skull" (Phil. Trans., 1873, Plates 6–8).

As compared with what is seen in the typical kind, the palato-quadrate arch is scarcely more ossified, but it is relatively, as well as really, very much larger. Instead of a single palatine bone running across under the right-angled ethmo-palatine cartilages, there are several separate pieces.*

On the left side (fig. 2, *pa.*) the large sub-falcate ridged palatine is single up to the handle or inner end; there, however, three at least small sesamoid centres are seen. On the right side the handle is entire, but the blade is composed of two unequal additional bones (*pa'*.); these are reversed in the figure.

The huge sigmoid pterygoid bone (*pg.*) is single from its fore end near the outer part of the palatine to the end of the quadrate condyle (*q.c.*); but the thick remnant of the dorsal end of the suspensorium—the "pedicle"—is invested with, and more or less ossified by, a group of bones. Two of these, on each side, are the familiar "metapterygoid" and "mesopterygoid" (Plate 8, figs. 2–4, *mt.pg.*, *ms.pg.*) of osseous Fishes.

The metapterygoid (Plate 8, fig. 2, *mt.pg.*) ossifies most of the pedicle (fig. 1, *pd.*), leaving the thick ovoidal condyle soft; but where it is passing inwards, at a right angle to the pterygoid bone, there in front of its outer end is a bone one-third its size: this is the mesopterygoid (*ms.pg.*).

The large out-turned, almost vertical quadrate region is largely fenestrate (Plate 8, figs. 1–4, and Plate 9, figs. 12–14); this fenestra is bounded above by a bridge of cartilage, which passes to the "otic process" (cut through in these larger figures of the details).

That bridge is largely ossified by an additional centre (*pd'*.) or supernumerary "metapterygoid." The fenestra is mostly occupied by three long splints—"intersuspensorial" membrane bones; one is sub-falcate and twice as long as the others; their form is sub-oval (Plate 9, figs. 12–14, *i.sp.*). The hinder part of the proper pterygoid is itself fenestrate on its inner face (Plate 9, fig. 13, *pg.*, *sp.*).

On the outside this arch is well invested by the three normal subcutaneous bones, viz.; the squamosal, quadrato-jugal, and maxillary. The first of these (*sq.*) is remarkable for the forwardly-bowed form and oblique position of its stem (Plate 8, figs. 1

---

* This specimen, the gift of Dr. Murie, came to me in a perfectly uninjured state; it had not been caught (and possibly bruised) in capture, but it had lived in the gardens of the Zoological Society for some time before it died. I often find specimens of Batrachia with some of these bones fractured in capture, but I have learned how to distinguish them from supernumerary bony centres formed naturally in the animal. There are several kinds that show additional bones, but none at all comparable to this species.

and 3, *sq.*), for the descending position of the post-orbital process, and for the bifurcation of the supratemporal plate, embracing the "pterotic" cartilage.

The quadrato-jugal (*q.j.*) grafts itself very little; the maxillary (*m.x.*) has a very distinct ascending facial process below the nasal; it does not, however, articulate directly with that bone, for there is a small squarish *pre-orbital* (*p.ob.*) between them.

Behind this (Plate 8, figs. 1 and 3, *l.*) there is an oblong bone, also attached to the descending crus of the nasal (*n.*); this is the "anterior suborbital" or lacrymal; this is a familiar bone in the Teleostei, and the other is common in one "Family" of that "Order," namely, the "Siluroidei."

But these are not all the *præter-normal* ossicles; there is no functional septomaxillary, as in *R. temporaria*, where it lines the nasal passage, but there are a number of generalised bony points close to where that bone appears, below, in the Common Frog. On the left side (Plate 8, fig. 2, and Plate 9, fig. 7, *p.mx.*) these are close to the *palatal junction* of the pre-maxillaries and maxillaries, some larger, and others lesser, "palato-maxillaries;" on the right side (Plate 8, fig. 2) there are two narrow bones, larger than the largest on the left side.

These do not make up the whole tale of the additional bones; the parasphenoid (Plate 8, fig. 2, *pa.s.*) has at its apex a separate "pro-parasphenoid (*pa.s'.*) as I showed in the larva of this species (Plate 3). This bone is about a quarter the size of the blade of the large bone, which is split where it underlies the new centre.

The vomers (fig. 2, *v.*) are large and quadrilobate, yet they neither meet at the midline nor reach near to the maxillaries; for the nasal region, although large really is small relatively to the huge face. The fourth or dentigerous lobe of each vomer is rounded, the two intermediate lobes that enclose the inner nostril are sharp, and the front lobe is crescentic.

Above (figs. 1 and 3, *n.*) the nasals are broad, roughly convex, in contact in their hinder part, do not reach the maxillaries externally, and together form a sub-pentagonal roof. Between them, behind, there is an emargination which passes into the notch between the fronto-parietals (*f.p.*) and exposes some of the girdle-bone (*eth.*), whose superorbital lobe (*s.ob.*) is also uncovered externally.

The frontals are not confluent with each other, but with the parietals; these are completely anchylosed together (fig. 1, *f.p.*).

The sides of this roof are almost parallel for two-thirds of the interorbital region; then the bones suddenly widen, then become pinched and bi-cristate, and then double their breadth over the inter-auditory region, where their outer margin is emarginate, and their surface sinuous over the double "canal." The two crests are divided by a fossa; they are first *temporal*, and then *sagittal*.

This compound roof ends, behind, in a transverse, dentate, squamous edge, leaving the narrow superoccipital region uncovered. Laterally (fig. 3), the bony roof and the bony floor are seen to be modelled over and around the endocranium, hiding half of it.

The middle and outer ears (Plate 8, figs. 1-4, and Plate 9, figs. 8-14) are in a high state of development; externally, the annulus (*a.ty.*) is almost an inch across, and is complete; its greatest width, below, is fully a quarter of an inch; it is thick-rimmed, unusually wide at the junction of its "horns," and altogether unique in size and finish.

The "columella" also is unique as to its development, having an osseous shaft in its extra-stapedial (Plate 9, figs. 8, 9, 10, *e.st.*, *e.st'.*), which is not seen in other Batrachia. The stapes and the dorsal part of the columella lie in a deep hollow, in the hinder part of which is the fenestra ovalis (Plate 9, fig. 8). The stapes (*st.*) is oval, obliquely truncate in front, and thick and convex externally.

Against it, and somewhat within, lies the thick, short inter-stapedial (*i.st.*), whose distal half is bony; it is quite segmented from the medio-stapedial (*m.st.*) which is long, phalangiform, thick proximally where it is not quite ossified, and narrow and arcuate further forwards. Then comes a short cartilaginous tract which is followed by the extra-stapedial (*e.st.*) whose proximal part is ossified as a styloform shaft bone (*e.st'.*), the cartilage within being also considerably ossified beyond the ectosteal tract.

On the inner side, near the dilated end the supra-stapedial (*s.st.*) is given off at a sharp angle, it passes upwards and backwards and is confluent with the cartilage lining the tegmen tympani; it is nearly as broad as the main bar, and is entirely unossified.

This division of the columella into *three* segments is of great interest, and can be understood only by comparison with the state of things seen in the upper hyoid region of Fishes in various groups.

In the Chimæroids (see HUBRECHT, fig. 2, *hy''.*) there is a "pharyngo-hyal" above the "epi-hyal" (*hy'.*), but this topmost piece, so constant in the branchial arches, proper, is not found in the hyoid of other Fishes. Much lower down, towards its distal end, the epi-hyal ("hyomandibular") is cut off as a short distinct cartilage in the Sturgeon and Paddle-Fish.

In osseous Fishes this cartilage acquires an additional bony sheath—the "symplectic" —but is not cut off.

I am now perfectly satisfied that this lower subdivision of the "epi-hyal" element is a *secondary* segmentation, such as is seen in the "pharyngo-branchials" of the Sturgeon, where each cartilage is sub-divided into two pieces.[*]

In illustration of this we see that the "hypo-hyal" of osseous Fishes has *two* bony centres; in the Menopoma it breaks up into *three* patches of cartilage.

Therefore we see that the middle ear of this Frog has parts that correspond to the following morphological elements of the upper part of the hyoid arch in the Fishes:—

1. Pharyngo-hyal = inter-stapedial.
2. Epi-hyal, subdivided into— *a*, hyo-mandibular = medio-stapedial; and *b*, symplectic = extra-stapedial.

---

[*] Mr. HOWES pointed out this to me, as shown in his excellent dissections at South Kensington.

The supra-stapedial is a secondary process of the distal part, and very inconstant; the segmentation of the extra-stapedial from the medio-stapedial is complete in *R. palustris* and *R. halecina* (Plate 5), and I doubt not is the same for a time in *R. pipiens*.[*]

The mandible in this species is chiefly remarkable for its great length (Plate 8, fig. 3). All its parts are normal; the coronoid process of the "articulare" is well developed.

The hyo-branchial plate (Plate 8, fig. 5) is also quite normal; it differs very slightly indeed from that of the Common Frog, and as in that kind the stylo-hyal end of the hyoid *articulates* with the floor of the tympanic cavity.

As compared with the skull of the "norma" we have many differences that are due to their size, such as the distance of the facial bars, laterally, from the nasal labyrinth, the small size of the cranial trough, and the huge suborbital spaces.

The endocranium is not so much hardened as might be expected in so large a kind; the bony masses behind are still distinct below, but not in the superoccipital region.

The investing bones generally are quite normal; the *roof* in the parietal region has its two sides anchylosed and crested: this is simply due to the extent and strength of the temporal muscles.

The cartilaginous foundations of the great palato-suspensorial arches is quite normal, only they are very large, relatively.

There are two groups of modified parts, namely, the various bony "centres" and the elements of the middle ear. The modifications may be classified as follows:—

1. Groups of small ossicles in a generalised state below the junction of the premaxillaries and maxillaries, instead of a functional septo-maxillary.

2. Ectosteal bones applied to the "palato-suspensorial arch" numerous, and in the palatine region unsymmetrical; several of those in the pterygoid region being referable to normal *ichthyic* centres.

3. Further additions of bony centres in the pedicle and quadrate, both *membrane-bones* and *endosteal* patches.

4. The quadrate region of the suspensorium fenestrate, below the division into "pedicle" and "otic process."

5. A *sub-* and a *pre-*orbital bone on each side.

6. Parietal region anchylosed and crested.

7. A distinct *pro-*parasphenoid.

8. Inter-stapedial semi-osseous.

9. Extra-stapedial half osseous, with distinct, but apposed, ectosteal and endosteal tracts.

10. Supra-stapedial confluent above.

---

[*] This *Acipenserine* subdivision of the epi-hyal element is to be seen also in the embryo of *Chelone viridis*.—"On the Skull of *Chelone viridis*," 'Challenger Memoirs.' ("Zoology," vol. i., part 5, plate 19, fig 8, 1880.)

d. Ethiopian species.

14. *Rana.—-—?* sp.—Adult female, 1 inch long. Lagos, W. Africa.*

The largest of the specimens examined by me was one inch long, and was a female with developed ovaries. Dr. GÜNTHER was doubtful as to the exact species of *Rana* to which it belonged; it was possibly immature, but evidently belonged to a small and very slender kind, with proportionally extremely long hind legs : its tongue was deeply bilobate, and one lobe was much shorter than the other in this specimen.†

My dissections show this kind to have a skull closely corresponding to that of a Common Frog of the first summer, or about four months after metamorphosis ; it is instructive any way, whether it be *young* or *arrested*, for it shows how close this Ethiopian species comes to our native kind, in all essentials of cranial structure. That it was a mature individual I have no doubt, and it came in the same bottle with adult specimens of a species of *Rappia*, the largest of which was only five-sixths of an inch long. The general form of the skull (Plate 13, figs. 7, 8) is half a long ellipse ; and the length is equal to the greatest breadth—a proportion which gives a rather long skull for a Batrachian.

The occipital condyles (Plate 13, figs. 7, 8, *oc.c.*) project very little, are not so wide apart as in some small species (*e.g., Rana pygmæa*, Plate 5, figs. 11, 12), and the emargination of the basal plate between them is slight ; the roof also (fig. 7) is not cut away much above.

The auditory capsules project moderately in front, but they have an arrested, smooth appearance ; their inferior surface is nearly as great as the superior ; the roof of cartilage is complete up to the post-orbital region, so that there are no lesser fontanelles behind the main space (*fo.*). The optic fenestra (II.) is small, and very near the foramen ovale (V.). The ex-occipitals and prootics (*e.o., pr.o.*) are but little advanced ; the former leave a wide basioccipital, and a still wider superoccipital, tract soft ; the former ride over the anterior ampulla (*a.s.c.*) above, and are seen outside the foramen ovale (V.) below. The interorbital region of the cranial "barge" is widish, and lessens very gently up to the antorbital region ; there is no ossification there ; and if this be, what I am satisfied it is, namely, a mature specimen, here is an instance of the entire absence of the "girdle-bone." I shall soon corroborate this view by showing that another small kind of Frog is equally devoid of this bony tract.

The ethmo-nasal region is nearly as wide as the occipito-auditory ; the roof and floor (*s.n., s.n.f.*) are broad, and nearly equal. The angles of the latter, and the pro-rhinals

* Collected by R. W. WALKER, Esq. ; specimens of the large *R. Bibronii* and *R. Grayi*, brought me from S. Africa by the Rev. AMBROSE WILSON, came too late for this memoir.

† This, according to Dr. GÜNTHER, is artificial, and due to contraction by the alcohol. This was, however, supposed by HALLOWELL to entitle it to a *distinct generic* name, viz.; *Heteroglossa* (see HALLOWELL, Proc. A. Philad., 1857, p. 64 ; and GÜNTHER, "Batrachia Salientia," p. 26). The latter author assures me that this is a normal species of *Rana*.

*fig. 8, *p.rh*.) are well developed, but there is a small prenasal rostrum; the bulbous end of septum nasi is somewhat emarginate. The fore part of the palato-suspensorial cartilage is thick, and forms a triangular pre-palatine; its narrow ethmo-palatine pedicle has on its postero-inferior surface a slight palatine ossification (*pa.*). This cartilage in its pterygoid, quadrate, and auditory regions, is slender; the pterygoid bone (*py.*) is, however, well-developed—not like that of a *young Frog*—and only leaves the upper and outer surface of the cartilage naked (fig. 7). Each pedicle (*pd.*) is small, and they are wide apart; so also is the quadrate region, but it ends in a large condyle (*q.c.*) that reaches as far backwards as to the middle of the stapes; this is quite an average retreat backwards. The Eustachian openings (*eu.*) are of medium size, oval, and transverse, and the space of the middle ear is well developed.

The *annulus* (*a.ty.*) is large, and its horns nearly meet; the stapes (fig. 11, *st.*) is large and oval, with a moderate anterior scooping for the columella. This rod is well ossified in its main part, and has no appearance of belonging to a young individual; its interstapedial segment (*i.st.*) is small, pisiform, and lies between the large unossified lobes of the long, arcuate medio-stapedial (*m.st.*, *m.st'*.). The extra-stapedial (*e.st.*) is a small, long-oval lobe of cartilage, and has no supra-stapedial process. The mandible (fig. 9) is quite normal. The styloid end of the hyoid bar (*st.h.*) is articulated with the ear-capsule; the bar itself (fig. 10, *c.hy.*) is narrow, and has no hypo-hyal lobe; the two side lobes of the basal plate (*b.h.br.*) are small; the thyro-hyals (*t.hy.*) are normal.

The narial valves (fig. 12, *n.l*.*, n.l*.*) are well developed and perfectly normal.

The investing bones lend no support to the opinion that this is anything but a mature individual: they are well developed; those on the upper surface differ from their counterparts in a full-grown Common Frog, by the nasals (*n.*) being relatively much larger and nearer together, and the fronto-parietals (*f.p.*) completely anchylosed along the middle—all save their pointed front ends, where a little of the fontanelle (*fo.*) is exposed. This common sheet of bone well covers the roof behind, and although the bone is *thin*, it is highly developed as bone tissue.

The nasals (*n.*) touch the septum nasi (*s.n.*) and curving elegantly outwards in front, overlie the large sub-tubular external nostrils (*e.n.*) which are very wide apart.

The fore part of the endocranium is well wrapped in the nasals, the pre-maxillaries (*px.*) and the maxillaries (*mx.*); these outer bones are but little developed in the palatine region, but are large and foliaceous on the outside.

The jugal part of the maxillary is joined to a quadrato-jugal (*q.j.*) which is well grafted on to the quadrate: a good sign of maturity.

The upper part of the squamosal (*sq.*) is narrow, but all its regions are well developed.

The parasphenoid (fig. 8, *pa.s.*) is well developed and large, but is not so elegant as in the Common Frog, its angles behind being attenuated and not outspread; the vomers (*v.*) are small crescents, widest in front and forming a mere hook behind:

they are *toothless*—a state of things commonly seen in small types of tooth-bearing Batrachia.

My specimen of this kind was equal in size to many other dwarfs in this group of Vertebrata, and the evidence in favour of its maturity greatly outweighs that which seems to tell of immaturity. Taking its skull then as that of an adult, and comparing it with the "norma," I find the following discrepancies, viz. :—

1. No septo-maxillaries.
2. Fronto-parietals of each side coalesced.
3. Quadrato-jugal ossifies part of suspensorium.
4. No teeth on vomers.
5. No "girdle-bone," even in rudiment.
6. Inter-stapedial very small.
7. No supra-stapedial, and extra-stapedial arrested.
8. Lesser fontanelles absent.

Skulls of "Ranidæ" not comprised in the typical "genus."

15. (A) *Tomopterna breviceps*.—Half-grown female; length, 1¼ inch. S. India.

This and the next kind are both put into the same genus in GÜNTHER's 'Reptiles of British India' (p. 411); but more recently the author has put this short-headed, high-faced species into the genus *Tomopterna*, removing it from *Pyxicephalus*.

This is as it should be. The two (this and *P. rufescens*) differ quite enough to entitle them to be placed in distinct generic groups; this species comes nearest to the *Ranæ*, the other is a more aberrant form.

The outline of this skull (Plate 14, figs. 5, 6) is half a short ellipse, and the width is to the length as 8 to 6¾; the general outline is very regular. A full grown individual would have shown a more irregular outline and denser bony centres, but the skull is fairly finished at this stage; *size* and *strength* are all that is gained afterwards.

The occipital condyles (Plate 14, figs. 5, 6, *oc.c.*) are rather wide apart, and project but little. The auditory capsules carry the hind skull out considerably, opposite the quadrate hinge; and this hinge is opposite the middle of the stapes (*q.c., eb.*) as in the last.

The suborbital spaces are very large, and each forms two-thirds of an almost regular circle. The interorbital region of the cranium only lessens very gradually forwards, and it swells out gently in the middle. The cartilaginous roof of the skull is complete up to the orbital region; there is only the main fontanelle.

The whole fore skull is extremely like that of the "norma," but the prenasal (fig. 9, *p.n.*) is a more distinct bud, and the pro-rhinals (*p.rh.*) are larger hooks; also the outer angles of the subnasal laminæ (*s.n.l.*) are simpler.

The hinder bony centres (*c.o.*, *pr.o.*) are just such as would be found in a Common Frog two-thirds grown; now, a considerable amount of roof, floor, and periotic capsules are soft. So also the "girdle-bone" (*eth.*) is not of great extent, it hardly reaches half way to the large optic fenestra (II.); from this space to the foramen ovale (V.) the wall is also unossified. The palato-suspensorial arch, and its ectosteal plates (*pa.*, *pg.*), are quite normal; the condyles of the quadrate are very large, and the cartilage at that part is but little affected by the quadrato-jugals (*q.j.*). The pedicles are wide apart, and well developed; the Eustachian passages (*eu.*) are large and oblique. The annulus (*a.ty.*) is moderate and complete. The stapes (Plate 14, fig. 10, *st.*) is large, thick, and oval; the medio-stapedial (*m.st.*) fits inside it by a decurved unossified lobe; but the inter-stapedial is not distinct. The medio-stapedial is therefore large behind; it is straight, and joins on to the narrow stalk of a broadly spatulate extra-stapedial (*e.st.*), which becomes double on its inner face, giving off a strong ligulate supra-stapedial (*s.st.*) which is confluent with the "tegmen" above. The stylo-hyal part of the hyoid band (*st.h.*) articulates with the floor of the tympanium, and turning round, borders the Eustachian opening (*eu.*).

The investing bones, one and all, are extremely like those of the type; there is a septo-maxillary on each side, but it is small.

The mandible shows in the dentary (fig. 7, *d.*) very little of the *crest*; yet in old individuals the mandible is high at that part, and also somewhat hooked at the synchondrosis, like an old male Salmon.

The hyo-branchial plate (fig. 8) is normal, but its lateral lobes are badly developed.

15 (continued).—(B) *Tomopterna breviceps*.—Adult female, 2 inches long. Ceylon.

The distinction between the skull of the half-grown young of this species and the young of a Common Frog of the same age, is slight as compared with what is seen in their adult condition (Plate 15, figs. 1–4; and Phil. Trans., 1871, Plate 9).

The facial outline now forms half a rather short ellipse, and the greatest breadth is nearly one-fourth more than the length; moreover, this is one of the highest (or deepest) of the skulls in the whole Order.

Indeed, this *Frog* is an isomorph in respect of its short deep head and its thorough want of neck of the most remarkable *Toads* of the same (the Oriental) territory, e.g., *Callula*, *Diplopelma*, *Cacopus*.

This, again, is an instance of what is seen in the main geographical territories, viz.: that some particular modification characterises the members of very different "Families" of the same Order, as if the Anurous type had, in each territory, broken up into groups isomorphic of, but not immediately related (*genetically*) to, those of other territories. Thus from a generalised root-form, in territories wide apart, there may have sprung, in each place, independently, *Frogs* with teeth and *Toads* without teeth;

types with sharp, or broad toes; with narrow or wide backs; with or without neck glands; with soft or stony skulls.

For this tribe is notable, perhaps above all others, for ready response to surroundings; here is a swimming creature who learns to crawl, to walk, to leap, to climb, and to whom (in a rare instance) even a sort of flight is not denied. Moreover, the slow metamorphosis of the larva may take place in these types in the water, in the damp floor of the jungle, in the foliage of trees, or in a large bag, or in little pockets, on the back of the mother.

This Frog, which is about equal in size to our native species but much stouter, has its skull even less ossified than the common kind; this is true both of the endocranial "centres" and the investing plates.

In this it differs from most of the species of *Rana* of its own territory, but agrees with them in other things, rather than with that familiar species, whose distribution reaches even to us. The nasal and auditory regions are about equal in axial extent; the orbital region is one-third longer; the hind skull is of extraordinary breadth.

The occipital condyles (*oc.c.*) are large, reniform, and postero-inferior; they are separated by a shallow notch less than their own breadth, and the foramen magnum (*f.m.*) is large and oblique, the roof retreating forwards. The double canal (*a.s.c.*, *p.s.c.*) is large and projects beyond the less evident horizontal canal (*h.s.c.*); the capsule projects as an obliquely oblong tract. The ex-occipitals (*e.o.*) do not reach to the stapes, right and left, and only slightly overlap the parasphenoid; the prootics (*pr.o.*) guard the outside of the foramina ovalia (V.) and mount up over the two front ampullae; thus there is a very large cruciform tract of cartilage on the floor of the hind skull. There are no lesser fontanelles, and the main space is only half as long as the cranial roof, and is long and emarginate behind; it is partly uncovered in front (fig. 1. *fo.*). The interorbital region is narrow, lessening steadily forwards; the roof is enlarged a little in front by the small oblong superorbital tracts of cartilage (*s.ob.*). The long, oval, optic fenestra (II.) lies in the middle of a large unossified tract, which leaves the girdle-bone (*eth.*) only two-fifths of the orbital space; that is, however, compensated by the bone running almost to the pro-rhinals, below, and along the hinder third of the septum nasi (*s.n.*), above. The nasal roof is only two-thirds as large as the floor (*s.n.*, *s.n.l.*), the front of it is elegantly sinuous, for the thick septum projects as a small subconical prenasal rostrum. The angles of the floor (*s.n.l.*) project well into the fore part of the wide face, and the pro-rhinals (*p.rh.*) are large, long, and uncinate. The rising wall (*al.n.*) grows up behind the nostril from the surface of the angle of the floor, but does not form a perfect ring by coalescing with the roof as in the larger Oriental Frogs.

The narial valves (*n.l*.*a.l*.) are large and normal.

The palato-suspensorial arches are slender behind, but in front they are thick at their root (*v.pa.*), and have a large pre-palatine spike (in front of *pa.*), and continue wide inside the maxillary until they end as a lobe (*pt.pa.*) from which the narrower pterygoid cartilage grows. The palate pedicles (*pd.*) are sharp and well formed, and reach almost to the prootic and parasphenoid. The angle of the fork is somewhat rounded, so that the large Eustachian opening (*eu.*) is sub-triangular. The bony plates (*p.e.*, *p.y.*) are normal, but thin and lathy.

The quadrate region or outer fork is very long, and retreats as far back as the middle of the stapes: this is one of the deepest cheeks to be found in the group. The cartilage is very slightly ossified by the quadrato-jugal (*q.j.*); the condyles (*q.c.*) have the peculiarly elegant form seen in the Oriental *Ranæ*, viz.: a very large, long, postero-internal trochlea, and a rounded, small, antero-external convexity.

The annulus (fig. 3, *a.ty.*) is rather large and perfect, as in the congeners of this species; the stapes (*st.*) is large, oval, and umbonate. The medio-stapedial (*m.st.*) has no proximal intervening segment cut off, the extra-stapedial (*e.st.*) is spatulate, and the supra-stapedial is confluent above.

The mandible (fig. 3) is normal, the condyle (*ar.c.*) is long and subreniform. The stylo-hyal (fig. 2, *st.h.*) is narrow at first and confluent; it widens gradually up to the hypo-hyal bend below (fig. 4, *c.hy.*, *h.hy.*); over the curve there is now a small thin extra-hyal (*e.hy.* for *ex.hy.*). The notch in front of the basal plate is wide, and the whole structure is normal; the front lateral lobes are now fan-shaped, and the hinder ligulate. The thyro-hyals (*t.hy.*) are large and bent outwards. The investing bones are very thin and splintery; the fronto-parietals (*f.p.*) are feeble and arrested in front; the nasals (*n.*) are feeble and ragged, like the newly-formed osseous centres of a young specimen; the premaxillaries (*p.x.*) are wide and well developed as to their processes; there is a pair of small seed-like septo-maxillaries (*s.m.x.*). The maxillaries (*m.x.*) are high, but thin; they are notched in front where they overlie the angle of the nasal floor, and have a bilobate ascending plate; they stretch along more than half of the temporal space behind. The dentiform quadrato-jugals (*q.j.*) are only slightly grafted on to the quadrate; the squamosals (*sq.*) lie well over the narrow tegmen, and have a shortish diamond-shaped postorbital process. The descending bar (fig. 3, *sq.*), is of great length, and widens gradually downwards.

The parasphenoid (fig. 2, *pa.s.*) is quite *Ranine*, and its basi-temporal plates are very large and bilobate externally. The vomers (*v.*) are *Cystignathine;* of the three spurs the foremost is twice the size of those which fence in the inner nostril (*i.n.*); the body is oblique, arcuate, almost reaches the middle, and is armed with an almost straight crest of teeth; the two crests form somewhat more than a right angle, but their sharp outer end is strongly turned forwards. As compared with that of our Native Frog, this skull is—

1. Much broader and deeper.
2. The hinder centres of the endocranium are smaller, and the girdle-bone is larger.
3. There is only the main fontanelle, which is rather small.
4. The parotic wings are extremely outstretched.
5. The interorbital region is very narrow in front.
6. There are distinct superorbital "eaves."

7. There is a short prenasal rostrum.
8. The pro-rhinals are unusually long and folded over.
9. There is no inter-stapedial, and the supra-stapedial is confluent above.
10. The suspensorium of the jaw is of great depth.
11. There is a small extra-hyal on each side.
12. Most of the investing bones retain their early condition as thin ragged laminæ.

On the whole this instructive skull shows that *Tomopterna* is a Frog close akin to the species of *Rana*, but leaning towards the Cystignathidæ; but it does not lie directly between the two families, for it has been modified, *as a Frog*, much as those neckless types *Cacopus, Diplopelma*, and *Callula* have been modified, *as Toads*.

Here we have a problem relating to the influence of *territory*; that which gave the *apoplectic look* to the toothless kinds, doubtless thrust together the head and shoulders of this tooth-bearing type.

16. *Pyxicephalus rufescens.*—Adult male; 1 inch 5 lines long. India.

This a much slenderer, smaller, and more warty kind of Frog than the last. Its *toes* are much more perfectly webbed in my specimen than in several specimens of *Tomopterna breviceps* in my collection. Dr. GÜNTHER says that the interdigital membrane is equal in both kinds (ibid., p. 412).

The outline of the head is so elongated and pointed as to be almost triangular—a great contrast to the last.

Dr. GÜNTHER says that the eyes are much smaller than in the last kind (p. 412); this is a correlate of the narrowing of the skull forwards.

The length is to the greatest breadth as 8 to $8\frac{3}{4}$; for although narrow in front, this skull is very wide behind.

The occipital condyles (Plate 14, figs. 1 and 2, *oc.c.*) are large, reniform, inferior, and wide apart; an evenly emarginate tract of cartilage, two-thirds their own width, separates them. This basioccipital cartilage is twice as broad as the superoccipital synchondrosis, and the bony masses formed by the extensively spread prootics and ex-occipitals (*pr.o., e.o.*) are much nearer together above than below; there they deviate more and more from the foramen magnum to the foramina ovalia (V.), which they reach below.

Above (fig. 1), there is a large transversely oval space unossified, and with small secondary fontanelles; the prootics meet in front of this, and are far extended in the roof. Laterally, the prootics and ex-occipitals are confluent, but only affect the posterior, and the arch of the anterior, canals (fig. 1, *au.*); the rest of the divergent ear-mass is soft: below, an oval floor is left soft to the vestibule, externally (fig. 2, *eb.*).

The actual cranial cavity is three times as wide in the post-orbital as in the ant-orbital region; far from being an extremely wide skull, it lessens forwards, to become one of the narrowest known. Its lateral margins are sinuous, the middle bellying a little. The narrow part opens out elegantly into the ethmoidal wings; up to them the girdal-bone (*eth.*) reaches, and no further. Above, this bone reaches half way to the extended prootic tract; below, two-fifths of the walls are soft, and this unossified tract has in its middle a very large optic fenestra (II.), and behind, it encloses most of the foramen ovale (V.).

The nasal region is quite unlike that of a typical Frog; above (fig. 1), the nasal roofs are very narrow, and merely form a crescentic selvedge to the septum nasi (*s.n.*), and on to the fore margin of the ethmo-palatine band.

The narrow roof curls round the front of the outer nostrils (*e.n.*), which are very near each other—only one-third the distance apart that we have just seen in the Frog from Lagos (Plate 13, fig. 7, *e.n.*); behind the opening the nasal wall (*n.w.*) is thick and crescentic.

This narrowing of the nasal end of the skull is made more remarkable by the development of a very distinct and pointed prenasal cartilage (*p.n.*).*

But below (fig. 2, *s.n.l.*), the trabeculae have united to form a very wide elegantly winged tract, which passes between the laminae of the maxillaries (*mx.*), externally: behind, it is narrowed, for each margin is cut away by a semi-circular notch, through which the internal nostrils (*i.n.*) pass. The "pro-rhinals" (*p.rh.*) are very long, slender, and bent back upon themselves.

These processes (Plate 15, fig. 5, *p.rh.*) are impacted between the two laminae of the premaxillary; they are equal to the prenasal in thickness.

The nasal valves (*n.v.*) are outside the nostrils; they are, in form, quite normal, but very small.

The whole of the palato-suspensorial arch is quite normal, and rather slender; the same is true of the bony plates (*pa.*, *py.*).

The pedicles (*pd.*) are wide apart, for the auditory masses are relatively large, and widely outspread. The condyles of the quadrate (*q.c.*) are a little further back than in the last, are opposite the end of the stapes, and are large and reniform; there is no grafting of the quadrato-jugal over them.

The Eustachian openings (fig. 2, *eu.*) are very large and are turned obliquely backwards, outside; the annulus is very large, and like that of the Bull-frogs (Eastern and Western). The stapes (fig. 4, *st.*) is large and like that of the "norma;" the columella is very generalised; there is no inter-stapedial segment at the scooped top of the club-shaped medio-stapedial (*m.st.*) whose interstapedial end (*i.st.*) is not ossified. By a short stem, the bony tract is connected, in front, with a small extra-stapedial (*e.st.*), which has no secondary process; here, the breadth of the "annulus," below, is more than twice as great as that of the circular extra-stapedial: this is the converse of what is seen in *Rana Kuhli* and *Dactylethra*.

The investing bones are about as strong as those of the skull of "the type," when

* The *right* dotted line from *p.n.* in fig. 1 points to the nasal process of the right premaxillary: behind its apex the first upper labial is hidden.

half-grown—or equal to the size of the full-grown individuals of this species. The parasphenoid (*pa.s.*) is very large, both in its main part and wings; this form is like that of the Common Frog, but the point is longer and sharper.

The vomers (*v.*) are typical; the septo-maxillaries extremely small *if present*, and the bones of the outer face normal, but slight. The nasals (*n.*) are long crescentic shells of bone; the fronto-parietals (*f.p.*) are large and out-spread over the hind skull and then become gradually fine sharp styles in front, leaving part of the girdle-bone and a little of the fontanelle (*fo.*) uncovered.

The "frontal suture" is permanent; the "sagittal" is filled in with bone: the two sides being continuous over the inter-auditory region.

The mandible (fig. 3) is quite normal.

The stylo-hyal end of the hyoid band (fig. 2, *st.h.*) is distinct and pointed as it passes behind the Eustachian opening; this band widens to double its breadth (fig. 3, *c.hy.*) in the lower half, and then turns suddenly round to join the basal plate (*h.hy.*, *b.h.br.*).

Round the outside of the broad distal half there is another band of cartilage as wide as the upper half of the main band.

This is an "extra-hyal" element (*ex.hy.*); it passes over the hypo-hyal loop as a short hook.

Other Oriental Batrachia show this, but not so distinctly, e.g. *Tomopterna*, *Callula*, and *Diplopelma*, and the Australian Tree-frogs have a rudiment of it: here it is most largely developed.

If this cartilage be compared with the pectinate "inter-branchial" of *Chimæra* (HUBRECHT, fig. 2), it will be seen to correspond with the base of that comb-like cartilage.

In Tadpoles the extra-branchials send pectinate processes inwards, but they show no separation of the extra-branchial bands from these "rays" of the septa, such as we see in Sharks, for the branchial apparatus of the Tadpole is as highly generalised, as that of the Shark is intensely specialised.

The "spiracular cartilage" belongs to the suspensorium (*of the 1st arch*), and becomes utilised as the annulus tympanicus; this rarer "extra-hyal" and the four extra-branchials all belong to one category.

This *infero-external* element of the hyoid arch, like the *supero-internal* (epi-hyal or "columella"), does not appear until after the metamorphosis of the Tadpole; that is utilised as part of the apparatus of the fast-improving ear; the extra-hyal merely serves as an additional platform for the thin fan-like muscles of the throat.

The rest of the hyo-branchial structure is quite normal; the lateral lobes are highly developed, and the thyro-hyals strong (fig. 3, *t.hy.*).

Besides the remarkable shape of this skull, so wide behind and so pinched in in front- necessarily modifying the form of the investing bones—we have the subjoined differences from the typical form, viz. :—

1. The septo-maxillaries are suppressed (?).
2. The fronto-parietals confluent behind.
3. The pro-rhinals doubled over as hooks.
4. A distinct sharp prenasal.
5. Very narrow nasal roofs.
6. No inter-stapedial segment.
7. No supra-stapedial band.
8. A large long uncinate "extra-hyal."

The folding over of the pro-rhinals, the length of the prenasal, the generalised state of the columella, and the additional hyoidean element, indicate a very *marginal position* for this type in the Family "Ranidæ."

Second Family. "CYSTIGNATHIDÆ."

First genus. *Pseudis*.

17. (A) Skull of *Pseudis paradoxa*.—First larva; total length, 10¼ inches; head and body, 3¼ inches; tail, 7 inches; greatest width of tail, 4 inches; depth of body, $2\frac{5}{8}$ inches; hind legs, ½ inch. S. America.

A side view of this, the *youngest* and *largest* Tadpole of *Pseudis* (natural size) (Plate 1, fig. 1), shows to what a magnitude the larva of a medium-sized Frog may grow; for the old individuals are but little larger than our native Grey Frog.

As to actual length, more than two-thirds of this larva is a temporary structure, and belongs to the tail; the bulbous fore part, below, is half pharyngeal and half abdominal.

Supposing a measurement by vertical lines across the length of the creature, the distance between the anal aperture and the occipital hinge is one-third the length of the essential animal; the single gill-opening (left side) is at the middle; and the eye-ball two-fifths from the front.

Altogether, the outlets and inlets to the essential animal are *five*: three azygous—the oral, anal, and pharyngeal; and two paired, the narial; these latter are above and behind the upper lip, and are small and rounded: they open into the palate almost vertically. The mouth is small but complicated (Plate 1, fig. 2), the upper labials (Plate 2, figs. 1, 2, *u.l.*) arching over the horseshoe-shaped lower labials (Plate 10, fig. 5, *l.l.*) and the small oral passage opening being between the halves of the horseshoe : the skin over these labials is formed into a horny plate, and the *plicæ* of the lips are developed into horny rasps.

Much fibrous tissue—some watery, some strong and tough—encloses the skull and the outworks; above, the head is marked by a median, and a pair of lateral, ridges; the facial cartilages are spread out on each side and lie on a plane but little below that of the cranial cavity (see Plate 2).

The "chondrocranium" at this stage is one of the best for comparison with that of the Lamprey, but it has already gone beyond the *suctorial* type in having both parosteal and ectosteal plates; it is profitable also for comparison with that of the larvæ of *Dactylethra* and *Pipa*.

The free growth of cartilage, here, gives us a character in this type which I have not found in others, but which GÖTTE describes in *Bombinator*, namely, that the notochord (*nc*.) is ensheathed in an azygous, tubular cartilage, as in the Salachians.[*]

Hence the parachordals are very narrow bands between this tube and the ear-sacs; and very short also; for the trabecular apices embrace the fore end of the notochord.

The cartilage, right and left, has formed a perfect arch above, and the flat roof runs up to the post-orbital region (Plate 2, figs. 1 and 3, *x.o.*).

Behind (Plate 2, fig. 3), the occipital condyles are forming; they look directly backwards, and the bony matter (*e.o.*) runs into them. The auditory capsules were not relatively large at first, but they are so largely developed outwards now as to seem of unusual size, being *winged* above and below.

By the study of the growth of the skull in many types I have satisfied myself that the upper wing ("tegmen tympani") has not the same morphological import as the lower, or floor of the tympanium, both of which are enormously developed.

The basal plate, formed by the trabeculæ and parachordals, grows round the base of the ovoidal ear-sacs, and appears outside, especially in front: hence this copious growth of cartilage, forming elegant *lower* wings to the ear-masses, and serving as a shelving tympanic floor (*f.ty.*). This undergrowth showing itself outside the ear-sacs, appears in many forms; it is a manifest striving of the basal plate to meet the arches that belong to it, this junction being impeded by the huge sense-capsules impacted in at this part.

But, as in the "Aglossa," the "tegmen tympani" (*t.ty.*) is a superficial growth of cartilage applying itself to the capsule, in the margin of which is imbedded the horizontal canal; this roof-plate is very large, still larger than the floor (Plate 2, figs. 1–3).

This tegmen is very thick as well as wide; it is widest in front, and there it overlaps, and is confluent with, a large process from the elbow of the suspensorium (*ot.p.*).

Here the tegmen cranii, and the tegmen tympani, of each side, remain soft, but the labyrinth is almost entirely enclosed in a generalised occipito-petrosal mass (*e.o., pr.o.*).

This is an undistinguishing spread of bony matter over and under these regions, a division into "periotic" and "ex-occipital" being attempted later in the growth of the head.

It is very solid bone, and leaving the azygous cartilage untouched below, and a similar breadth of roof above, it builds a side wall to the foramen magnum, forming a solid rough kind of masonry, and enclosing most of the "canals" and part of the "sacculus."

The cartilaginous stapes (Plate 2, fig. 3; Plate 10, fig. 5, *st.*) lies in its fenestra just under the middle of the "tegmen," and above the deep shelving floor (*f.ty.*)

[*] The outer sheath of the notochord is extremely thin in most of the Batrachia.

Already the passages for the 5th and 7th, and the 9th and 10th nerves (V., VII., IX., X.) are being enclosed in bone.

The trabeculæ and "intertrabecula" form the whole skull from the notochord to the upper labials, the hinder half being *cranial* and the fore half *nasal*.

The fontanelle (*fo.*) is not large, and it appears of a pyriform shape, because of the fronto-parietals (*f.p.*) which partly cover it; they are strong wedges of bone close together behind, where they are broad, and wide apart in front, where they are narrow.

There is an enlargement of each bone in the temporal region, and from thence the uncovered skull rapidly widens into the auditory masses.

In front of that point it is almost oblong, only gently widening towards the front, where it is continuous with the large facial bars.

The floor is flat—scarcely convex—and above the edge of the floor the sides are scooped, and shelve inwards, so that the flat top is narrow, especially in front, where the ".tegmen" (*t.cr.*) reappears.

The tegmen is continued in front of the closed cranial cavity, whose only fore outlet is through the olfactory foramina (I.); therefore between these holes there is a part answering to the "perpendicular ethmoid" (*p.e.*). This is continued forwards as a crest growing up from the line of junction of the trabeculæ; it is the "septum nasi" or fore part of the intertrabecula. Beyond this wall, with a slight interruption, the trabeculæ are united for a space, and then are free, and diverge at their end.

These "cornua" are convex above and concave below, with a thick outer and front edge (Plate 2, figs. 1, 2, *e.tr.*); and this part, ending in a flat facet, articulates with a like facet on the upper labial (*u.l.*).

The coalesced part of each cornu is somewhat constricted, and a similar lunate outline exists in the bar outside. Thus the inner nostril (*i.n.*) is made; it is finished in front by the short *pre-palatine ligament* (*pr.pa.*), both cartilages approximating.

Nearly the middle third of each trabecula is continuous with the outer band; this is in the ethmoidal region, which is very extensive.

There are in reality *two* bands—one outer very large, and a smaller band lying between this and the trabecula; this lesser part is the "post-palatine" (*pt.pa.*), and the huge bar outside is the suspensorium of the lower jaw (*p.pg., sp.*).

The inner bar is half the length and one-third the width of the outer; its hinder third is free, projecting backwards and hooked a little outwards between the trabecula and the suspensorium.

Its fore part is continuous with the inner edge of the suspensorium, where it bounds the inner nostril (*i.n.*), and ends in the inturned pre-palatine spike. The middle part, which is the longest, is the conjugational tract binding together the two main cartilages, outer and inner (*c.tr., sp.*).

The upper part of the ethmo-palatine is strongly crested on its inner side, and this crest meets the lateral ethmoidal wall at an acute angle behind the inner nostril (Plate 2, fig. 1, *p.pg., i.n.*).

The nasal capsules are membranous as yet; and this crescentic ethmoidal wall, which *articulates* with the "tegmen," bounds them behind.

The dorsal element of the mandible, or its "suspensorium," is the most extraordinary thing in this remarkable "chondrocranium." Each bar is equal in size to the combined trabeculæ cranii, and to each is suspended three other cartilages, and a *process* which becomes a free ray (*ot.p.*).

The dorsal end of this bar early coalesced with, or rather *grew from*, the trabecula, beneath the emerging trigeminal nerve (V.); that part is now narrow, but terete; an elegant crescentic margin to the subocular fenestra (Plate 2, figs. 1, 2, *s.o.f.*) is formed by this "pedicle" as it passes outwards and forwards, and becomes the main bar; that bar is becoming rapidly widened, so as to be five or six times the breadth of the dorsal end.

A rounded notch exists between the "pedicle" and the spiracular cartilage (*ot.p.*), and this cartilage is continuous with the tegmen tympani at its distal end, and with the "elbow" of the suspensorium at the proximal end.

The "elbow" of the suspensorium passes outwards into the cheek as a large projection, with a rounded outline; the bar is then bent in a falcate manner, so as to run into the face where the ethmo-palatine projects. In front of this, opposite the middle of the large conjugational tract, there is the pyriform, gently concave condyle for the cerato-hyal; it is just beneath the edge of the bar (Plate 2, figs. 1 and 2, *hy.f.*).

Over this part the *larval Batrachian* "orbitar process" grows upwards and inwards as a sessile, semi-oval leaf, with decurved edges and a swollen base; it is attached to the post-palatine crest by a short ligament, and is not confluent as in *Bufo vulgaris*.

The decurrent enlargement of its edge runs backwards insensibly into the main bar as a thickened margin; in front, it projects over the edge of the bar as a free point, and then runs along the rest of the bar as a selvedge (Plate 2, fig. 1, *or.p.*).

From thence the anterior fourth of the suspensorium is a many-sided flap, not sensibly lessened in width, and having on the outside a snag, and in front the convex sinuous condyle of the mandible (*q.*), which looks a little inwards, and is only a trifling distance behind the cornu trabeculæ.

That *ending*, however, is but the *beginning* of the proper mandible, which is a stout, transversely-directed, sigmoid bar (Plate 10, fig. 5, *mk.*); it is twisted and notched (like the human *ulna*) to roll upon the quadrate. To its flattened inner (distal) end is attached the lower labial (Plate 10, fig. 5, *l.l.*), a thick, short, arcuate bar, lessening in size downwards, where it is attached by a strong ligament to its fellow to form the *horseshoe*, or imperfect suctorial disk. The elongated angle of the upper labial (*u.l.*) overlaps the double jaw-piece (labio-mandibular), and is represented by a separate cartilage in the Lamprey, and in the Tadpoles of many kinds of Batrachia.

The hyoid (Plate 10, figs. 5 and 6, *c.hy.*) is a large irregular lozenge of cartilage, growing towards the mid-line below, and connected to its fellow by simple cartilage (*c.hy.*, *b.hy.*). Externally, or above (Plate 10, figs. 5 and 6, *c.hy.*, *st.h.*), it sends out a

DEVELOPMENT OF THE SKULL IN THE BATRACHIA.

stout angular styloid process, on the upper surface of which there is a large, oval, convex condyle, which rolls in the shallow cup of the suspensorium.

The sinuosities of the hyoid bar are filled in by the convexities of the branchial apparatus. In the middle, behind the soft basi-hyal, there is a large pyriform basi-branchial (Plate 10, fig. 6, $b.br.$), composed of hyaline cartilage; a short process on the postero-inferior surface of this is the rudiment of a second basi-branchial segment.

Outside and behind this median piece there is a pair of flat lozenge-shaped cartilages, the hypo-branchials ($h.br.$); these grow outwards and are connected with cartilages above (inside) and below (outside); these latter are the branchial pouches.

Two of the small upper cartilages are distinct from the hypo-branchials, but are partly confluent with the large outer bars ($c.br^1.$, $c.br^2.$, $c.br.$); the two hinder rudiments ($c.br^{3, 4}.$) are continuous with the hypo-branchial plate, and the space between the two is filled in with cartilage.

The first and fourth outer bars ($c.br^1.$, $c.br^4.$) are pouch-like, the others are thin broad bands.

The rudimentary inner arches ($c.br.$) are less differentiated than in the larva of the species of *Rana* and of other kinds (*Cyclorhamphus*, *Calyptocephalus*, *Cystignathus*, &c.) that I have worked out.

These parts are at their fullest development at this stage.

I have already mentioned the roof-bones in relation to the fontanelle; the parasphenoid is a dagger with a guard, but without a handle.

It seems small, yet it occupies the same place, and has precisely the same relations as in the adult (Plate 2, fig. 2; and Plate 10, fig. 2, $pa.s.$); it is *one-third longer* than in the old male. Even now it is split in front: a character which is retained throughout life.

17 (continued).—(B) Second Tadpole of *Pseudis paradoxa*.—7 inches long; tail, $4\frac{2}{3}$ inches; greatest width of tail, 2 inches; hind legs, 3 inches long; fore legs hidden.

In this stage the legs are six times as long as in the last, and the tail two-thirds the length and half the width; here the chondrocranium is but little more than half as broad across the suspensoria and only two-thirds the length.[a]

There are many things to be noticed in this stage besides its lessened size and more oblong general shape.

And first, the *Selachian* character of a huge notochord enclosed in a tubular azygous cartilage is now as difficult to find as in most Tadpoles; it has become part of the basal plate by coalescence with the "parachordals" and trabeculæ, and the gelatinous axis is

---

[a] If I had only found these discrepancies in size in two or three, I should have thought it accidental; but my specimens are too numerous, and run over too many stages, for there to be any mistake: the skull in the *third* stage (C) is but little more than one-third the length of that of the youngest (A).

very much shrunken. The wild and general growth of bony matter, without division into periotic and occipital regions, is now still more remarkable, and, above, the right and left tracts are rapidly coalescing, so that there is no superoccipital cartilage (Plate 11, fig. 1). This bony matter reaches forwards to the great fontanelle (*fo.*) and laterally up to the inner margin of the horizontal canal (inside *t.ty.*).

Below (fig. 2) there is a wide basioccipital tract (*iv.*) unossified, and the bony matter only skirts the floor of the vestibule; it runs, however, well round the pre- and post-auditory nerve-passages (V., IX., X.).

The steeply-sloping floor of the tympanic cavity (Plate 11, fig. 3, *f.ty.*) has become much reduced in size, and thus the oval stapes (*st.*) is seen clearly from below.

The fontanelle (*fo.*) has the same relative size, and so have the fronto-parietal bones (*f.p.*); but the upper part of the skull in the ethmoidal region has altered greatly; ready to alter still more as the tail, and indeed the whole creature, keeps lessening in size.

The ethmoidal roof (Plate 11, fig. 1, *p.e.*) has grown over the growing septum nasi, but the nasal roofs are still membranous, or only composed of soft cartilage.

The back wall of the nasal fossæ is not so distinctly articulated to the edge of the tegmen cranii; but, gently shelving down, the median part passes into a concavity which lies between the roof and the post-palatine (*pt.pa.*). The cornua trabeculæ (*c.tr.*) have largely united together, but a small hole some distance behind the notch may be still seen; this soon fills in, it represents the large open space in the "Urodeles."

The post-palatines (*pt.pa.*) have been, as it were, moulded into a more solid, but altogether a rounder and smoother, structure; the sharp crest has become a neat longitudinal ridge, and the hinder process is not a flap but a geniculate projection into the fore part of the suborbital fenestra.

Laterally (Plate 11, fig. 3, *pt.pa.*), it appears as a gently concave plate, like a "postzygapophysis;" the orbitar process playing against its scooped face. The fore part of the palatine has chondrified the pre-palatine ligament, and the apex of this bar abuts against the trabecular cornu (*pr.pa., c.tr.*); thus the internal nostril (*i.n.*) is fairly enclosed by cartilage.

Beneath (Plate 11, fig. 2), the trabecular, palatine, and suspensorial regions all pass gently into each other; this, however, is a very temporary condition.

Besides the general lessening of the cartilage, the suspensorium is not now so bent outwards, and the "pedicle" has bent itself into a sigmoid form for want of room (fig. 2, *pd.*).

The space between the "elbow" of the suspensorium and the spicular cartilage or otic process (*ot.p.*) is larger, and that band seems to belong equally to the tegmen tympani and the suspensorium; the ridged edge of the orbitar process (*or.p.*) runs along to the tegmen, strengthening the whole band; this "extra-suspensorial" tract has no counterpart in the Lamprey.

The quadrate region (*q.*) stands out further from the trabecular cornu, and its condyle looks more directly forwards, and not so much inwards; the orbitar process is more elongated.

The lower arches and labials (Plate 11, figs. 3–5, *mk.*, *l.l.*, *c.hy.*, *b.hy.*, *b.br.*, *h.br.*, *c.br.*, *ex.br.*) are but little changed.

17 (continued).—(C) Tadpole of *Pseudis paradoxa*, with all the legs large and free; tail, 5 inches long; its greatest width $1\frac{1}{2}$ inch.

The chondrocranium (Plate 11, figs. 6, 7; and Plate 12, fig. 1), although relatively wider on account of the throwing out of the suspensorial bands, is only half the length it had, even in the second stage (B), and a third the length of that of the first (A).

This temporary transitional form is very extraordinary; I have caught no stage quite like it in any other species.

The third Tadpole of *Rana clamata* (Plate 4, figs. 5, 6) comes nearest to it. The ethmo-nasal region, which was one-half the length of the skull in the last stage, is now less than one-third; thus the orbital and auditory regions are now very large, relatively, and with them the cranial cavity is equally increased in size (relatively).

The occipital condyles (Plate 11, figs. 6, 7, *oc.c.*) are now wider apart; but the ossification here is but little increased in extent. The bony substance has united over the occipital roof, and a definite tract of cartilage lies in front of the bone. There are no "secondary fontanelles," and only a wedge-shaped tract of the main fontanelle (*fo.*) is uncovered by bone (fig. 6).

The notochord (*nc.*) is still present as a fine thread, and the basioccipital synchondrosis is unaltered; moreover, the floor of the vestibule below, and the outer part of the capsule with the tegmen tympani, continue unossified.

The semi-circular canals stand out strongly, and along the ridge of the posterior canal and part of the anterior (*pr.o.*, *au.*, *cp.*), the bone has undergone dehiscence, exposing the cartilage within, and partly separating the prootic from the ex-occipital on each side (fig. 6, *pr.o.*, *e.o.*).

In like manner, over the skull, the large frontals are partly severed from the small parietals (*f.*, *p.*); the parasphenoid (*pa.s.*), below, has become smaller (actually), and more elegantly formed.

The unossified auditory floor (fig. 7) is now a more transverse, sub-oval mass, and the projecting, descending lip of the floor is very small (Plate 11, figs. 7 and 8; and Plate 12, fig. 1, *f.ty.*). The upper part of the auditory capsule projects beyond the vestibular pouch, very considerably.

The cranial cavity lessens in width gently to the fore end; the "tegmen" is of moderate extent over it, and part of it passes into the nasal region, as the roof of the ethmoid.

The optic nerve (Plate 11, fig. 7, II.; and Plate 12, fig. 1, II.) passes out of the hinder part of a large fenestra; the roof-bones have a down-turned orbital edge.

The ethmoidal wall behind the nasal sac was articulated to the " tegmen cranii" (Plate 2, fig. 1); that joint was lost in the last (Plate 11, fig. 1); now a new articulation has appeared, namely, of the post-palatine crest with the ethmoidal wall (Plate 11, fig. 6, *pi.pa.*); this, however, is merely at its upper part; the bar is continuous with the skull below (Plate 11, fig. 7, *p.pg.*).

The *direction* of the endoskeletal palatine ("post-palatine") was directly forwards (A); the pre-palatine point being slightly turned inwards to join the trabecula by a ligament (Plate 2, fig. 1, *pt.pa., pr.pa.*).

Afterwards (B, Plate 11, figs. 1–3), the pre-palatine point was well developed, and turned directly inwards to join the trabecula, but the main piece was parallel with the skull; the sharp crest had become an even thick ridge.

Now (C, Plate 11, fig. 6; and Plate 12, fig. 1, *pt.pa.*), this part has behaved like a railway signal: it has turned outwards, almost directly, but is bent a little forwards; the pre-palatine has again become reduced to a mere point carrying the ligament that runs to the cornu trabeculae.

The flat, but crested, bar (A, Plate 2, fig. 1, *pt.pa.*) has become a massive, rounded rib of cartilage, standing high above the thin pterygoid plate that unites it with the suspensorium (Plate 11, fig. 6, *p.pg.*); thus that outer bar now stands off a good distance from the skull.

The tegmen cranii has grown well over the perpendicular ethmoid, and the septum nasi has become surmounted by a pair of thick crescentic folds, lying back to back (Plate 11, fig. 6; and Plate 12, fig. 1, *al.sp.*); from these the cartilage grows to some little distance over the nasal sacs, but I cannot find, either now or afterwards, any such clearly distinct nasal roof as is seen in the species of *Rana*.

The anterior part of the skull has become reduced in size and modified to a much greater extent than the outer bars; thus the quadrate hinges (*q.*) are now a sensible distance in front of the cornua trabeculae.

These latter parts are now very small and bifurcated: the inner lobe is a short hook, (*c.tr.*) pointed inwards; this will be the pro-rhinal, it lies in front of the nasal sac; the outer lobe is a long "ear" of cartilage (*s.n.l.*) directed outwards and little forwards, and attached by ligament to the pre-palatine spur (*pr.pa.*).

The position of the pre-palatine ligament shows the amount of lessening the trabecular cornua have undergone, for the ligament and this *outer* cornu, together, form the fore boundary of the internal nostril (*i.n.*). The last stage (Plate 11, figs. 1, 2, *c.tr.*) shows how much of the cornu there was then in front of that passage.

The ethmo-palatines meet below (fig. 11, *p.pg.*); there was no such appearance in the last stage, before these bars had turned round (fig. 2); here, if the segmental nick (fig. 6), which is seen above, had extended between the post-palatine and the trabecula, it would have severed it from the skull, beneath which the right and left bars are confluent. The sub-cranial confluence is peculiarly Petromyzine.

The huge suspensorium (*sp., q.*) is being folded up that it may be changed; huge indeed now, it will soon become less than one-fourth its present size. Even now it is less than a fourth the size of that of the young Tadpole (A, Plate 2).

Relatively to the cranium, it is immense, yet, and retains all the larval characters, being still confluent with the chondrocranium at three places.

The pedicle (*pd.*) is now a very narrow band, and like the pterygo-palatine, is at right angles with the main bar.

The "elbow" of the suspensorium has developed a new "otic process" (*ot.p.*), and the spiracular band has become a mere thread (Plate 11, figs. 6, 7, *sp.c.*).

As in the "Urodeles," the permanent "otic process" (*ot.p.*) mounts up against the fore edge of the tegmen tympani and the swellings caused by the ampullae: it reaches inwards, as in them, to the anterior ampulla, and even a little further.

Above (fig. 6), it is a thick, rounded mass, but below (fig. 7), it is flat; it has, as it were, been thrust back against the auditory mass, and cleaves to, and lies outside, as well as in front of it.

The greatly developed ridge on the edge of the suspensorium, instead of passing in a gentle arc forwards to become the thickened edge of the orbitar process, turns suddenly as a round loop, and then, at an acute angle, bends back again round the front margin of the orbitar process, scarcely dying out as it approaches the condyle of the quadrate, in the front of the face; all this growth is ready to vanish away.

This swollen upper selvedge is thickest behind, then narrows up to the apex of the orbitar process, and keeps its breadth until it dies out in front.

The quadrate region and condyle (Plate 11, figs. 6, 7; and Plate 12 fig. 1, *q.*) is very broad and also thick, especially at its outer edge.

In conformity with the divergence of the quadrate bars the mandibles (Plate 12, fig. 1, *mk.*) are considerably longer; but the lower labials (*l.l.*) are full sized, as yet, and the condyle for the hyoid (*hy.f.*) is perfect.

So also is the bar itself (*c.hy.*), but one can see that the styloid region (*st.h.*) is elongating.

In the Common Frog and Toad there is no upper hyoid element until about three months after the loss of the tail; in *Pseudis* the tail has lost only two-sevenths in length, although much narrower, when that element (the epi-hyal or "columella") appears. Here, at any rate, this rod appears much earlier, *relatively*, than in the common kinds; it is possible, however, that the Tadpole, at this stage, may be *several years old*.

In these large Tadpoles it is as easy to show that the early "stapes" belongs to the periotic capsule as that this *late* segment does not. There are two cartilages, besides the stapes, in certain "Proteidea" (*e.g.*, in the *Menopome*), both inside the facial nerve. The upper piece sends its narrow proximal part to the stapes; the lower cartilage, manifestly part of the epi-hyal, becomes partly confluent with the hind margin of the suspensorium.

Here things take a very different course, for the cartilage (Plate 11, figs. 8, 9) is undivided, and forms a rudimentary epi-hyal.

This *new* cartilage (*co.*) lies inside the nerve, as it forks to join the glosso-pharyngeal (VII., IX.), and to supply the region of the first cleft, fore and aft.

This rudimentary columella appears as a solid bud of cartilage abutting against the fore edge of the large, almost circular, stapes (*st.*).

A section of these parts (Plate 11, fig. 9, *st., co.*) shows that, according to the rule in the Batrachia, the dorsal end of this cartilage is wedged in between the antero-superior margin of the stapes and the auditory capsule; it is already constricted into two regions; the massive "inter-stapedial" part (a distinct segment generally, but not in *Pseudis* (see Plate 10, fig. 4, *m.st.*), is already marked off from the main rod (Plate 11, fig. 9, *co.*). Below the stapes (Plate 11, fig. 8, *f.ty.*) the tympanic floor, once so large as a down-growth of the actual skull-base, is now a mere thickening of the bulbous vestibular region.

The præmaxillaries (Plate 12, fig. 1, *px.*) have been added to the bony plates, and they fit against a small inner (anterior) cartilage (*n.l.*); a crescentic valve to the nostril is also found outside this; the large primary upper labials have been removed; they were ready to be absorbed.

If the metamorphosis of this skull stayed here, we should have much to contemplate that is both striking and instructive; this stage, however, is scarcely mid-way to the actual end of this changing skull.

17 (continued).—(D) Fourth larva of *Pseudis*, rapidly acquiring the Frog-form, but with a tail still 3 inches long and 7 lines wide.

In this species the tail co-exists for a long while after many of the characters of the adult skull have appeared (Plate 12, figs. 2-7). At first sight this larva appeared to be not very different from the last, but it had lost the contracted *suctorial* mouth, and had gained the characteristic open gape of an adult Frog.

This had been done by the rapid enlargement of some parts, and the rapid lessening of others—an interstitial change which appears the more marvellous the more it is contemplated.

In the broad basioccipital region the notochord (Plate 12, figs. 2, 3, *nc.*) still persists as a small thread; and a three-rayed tract of cartilage has escaped the ex-occipital growths (*e.o.*); the lateral extensions of cartilage are the occipital condyles (*oc.c.*), which are large and wide apart.

The cranial cavity has become still wider than in the last; it lessens gently up to the ethmoidal region.

The parietals are now fairly distinct from the frontals (*p., f.*), and the hinder part of the "tegmen" has no uncovered cartilage, the confluent ex-occipitals having the narrow part of the subquadrate parietals lying on their fore margin.

These hinder bones send out a temporal angle behind the widest part of the frontals, and then the latter bones continue this down-turned edge (orbital plate) up to their narrow apex (fig. 4, *f.*).

DEVELOPMENT OF THE SKULL IN THE BATRACHIA.

Thus in a side view (fig. 4) the wall is about half of it bony, but this is due to the curling over of the roof; the "sphenethmoidal" wall itself is quite unossified. The orbito-sphenoidal region is largely fenestrate (*o.s.f.*), as in Lizards, but in a very different manner.

In this stage the fontanelle (fig. 2. *fo*.) is seen in front between the diverging ends of the frontals, and behind, as a lozenge-shaped space (*fo'.*) between the frontals and parietals.

Laterally, the ear-capsules are very large relatively (compare Plate 11, figs. 1-3; and Plate 12, figs. 2-4); and the tegmen tympani with the enclosed horizontal canal (*t.ty.*) projects much in front; this part is not ossified.

Below (fig. 3), the two unossified vestibules, embraced at their inner margin by the large grooved ex-occipitals, resemble a pair of symmetrically imbedded acorns, the rough "cup" being bony, and the smooth "fruit" cartilaginous.

Each of these capsules is fenestrate, and operculate, obliquely; the opercular stapes (*st.*) looking outwards and somewhat backwards; the rough "cup" is perforated in two places for the glosso-pharyngeal and vagus nerves (IX., X.). In front, the bony matter scarcely reaches the 5th nerve (V.); the optic nerve (II.) passes out of the hind part of the orbito-sphenoidal fenestra, and the olfactory escapes through the closing wall in front.

The fore part of the tegmen cranii is rather long; laterally, it is articulated to the ethmo-palatine (*e.pa.*); in front, it passes into the two narrow crescentic nasal roofs (fig. 4, *al.sp.*), which are partly covered by the small ear-shaped nasal bones (*n.*).

The nasal floor (fig. 3, *s.n.l.*) is of the normal breadth, and in front it is now a finished structure, elegantly crenate, with seven lobes.

The outer of these are the outer angles of the trabecular cornua, now finishing the nasal floor in front. Next to these, but on a higher plane, are the horns of the nasal crescents (*s.n.*, *al.sp.*). Near the mid-line we see a pair of short inturned pointed "pro-rhinals" (*p.rh.*); and now, at the middle, there is a little bud of cartilage, the prenasal (*p.n.*).

Besides the nasal roof-bones (*n.*), there are now, below, two oval dentigerous plates, placed transversely and some distance apart; they touch the inner nostril (*i.n.*) by their outer edge; these are the vomers (*v.*).

The semi-oval outline of the face is now nearly finished by the dentigerous pre-maxillaries and maxillaries (fig. 4, *p.x.*, *m.x.*); most of the maxillary bone is moulded on a cartilaginous bar, which is quite unlike anything seen in the last stage.

The solid ethmo-palatine bar (Plate 11, fig. 1; and Plate 12, fig. 1, *pt.pa.*) is now (Plate 12, fig. 4, *e.pa.*, *pt.pa.*) a slender rod, flatter below than above, and instead of being curved forwards outside, it is turned suddenly backwards.

The prepalatine spike (fig. 3, *pr.pa.*), which turned directly inwards in the second stage and touched the trabecula (Plate 11, figs. 1–3) now has its outer margin coincident with that of the whole subocular bar (pterygo-palatine), whilst the point looks directly forwards.

It forms now a pointed wing to the fore third of the crescentic palatine region; the upper process of which (*e.pa.*) is half the length of the post-palatine (*pt.pa.*) which ends opposite the middle of the optic fenestra (II., *o.x.f.*), whereas in the last stage it ended opposite the middle of the internal nostril (Plate 11, fig. 6).

The last fourth of the pterygo-palatine bar (Plate 12, figs. 2–4, *py.*), is the thin pterygoid foregrowth of the suspensorium, which was in the antorbital region in the last stage.

The parasphenoid (Plate 12, fig. 3, *pa.s.*) retains its shape, but its point is more split, and these two spikes end in a sulcus caused by the meeting of the ethmo-palatines (*e.pa.*) beneath the ethmoid: their commissure is elegantly bowed forwards.

A thin ectosteal plate (Plate 12, fig. 3, *py.*) has applied itself to the "chondro-pterygoid;" this latter part, although it is not segmented off from the post-palatine as in the genus *Bufo*, is yet quite distinct from it to the eye, it being suddenly compressed; the part in front retaining its thickness up to the meeting of the two regions.

The suspensorium (*e.pa.* to *q.c.*) has shrunk to one-fourth the size it had in the last stage; it has lost the large leafy orbitar process; and the quadrate condyle which reached to the frontal wall of the head, now reaches nearly as far back as the "fenestra ovalis."

The hinge for the hyoid has also gone, and that bar (*c.hy.*), which did hang opposite the ethmoid, is now tied by a ligament (hyo-suspensorial), so far backwards that its styloid apex is directly below the middle of the auditory capsule.

This position is attained by the *Axolotl* when it measures 4½ lines—that is, soon after hatching (see "Urodeles," Part I., Plate 22, fig. 3, *q.*, *c.hy.*).

In all this change of *size, form*, and *relation*, the enlarged "elbow" or otic process (*ot.p.*) has been the fixed point, and the quadrate has been thrust back by the reversed curve and increased length of the ethmo-palatine.

In the last stage (Plate 11, fig. 7, *pd.*) the pedicle was becoming a mere thread of cartilage, and the inner and posterior edge of the lateral bar was becoming solid and rounded in front of the groove where the facial nerve emerges.

Now (Plate 12, fig. 3, *pd.*), the attenuated pedicle has become fibrous proximally, and the outer part has developed itself into a thick pedate mass, the concave side of which fits against the convex fore face of the unossified wall of the vestibule. This "condyle of the pedicle" is a process growing from the *under*, as the otic process is an outgrowth from the *upper*, part of the suspensorium.

Over the outer surface, from which the orbitar process has vanished, a thin straggling tract of membranous bone now lies; this is the squamosal (*sq.*), with all the characters, as yet, of a "preoperculum."

The little bridge of cartilage which arched over the facial nerve (Plate 11, fig. 8, VII.) is now once more free (Plate 12, figs. 2, 4, 6, 7, *a.ty.*); it is now an elegant ear-shaped flap, notched below, and in front sending down a membranous tract, which will soon become cartilage.

The broad part of this rudimentary "annulus" lies on the swollen unossified tegmen tympani, covering the ampulla of the horizontal canal (figs. 6, 7, *h.s.c.*), and forming a bridge over the 7th nerve and the tympano-Eustachian cleft, which opens inside, between the suspensorium and the condyle of the pedicle (Plate 12, fig. 3).

In the recess beneath this double roof the epi-hyal element (columella) has now become almost normal in form, but is not yet ossified (figs. 4, 6, 7, *co.*).

Although relatively so much smaller, it is no hard task to harmonise this element with the hyo-mandibular of a Skate, which is developed quite independently of the cornto-hyal, and perchance a little after it in point of time.

There is but one piece of cartilage, for the inter-stapedial segment is not distinct; this part is broad, oblique, and wedges in between the ear-sac and the stapes (*st.*), now a large lozenge-shaped cartilage.

The extra-stapedial region is bent downwards, it is ligulate, and retains the breadth of the medio-stapedial region; there is no supra-stapedial process.

The mandible has more than doubled the length it had in the last stage (Plate 12, fig. 1, *mk.*); the angular process is reduced greatly, and the loosely swinging hinge is quite unlike that of the larva.

The inferior labial has coalesced with the distal end of MECKEL's cartilage (*mk.*), but the dentary (*d.*) has not yet converted it into a "mento-Meckelian" bone. The proximal broad end of the lower jaw lies in a thin bony trough, the "articulare" (*ar.*); this is developed most on the inside.

The cerato-hyal (*c.hy.*) is now attached by the "hyo-suspensorial ligament" (fig. 4, *h.s.l.*) to the inside of the suspensorium at its lower third; its styloid region (fig. 4, *st.h.*) is somewhat uncinate, and its hypo-hyal region (*h.hy.*) decurved; the basi-hyal has vanished.

A single rudiment (fig. 5, *ex.br.*) on each side remains of the four inner and four outer branchials; this plate embraces the basi-branchial (*b.br.*) which is wider than long, and has a crescentic emargination in front.

The pair of hypo-branchials (*h.br.*) are not much altered, but their hinder part, which sends out rudiments of a *third* and *fourth* cerato-branchial (Plate 11, fig. 4, *c.br.*, 3, 4), has now become a solid "horn," ready to ossify as a "thyro-hyal" (*t.hy.*).

This is a very instructive stage of the hyo-branchial apparatus; if it be compared with the larval structures (Plate 10, fig. 6; and Plate 11, fig. 4), and with those of an adult Frog (see in *Rana pipiens*, Plate 8, fig. 5), we shall see what becomes of the copious cartilaginous growths of the larval branchial arches.

Before leaving this stage I have to remark upon the evidence these changing larval skulls supply of the existence of a pre-oral visceral arch in the ethmoidal region.

Years ago my study of the development of the skull in Birds and osseous Fishes, besides what I saw in various kinds of adult Birds, left me in a state of mind that scarcely admitted a doubt upon the subject.

The skull of the Tadpole, evidently a most *generalised* structure, covered the whole question with a cloud of doubt.

Since then, whilst wavering, I have collected copious evidence of what seems to me to be a true *second pre-oral rudiment* in the antorbital region, and a first or terminal rudiment in the front of the face.

I have already shown ("Batrachia,' Part II., and "Urodeles," Part I., and in my papers on the "Selachian and Avian Skulls") the very constant occurrence of an ethmo-palatine in several large natural groups of Vertebrata.

Here, in the larva of *Pseudis*, I have shown this part partly segmented off, and when not segmented off, it is yet very distinct, and not to be misunderstood; it assumes, in fact, *four* positions, each of which is like what is seen in the members, generally, of some large group or groups.

This large and most remarkable conjugational bar is parallel with the skull in the two first stages (A and B, Plates 2 and 11).

So it is in the Siluroid Fishes (*e.g., Clarias capensis*), where the palatine bone formed by the ossification of an autogenous cartilage (in these and other Teleostei) remains distinct, but has no "ethmo-palatine" or ascending process, as in the Salmon.

In the third stage (C) it is bent outwards and forwards; this is the natural form and position of its independent homologue in a large number of the "Urodeles."

It afterwards turns round; but in doing this there is a time in which it is directly transverse, or at a right angle to the skull, as in *Menobranchus*.

In *Proteus anguinus*, where it is not drawn forwards by a chondrified nasal capsule, it turns backwards, and so it does also in *Notidanus*, in the Skates generally, and in many Birds.

In the Salmon, as in the adult Toad, the ethmo-palatine has its three regions well developed—an ascending, an anterior, and a posterior part; all these are well seen in the stage of *Pseudis* just described.

Such a rudiment of the ethmo-palatine as exists in the adult of many Urodeles, where it has coalesced with the back and lower part of the nasal capsule, is seen also in the "Sauropsida" (*e.g., Chamaeleo, Dromæus, Casuarius,* and *Struthio*); whilst in many Birds it is separately ossified, and forms, as the "os uncinatum," a most characteristic *endoskeletal* bone in the fore-palate.

The foremost pleural rudiment is less widely distributed as the "pro-rhinal," or "recurrent trabecular cornu." I should not be surprised if it turned out to be the serial homologue of the mandibular suspensorium, of the epi-hyal, and of the epi-branchial elements.*

---

* I have already mentioned (p. 18) that Professor HUXLEY and Mr. BALFOUR do not take the same views of these parts as Professor MILNES MARSHALL and I do.

17 (continued).—(E) Skull of adult *Pseudis paradoxa*.—Old male; $2\frac{1}{4}$ inches long. Surinam. (HYRTL's prepn. in Mus. Coll. Surg., Eng.)

The specimen whose skull is here figured was evidently a very old individual; it is therefore of great value in this series.*

It is at once seen to be a very generalised type (Plate 10, figs. 1–4), sharply separated from a normal Frog, such as *Rana temporaria*.

The skull is one-half the length and a little more than half the breadth of that of the perfect larva.

The skull of the nearly metamorphosed larva (D) is *five-sixths* the length of that of the adult; so that after lessening *much*, it enlarges again, a *little*.

About the size of the skull of a large Common Frog, it is more elegant in shape, being a very perfect semi-ellipse in outline.

The well-bent bow formed by the two series of cheek bones has its "horns" meeting at a sharpish angle, for the parts in front are modelled on three trabecular outgrowths, the "pro-rhinals" and a small "prenasal."

The complex structures that run across, behind, from one cheek to the other, are also very beautiful in their construction (Plate 10, figs. 1–3).

The hind part of the skull is densely ossified from the occipital condyles to the optic foramen (oc.c., II.), and in front the ethmoidal girdle-bone (*eth*.) occupies the hind half of the nasal capsule below, and all but the edge in front; the parts in front of that, and the anterior sphenoid (*o.s*.), are unossified.

The occipital condyles are rather wide apart, being separated by a large semicircular emargination; they are hemispherical and subpedunculate, as in those Urodeles that have an *intercalary* "odontoid" vertebra.

Below (fig. 2), the auditory masses seem to stand out in a directly transverse line; but above (fig. 1), they are seen to turn forwards as well as outwards, and the well-marked semicircular canals (between *pr.o*. and *e.o*.) are some distance from the edge, which is flanked by the squamosal (*sq*.): the intervening part is the ossified "tegmen tympani."

The gently rounded floor of the vestibule (fig. 2) shows no trace now of that *Selachian* development of the basal plate, outside the capsules, which formed the *tympanic floor*.

The main nerve-passages (fig. 2; II., V., IX., X.) are well seen from below; the two hinder passages are separated by a bony bar: the "foramen ovale" is enclosed in bone, and the optic *fenestra* is partly margined, also, by the extended alisphenoidal wing of the prootic (*pr.o*.).

The foramen magnum has less *roof* than *floor*; neither of these plates has any divisional line as in a normal Batrachian.

* It is a most fortunate thing for me that our College has possessed itself of this invaluable specimen; it is not an easy thing to get an adult.

Nor has this generalised growth (both of cartilage and bone) allowed any deficiency of roof below the parietal region; there is only one, the main, fontanelle: this space is barely covered by the shortish frontal end of the roof-bones (*f.p.*).

The roof-bones are completely *re-united*, and the frontal suture is partly closed, as well as the whole of the sagittal; the skull is sub-sulcate towards the mid-line (fig. 1.).

These roof-bones leave much of the endocranium uncovered behind as well as in front; a little temporal wing, and a moderate orbital plate is formed by them.

The parasphenoid retains the form it had in the large larva (Plate 2, fig. 2, *pa.s.*); and it is very characteristic of *Pseudis*, being a dagger, gapped at the point, with a wide elegant "guard," and no "handle."

The large girdle-bone (*eth.*) extends its osseous substance half across the cheeks, half along the internasal region, and it trespasses on the orbito-sphenoidal headland.

Together, these elements make the interorbital part of the cranium (its anterior two-thirds) a boat-like structure, gently narrowing forwards.

Then, on a sudden, it widens again, passing out into the palatal and nasal structures.

Above (fig. 1, *n.*), the elegant conchoidal nasals cover much of the roof, and run down over the ethmo-palatine wings, but they do not hide the middle over the front extension of the creeping ossification from the girdle-bone; above (fig. 1), the soft prenasal (*p.n.*) finishes the face.

The ectosteal palatine (fig. 2, *pa.*) binds under the half-ossified ethmo-palatine; it is falciform, with the narrow handle inwards, far from that of its fellow of the other side.

Hiding the semi-osseous sub-nasal plate, except in front and in the middle, are the two vomers (Plate 2, *v.*), they are sub-reniform, with a lateral spike; behind, they are massive and dentigerous: this lenticular part is like what is seen in several Frogs (both "Platydactyla" and "Oxydactyla") from the *Notogœa*.

The internal nostrils (fig. 2, *i.n.*) are large and round; the external nostrils (fig. 1, *e.n.*) are moderate in size and are guarded by two valvular cartilages, the inner and outer upper labials.

The arc of cartilage running from the palatine to the pterygoid becomes thin, but I cannot find that it is segmented; the latter bone (*py.*) is pointed in front, then becomes very broad, and this wide part forks into the sheath of the pedicle (*pd.*) and a posterior process.

The posterior process becomes sub-vertical, binding the inside of the suspensorium to the condyle of the quadrate (*q.c.*).

The inner fork arches inwards and enclothes the stunted pedicle (*pd.*) up to the condyle, which glides on a pad of cartilage on the fore-face of the auditory capsule (fig. 2.).

In the rounded angle between these two forks we see the large, oval, oblique, Eustachian opening (*eu.*); this is enclosed behind, by the geniculate stylo-hyal (*st.h.*).

The suspensorium forms a very obtuse angle with the axis of the skull, and its condyle reaches to a transverse line that cuts the neck of the occipital condyles.

The hinge (*q.c.*) is a bilobate condyle, the inner lobe being the larger; the substance of the cartilage above the condyle is ossified as a *quadrate bone*, by the engrafting on it of the "quadrato-jugal" (*q.j.*).

A tract of cartilage can be seen outside the pterygoid (fig. 3, *pg.*), and this may be traced into the pedicle, and to the *borrowed* quadrate centre.

The backward position of the suspensorium is the cause of the very tilted position of the squamosal (fig. 3, *sq.*), the lower part of which, partly hidden by the ear-drum, runs backwards as well as downwards; the supratemporal part (fig. 3, *sq.*) is very long and sigmoid, for it sends forwards a sharp out-bent postorbital process.

The premaxillaries (*p.x.*) are large and well formed, having a well-defined dentary margin, a triangular palatine process, and a high nasal process, capped, inside, by the inner upper labial.

The dentary edge of the face is finished by the maxillary, the teeth ending opposite the Eustachian tubes, and the bone opposite the middle of the auditory capsules.

There are small "septo-maxillaries (*s.mx.*)," and the maxillaries (*mx.*), notched in front, run well up to the premaxillaries and the nasals; the jugal part is high.

The quadrato-jugals (*q.j.*) are short, high, and are grafted upon the quadrate region of the suspensorium, as aforementioned.

The cartilaginous mandible is placed obliquely in its "articular" trough (*ar.*), that bone lying mainly below, and on the inside; the dentary (*d.*) has formed a "mento-Meckelian" bone (fig. 3, *m.mk.*), by ossification of the end of the rod; *once* the free inferior labial!

The stapes (Plate 10, fig. 4, *st.*) is thick and reniform; for the antero-superior edge is sinuous to admit of the large dorsal part of the "columella," between it and the capsule.

The inter-stapedial end of the columella is thin and clawed below; the upper part is large and blunt; by these it holds the stapes, as it were.

The "medio-stapedial" bone (*m.st.*) is dilated where it runs into these spurs, and then runs as a straight rod up to the extra-stapedial (*e.st.*). This latter part is at first no thicker than the end of the bone from which it arises; it turns downwards, and soon enlarges into a tranversely oval disc.

There is no "supra-stapedial," even as a membranous band, and the interstapedial also not being segmented off, this columella is much below the normal condition.

The "annulus tympanicus" (*a.ty.*) is large, and its horns nearly meet above.

The stylo-hyal end of the cerato-hyal band (*st.h.*) has coalesced with the auditory capsule, a moderate space from the front of the fenestra ovalis; it then turns directly outwards, finishing the Eustachian rim (*eu.*), and from thence passes downwards and forwards to the basal plate, which differs but little from the norma.[*]

Second genus. *Gomphobates*.

18. *Gomphobates* (*Leiuperus*).——? sp.—Adult (?) ; 10 lines long. River Plate.

This small Frog was, apparently, adult, and yet so arrested in certain respects, as well as so very small, that it answered to the young of the Common species taken about the end of July. The chondrocranium of this kind is so very instructive that the figures are only made to show this, and certain of the bony plates attached to it.

The general outline of the skull (Plate 13, figs. 1, 2) is semi-oval and the breadth is a little more than the length. The notochord (*nc.*) is still large, and lies upon the united moieties of the basal plate. The occipital condyles (*oc.c.*) are normal; and the emargination between them moderate; but the roof has a much larger crescentic notch than usual over the foramen magnum (*f.m.*). The auditory capsules are like those of a young Common Frog in general condition, but they have a peculiar character seen in metamorphosing larvæ of *Pseudis* (Plate 11), viz.: a floor to the tympanic cavity projecting beyond the fenestra ovalis and stapes (fig. 1, *st.*). The tegmen tympani is of very small extent, and only exists in front.

The 9th and 10th nerves (IX., X.) are surrounded by a very limited ex-occipital (*e.o.*), and the prootic (*pr.o.*) only forms an oval patch behind the foramen ovale (V.); this is entirely on the lower aspect of the skull; the ex-occipitals just reach the upper surface between the ear sac and the occipital arch.

The interorbital region is rather broad, a little bulging, and considerably narrower in front than behind; it is well walled in with cartilage, the optic fenestra (II.) being of the average size; and the top wall grows over on to the roof for some distance, leaving *one* oval fontanelle (*fo.*) about half as large as the whole roof. The anterior part of the tegmen cranii is well developed, and the hind part is nearly as long as the fontanelle. The ethmoidal wings end abruptly above, and *articulate by a flat facet* with the ethmo-palatine (*e.pa.*), but are quite continuous with it below; there is no rudiment even of the girdle-bone.

The figures given of the upper and lower aspects of the ethmo-nasal region, might

---

[*] The skull of the adult *Pseudis* differs from the "norma" in many things, viz.:—
1. Intense ossification of endocranium.
2. Sub-pedunculate occipital condyles.
3. No secondary fontanelles.
4. A prenasal rostrum.
5. Solid dentigerous lobe to vomers.
6. No proximal segment or distal process to columella.
7. Hind edge of parasphenoid emarginate.
8. Extreme solidity of pedicles.
9. Ossification of quadrate region.

be taken as *typical diagrams* of those parts in the "Anura" generally; for although coalesced, they show the outlines of all the parts that go to make up the whole (figs. 1, 2).

The tegmen cranii in the ethmoidal region is continuous with the top of the large high intertrabecular crest, the outline of which is clearly seen, both above and below, as a gradually narrowing tract of cartilage, convex above and below, and ending in a free, blunt spike. Here we have the perpendicular ethmoid passing into the septum nasi and ending in the prenasal rostrum (*p.n.*).

In front of the ethmo-palatine bars the trabeculæ have a large crescentic notch on each side; here the inner nostrils (*i.n.*) are situated, yet at this part the latter are not much narrower than in the interorbital region.

In front, they spread into wings three-fourths the extent of the auditory masses, behind; and each wing is divided by a large rounded notch into a large hind, and a lesser fore, lobe.

The hind lobe, or angle of the subnasal lamina (*s.n.l.*) forms a rounded hook in front, which turns inwards; on the upper surface (fig. 1) each plate has a transverse crest, which runs inwards and articulates with the out-turned, anterior horn of the nasal roof-cartilage (*na.*). The anterior lobes of the primary trabecular cornua are nearly as long, and one-third the width of the hind lobes; they are a little arcuate, finger-shaped, and diverge at more than a right angle; these are the pro-rhinals (*p.rh.*). We have, thus, a five-fingered end to the chondrocranium, and these five lobes are all developments of the *pre-pituitary on-growths of the three basi-cranial cartilages* that in the post-pituitary region embrace and enclose the notochord.*

Over the nasal sacs, the cartilaginous roof (*na.*) is seen to be composed of a pair of ear-shaped shells, lying back to back, not touching each other, but confluent with the edges of the septum nasi.

Behind, they have also united with the ethmo-palatines (*e.pa.*), and in front with the transverse ridge on the outer trabecular lobe (*s.n.l.*). The inner margin of each is semi-circular, the outer is sinuous, widest in the middle, and each end is developed into a cornu; the front horn is blunt and the hinder horn sharp. On the front horn there is seen the second upper labial (*u.l².*) (the *first* has been left unfigured with the pre-maxillary); between this and the front horn is the outer nostril (*e.n.*).

---

* Yet the *true skeletal axis* ended behind the pituitary space and body, and the *organic fore end* of the neural axis ended just in front of that part, through the overbending of the vesicles of the brain. The *continuous*, or *distinct*, conjugating bar by which the fore-growth of the mandibular pier becomes attached to the ethmoid, bears an endoskeletal relation to the outer bones in the *malar region*.

The angle of the "subnasal lamina" bears the same relation to the *maxillary region*, and the pro-rhinals to the *inter-maxillary region*, and these pre-oral endoskeletal and exoskeletal structures, combined, make up one antagonistic upper jaw, to work against the *post-oral* lower jaw.

Moreover, there are, besides these three pre-oral *rudiments* (or *imitative arch-pieces*), the "post-palatine" cartilage of the Axolotl, the "transpalatine" of Singing Birds, and the "epipterygoid" of Crocodiles, Chelonians, and Lizards, to be accounted for. These three latter, I suspect, are different modes of the same element.

The palato-suspensorial arches and their relation to the nasal region are like those of the *fourth* of my stages of *Pseudis* (see Plate 12). The "ethmo-palatine" (*e.pa.*) is a thick transverse bar, whose flattened upper end articulates with the ethmoidal wall. The "pre-palatine" (*pr.pa.*) is split into two laminæ, the outer one is rounded, and the inner and upper lobe is sharp; the outer runs into a second spike, turning inwards on the transverse bar (fig. 2). The "post-palatine" (*pt.pa.*) now forms a crescentic crest on the top of the far-extended pterygoid cartilage (*pg.*), where it turns inwards, in front.

The former was free behind; and its new and fixed position, as compared with its original condition, is well seen by comparing my *third* and *fourth* stages of *Pseudis* (Plates 11, 12).

The rest of the arch keeps its even size as a rather wide band, until it forks; the short inner fork is the pedicle (*pd.*), the longer outer fork is the quadrate and its hinge (*q., q.c.*), which reaches to a line crossing the middle of the stapes (*st.*). The palatine and pterygoid bones (*pa., pg.*) are normal: the former looks like the bone of an *old* individual; the latter is thin and arrested.

The Eustachian openings (*eu.*) are large and turned obliquely outwards and forwards; the stapes (figs. 5, 6, *st.*) is an elegant oval, and has wedged inside its front margin a much smaller oval cartilage quite distinct—the interstapedial hind part of the columella. The main part of the columella (*co.*) is not developed into regions; it is long, sinuous, and finger-shaped, with a crest near the stapes: but for the separate interstapedial, this columella would have corresponded with that of *Pseudis* at its *fourth* stage (Plate 12, fig. 6). The narrow "annulus" (*a.ty.*) is arrested; it is a knee-shaped strap of cartilage bent forwards.

The mandibular and hyo-branchial plate (figs. 3, 4) correspond with those of a very young Frog; the mento-Meckelian is not finished, and the cerato-hyal (*c.hy.*) runs short of the ear-capsule; there is no hypo-hyal lobe, and only the hinder part of lateral lobes.

Nearly all the investing bones have the same arrested character; the parasphenoid (fig. 2, *pa.s.*) is, however, very large; the vomers (*v.*) are small, angulated crescents, without teeth.

The fronto-parietals fail to cover more than half of the great fontanelle; this corresponds to what has been found in other species of this genus.[*]

The deflection from the *norma* seen in this small skull is very largely due to arrest, on account of the small size of the species, and perhaps also to a somewhat immature state. These deficiencies, as compared with the skull of an adult of the common kind are:—

---

[*] Professor E. D. Cope—"Eleventh Contribution to Herpetology of Tropical America," Amer. Phil. Soc., June 20th, 1879, p. 264—says that *Leiuperus* (a synonym of *Gomphobates*) has an open fontanelle.

1. The small size of the ex-occipitals and prootics.
2. No girdle-bone.
3. The fontanelle single and largely uncovered.
4. The evident outlines of the elements composing the nasal region.
5. The presence of a prenasal rostrum.
6. The articulation of the ethmo-palatines with the ethmoid above.
7. The rudimentary condition of the annulus and columella.
8. The arrested state of the mandible and simple condition of the hyo-branchial structures.
9. The absence of teeth on the vomers.

Third genus. *Cystignathus.*

19. (A) *Cystignathus ocellatus* (?).—First larva. Brazils.*

First larva; 2⅙ inches long; tail 1½ inch; hind legs not visible.

This is the most immature of the Batrachian larvæ of which I have as yet given illustrations in this memoir; I have already referred the reader to much earlier stages, figured and described in my published papers.

These skulls are relatively very solid, and what strikes the eye at once is their *oblong* form and the straightness and width of the cornua trabeculæ. The chondrocranium (Plate 17, figs. 1, 2) is quite complete, and it has passed from the simple primary condition by having had three osseous tracts applied to its surface (*f.p., pa.s.*).

Except for the projecting cornua and labials, the skull up to the quadrate condyles (*q.c.*) is very evenly oblong, and only one-tenth longer than broad. The occipital condyles (*oc.c.*) project but little beyond the auditory swellings; the halves of the basal plate are separated by a considerable notochord (*nc.*), and the cartilage which unites the basal plates with the periotic capsules is young and crowded with corpuscles.

The tegmen cranii, both fore and aft, is also made of young cartilage, and its boundaries are traceable; it leaves *one* large fontanelle (*fo.*), which is pyriform in outline, with the stalk behind. Up to the middle of the internal nostrils (*i.n.*) the intertrabecular plate has conjugated the trabeculæ (*tr.*); from thence the massive cornua (*c.tr.*) diverge gently. Behind, the canals of the ear (fig. 1, *a.s.c., h.s.c., p.s.c.*)

* These were taken with the species here named, *and presumably belonging to it*, from Rodsio (a tributary of Rio des Macacus) above the Falls, Brazils, May, 1865; they are the gift of Prof. A. Agassiz. I shall treat of these as belonging to the type named, with the above explanation, as also the skull of another, much smaller larva, with no title, but only the locality from which it was taken, namely, Lake Jannarg, Manaoo, Brazils. Whichever genus the larger Tadpoles belong to, to that also the little larva belongs; these two species have larval skulls very distinct from what is seen in *Rana, Calyptocephalus, Cyclorhamphus,* the Hylidæ, or the species of *Bufo;* they are most like those of the earlier stages of *Pseudis.* Merely as a study of a variety of the larval Batrachian skull they are very valuable; and I trust that some friend will verify their title for me, and if there be any error, correct it.

shine through the thin walls of the sac, which is a small spheroid, growing into a free edge on the outside, above. That "cave" of cartilage is the tegmen tympani (*t.ty.*), behind, and the confluent spiracular cartilage (*sp.c.*), in front; this ear-shaped flap just touches the round elbow of the suspensorium—its otic region (*ot.p.*). The infero-lateral face of the capsule is splitting to form the "fenestra ovalis" (*f.o.*), but the *stapes* is not yet formed. Contrary to what was seen in the last two kinds, the inter-orbital or fore half of the cranium proper enlarges from behind forwards; it also bulges a little at the sides.

There are scarcely any "ethmoidal wings" (*al.e.*) for the post-palatine rudiment (*pt.pa.*) lies close to the skull, and the pterygo-palatine band (*p.pg.*) is very short; it turns forwards and outwards, and is surmounted by the thick, oblong post-palatine. In front of the internal nostrils (*i.n.*) the trabeculæ (*c.tr.*) do not grow towards the pre-palatine rudiment (*pr.pa.*) as in the last two kinds; they only, there, suddenly become wider. The sides of the ethmoid (*al.e.*) can be seen to be *older* than the tegmen (fig. 1, *tr.*) between and over them, and this latter part covers the rudimentary "mesethmoid."

The gently diverging cornua (*c.tr.*) increase a little in size up to their fore end; but they do more than this, they are convex above, turn gently downwards, and at their outer edge curl over; this fold thickens up to the end, where it is very solid; it there articulates with *both* the upper labials of that side (*u.l¹.u.l².*); this is very similar to what is seen in *Pseudis* (Plates 2 and 11), where the upper labials are not divided.

The falcate hinder part of the suspensorium (*sp.*) unites by its hooked end (*pd.*) with the basis cranii, and this "pedicle" passes in between the orbito-nasal and vidian nerves (V′., VII′.). The sub-oval condyle for the cerato-hyal (*hy.f.*) is inferior, and projects from the root of the large orbitar process (*or.p.*) which nearly touches the skull-wall; it is only separated by a thinnish tract of fibrous tissue. Its ribbed edge is decurrent along the suspensorium, fore and aft, reaching nearly to the condyle (*q.*, *q.c.*) and nearly to the ear. The pre-palatine rudiment (*pr.pa.*) is an upturned "ear" of cartilage, giving attachment to the ligament that bounds the inner nostril in front.

The quadrate condyle (*q.c.*) is saddle-shaped, and looks somewhat inwards, as well as forwards and downwards; it has a round knob of the suspensorium outside it. The upper labials (*u.l¹.u.l².*) are formed (*as cartilages*) separately; the inner pair are sub-crescentic, the outer pedate,—the "toe" being the postero-external angle.

The mandibles and lower labials (fig. 3, *mk.*, *l.l.*) are of the same form as in *Pseudis*; they are short, massive, and obliquely articulated together.

The other inferior arches (fig. 4) have their own characters.

The cerato-hyal (*c.hy.*) is remarkable for the hollowing out of its upper part, which is a curved shell of cartilage; the condyle (*hy.c.*) is oval, and there is a spike in front of it. The basi-hyal (*b.hy.*) is composed of simple cartilage, the basi-branchial (*b.br.*) of hyaline, and it has a budding second joint. The large hypo-branchials (*h.br.*) are wide in front, and gradually narrower behind, where they are notched.

The extra-branchials ($ex.br.^{1-4}$) are quite normal, and each has its own cerato-branchial rudiment ($c.br.^{1-4}$) coalesced below with it.

The fronto-parietals (fig. 1, $f.p.$) are small blades of bone, pointed in front, broader behind, and resting on the roof-edge of the skull.

The parasphenoid (fig. 2, $pa.s.$) is very noticeable; it is U-shaped, and half cleft into two pieces; this character attains in this, and in its congeneric larva, to an extent not seen by me in any other type of the "Ichthyopsida;" a front notch is seen in every stage of *Pseudis*, but it is small.

Most of the cranial nerves (fig. 2, II., $V^1.$, $V^{2,3}.$, VII., VII'., IX., X.) are shown issuing from their foramina, as in the last kind of larva.

19 (continued).—(B) *Cystignathus ocellatus* (?).—Second larva; $3\frac{1}{4}$ inches long; tail, $2\frac{1}{4}$ inches; hind legs, 1 line long.   Brazils.

In development, this Tadpole corresponds very exactly with my huge *youngest* larva of *Pseudis* (Plates 1 and 2); the hinds legs are *relatively* of the same size and are at the same stage.

The chondrocranium (Plate 17, figs. 5, 6) is one-half larger than the last, and certain changes have taken place in it of importance. The notochord ($nc.$) is shrinking, and the basal plate closing in upon it; the occipital condyles ($oc.c.$) are better formed, and the cranial cartilage has fused more completely with the auditory capsules. A discoidal exoccipital encloses the 9th and 10th nerve ($e.o.$, IX., X.), and the prootic ($pr.o.$) is beginning to creep up the front of the auditory capsule behind the foramen ovale.

The canals of the ear ($a.s.c.$, $h.s.c.$, $p.s.c.$) are very clearly seen above; below, the capsules are getting flattened sides, which lessen backwards; in the side, under the large leafy tegmen tympani ($t.ty.$) the fenestra ovalis has become almost vertical, and a nucleus of cartilage is now to be seen in the soft plug of tissue that fills it; this is the stapes (fig. 6, $st.$). The tegmen tympani is elegantly angulated behind; in front it is continuous with the spiracular cartilage ($sp.c.$), a large sub-falcate flap, which has coalesced also now with the budding "otic process" ($ot.p.$). The interorbital region of the skull is more perfect, but of the same shape; the pear-shaped fontanelle ($fo.$) is widest, now, further back, and the front tegmen is more solid, and has united with the trabecular edges and hidden median cartilage to finish the ethmoidal region. The olfactory nerves (I.) are seen emerging, but the nasal roofs are still membranous, and were removed in the preparation. Yet the intertrabeculæ is filling in the space between the trabeculæ up to the membranous margin of the internal nostrils ($i.n.$). That membrane, however, has in it no spur growing from the trabecular cornu ($c.tr.$), which is narrowed further forwards than in the last stage; for the rest, these, and the upper labials ($u.l.^1.u.l^2.$), are only larger. The pedicle of the palato-suspensorial arch ($pd.$) is less transverse; the orbitar process ($or.p.$) is further out; the post-palatine rudiment ($pt.pa.$) is now fairly differentiated from the ethmoidal wing ($al.e.$), now manifestly seen to be an *external crest* of the trabecula, a continuation of the cranial wall, but in front of the cranial cavity.

For the rest, the quadrate region and its condyle (*q.*), the mandibles (fig. 7, *mk.*) and the lower labials (*l.l.*) have merely increased in size since the last stage. The same may be said of the fronto-parietals (*f.p.*); they are a little straighter, however. But the parasphenoid (fig. 6, *pa.s.*) has been transformed into the normal shape, or nearly; the notch is now relatively much smaller; its shaft is long and wide also; the basi-temporal wings are now developed, they are pointed, raggedly lobate, and meet in a triangular "handle."

19 (continued).—(C) *Cystignathus ocellatus* var.—Adult male: $5\frac{1}{2}$ inches long. Dominica.

This large sub-typical Frog has a skull (Plate 16, figs. 1-5), whose length is to its breadth as 7 to $8\frac{1}{2}$; it is a very regular half-oval in outline, and the condyles for the lower jaw reach, very nearly, as far back as those for the *atlas*. Being of great breadth, the auditory regions are extended like outspread arms, the parotic processes being as wide as the capsules proper, outside the horizontal canals.

The extent of the ossification of the endocranium is normally Ranine, and there is clear tract of cartilage above and below the foramen magnum (*f.m.*), and the ex-occipitals and prootics (*e.o.*, *pr.o.*) are distinct.

The occipital condyles (*oc.c.*) are large, reniform, postero-inferior, and separated by a space less than their own size; the occipital arch itself is of great width, and well marked off from the auditory capsule, especially above. This is due to the great size of the epiotic eminence (*ep.*), caused by the arch of the posterior canal; this eminence is unossified. So also is the outer half, below, and the outer third, above, the prootic not reaching, even there, as far as to the squamosal (*sq.*). Thus the whole of the extensive tegmen tympani (*t.ty.*) is unossified.

Besides the high epiotic region another ichthyic character appears (seen to a less degree in *Rana pipiens*, Plate 8), viz.: an extensive "pterotic" crest (*pt.o.*); this is a backward and outward sigmoid process of the tegmen tympani, which doubles the extent of the margin and overhangs the large annulus tympanicus (fig. 3, *a.ty.*), behind.

The prootics (*pr.o.*) reach as far as to the middle of the space between the 5th and 2nd nerves (V., II.); the latter lie in a fenestra which reaches half-way from the prootic to the girdle-bone (*eth.*). The interorbital region widens steadily from behind forward, it is rather hollow above and below, for the side view (fig. 3) shows a concave outline both to the roof-bones and the parasphenoid (*f.p.*, *pa.s.*).

This enlargement forward is due rather to the growth of an "eave" over the fore part of the orbit than to any great increase in the size of the cranial trough within; but for that superorbital expansion the cranial trough would be seen as somewhat spindle-shaped, bellying in the middle and narrower at either end. The outer edge of the superorbital expansion is unossified, and in its fore edge the orbito-nasal nerve (fig. 3)

DEVELOPMENT OF THE SKULL IN THE BATRACHIA.

enters. The girdle-bone (*eth.*) is three-fifths the size of the orbital region and stops exactly at the fore edge of the true ethmoidal region ; it does the same right and left.

In the smaller species (Plate 16, fig. 6) a pair of secondary fontanelles exist, and I have no doubt of their existence in the large kind also. The great fontanelle is not very large and is covered in well by the roof-bones (*f.p.*).

That which strikes the eye at once in this skull is the tumid shape of the whole nasal region, which is large in every direction, along, across, and from top to bottom. It is well roofed by cartilage (*al.n.*), but the floor (fig. 2) is still more extensive; in front, the snout is broad and gently convex, not produced, but rounded in the middle and rising well over the openings (*e.n.*). Below (fig. 2), the growth of cartilage is unusually extensive, the angles of the subnasal laminæ (on each side of *s.n.*) are large and falcate; the pro-rhinals (*p.rh.*) are long, wide, and somewhat decurved bands; the wall of the nasal capsule (fig. 3, *n.w.*) is well developed. But the greatest modification is seen in the large size of the ethmo-palatine arms (*e.pa.*), which however have no definite pre-palatine spike (*pr.pa.*). The bony bars correlated with these arms are also very large; they are falcate, with the point growing back under the post-palatine cartilage (fig. 2, *pa.*).

Here we have also what is seen in several Toads, viz.: an additional semi-distinct bony bar, sharp and denticulated, stuck to the lower surface of the main bone at its middle.

The pterygoids (*pg.*) are slenderer than the palatines; they have used up the feeble cartilaginous tract to which they were applied; they fork in front of the Eustachian opening (fig. 2, *eu.*) at a right angle, and thus that large space is itself triangular. The part covering the pedicle (*pd.*), and faced with unaltered cartilage, is very long and pedate, for this point is not far outside the foramen ovale (V.). The hind part of the pterygoid faces the inside of the strong suspensorium, whose condyle (*q.c.*) is a large and well made trochlea. There is some deposit of bone in the quadrate region from the quadrato-jugal (*q., q.j.*).

The annulus (figs. 1 and 3, *a.ty.*) is large and perfect; the stapes (figs. 1, 2, 5, *st.*) is oval, umbonate, and cut away in front; the columella has a condyloid cartilaginous proximal end without a segment; and the strong bony bar (*m.st.*) has two projecting "trochanters." The extra-stapedial (*c.st.*) is tongue-shaped, and its ascending process (*s.st.*) is confluent above.

The mandible (fig. 3) is strong; the condyle (*ar.c.*) is large and reniform; the coronoid crest (*cr.*) is not high; the dentary (*d.*) is only half as long as the ramus; the mento-Meckelian (*m.mk.*) is well developed.

The stylo-hyal (*st.h.*) is confluent above, the rest of the band (fig. 4, *c.hy.*) is narrow up to the hypo-hyal lobe (*h.hy.*); the notch is very large, leaving a basal plate (*b.h.br.*) of small extent at the middle; the fore side lobes are very large, elegant, and reniform; the hind lobes are narrow arcuate bands; the thyro-hyals (*t.hy*) are rather straight and of moderate size; between them there is a lozenge-shaped ossification of the basal plate (*b.br.*).

The roof bones (*f.p.*) are strong, well developed behind, where they rise over the arches of the canals (figs. 1 and 3), have a small temporal wing, and are short and rounded in front, where they leave the girdle-bone uncovered, scarcely covering the great fontanelle.

Also between the temples there is a small gap which answers to the "parietal fontanelle" of a Lizard; the orbital flange of each bone (fig. 3, *f.p.*) is thickish, and very irregular in outline.

Answering to the large size of the cartilaginous ethmo-palatines the nasals (*n.*) are huge bones, the left much the larger of the two, and sending an angular process over the fronto-parietal. Their fore part is pointed and divergent, their facial or descending region thick ribbed and projecting; their top is strongly convex.

The premaxillaries and maxillaries (*px., mx.*) are quite Ranine; they are strong and well developed. The septo-maxillaries (*s.mx.*) are small and inconspicuous.

The quadrato-jugals (*q.j.*) are grafted on to the quadrate; they are strong, arcuate, tusk-like bones.

Also the top of the squamosal (*sq.*) is like a tusk, being curved, and turning inwards in front as a sharp fang; the supratemporal part is of small extent, and the descending plate long and broad.

The parasphenoid (fig. 2, *pa.s.*) is perfectly normal; it is large and swelling, has a long point in front, its wings bent backwards, and its handle triangular.

The vomers (fig. 2, *v.*) are large, many-cornered bones; they bind crescentically round the inner nostrils (*i.n.*), which are large, circular, and very wide apart. The front process of the bone is small, not reaching nearly to the end of the subnasal angle; the terminal plate almost touches its fellow, is a fan-shaped tract, and ends in a thick crescentic ridge, turned obliquely outwards and backwards, and armed with an arched row of sharp teeth. The nasal valve-cartilages (*n.l¹.n.l².*) are well developed.

This skull differs from that of a true *Rana* in :—

1. The epiotic growth.
2. The pterotic growth.
3. Its widening roof forwards.
4. The huge ethmo-palatine.
5. The additional palatine bony crest.
6. Absence of an inter-stapedial.
7. The small extent of, and bony tract in, the basal plate.
8. The large size of the nasals, and their want of symmetry.

20. *Cystignathus*——? sp.—Larva; 1 inch long; tail, $\frac{3}{8}$ inch; hind legs, 1 line.   Lake Januarg, Manaoo, Brazils.

This larva of a Neotropical Frog belongs, evidently, to *the same genus* as the last; it is less than a third the length of the large Tadpole, whose skull has just been described, and in development it is intermediate between the last two.

The difference between this skull (figs. 8, 9) and that of the lesser of the two large Tadpoles is slight, indeed, but there are noticeable modifications. The occipito-auditory regions are more developed, and the parts more fused; the fenestra ovalis (*f.o.*) is larger and wider open, and more *lateral* in position; but there is no stapes. The interorbital part of the skull is of the same shape, and so is the great fontanelle (*fo.*). The ethmoidal region is in the same unfinished condition, but the cornua trabeculae (*c.tr.*) are a little shorter, and the thick edge below is less solid, but it creeps round to the inner edge. Also at their root the cornua have a small snag for attachment of the membrane in front of the internal nostril (*i.n.*). The ribbed edge of the orbitar process (*or.p.*) is extended further back, as a strong decurrent enlargement, nearly to the apex of the pedicle (*pd.*).

The pre- and post-palatine rudiments (the snag inside *q.*, and *pt.pa.*) are much alike in both species; so are the mandibles (*mk.*), the lower labials (*l.l.*), and the cerato-hyal (*c.hy.*). But the upper labials (*u.l'.u.l².*) differ; the right and left inner pair are united in front, and all four are rounder and broader; here we get nearer the Lamprey.

The fronto-parietals (fig. 8, *f.p.*) are large, and moreover they have a sigmoid, and not a straight, form.

But the parasphenoid (fig. 9, *pa.s.*) is the true generic diagnostic; it is more split than in the younger Tadpole of the other species; the notch and the two sharp blades being one-fourth longer than the united part, which has more distinct lateral processes.

If we reflect upon the short time in which this bone changed from what is figured in fig. 2, to what is shown in fig. 6, it will occur to anyone that there is strong reason to believe that, *for a short time*, each "trabecula" had its own investing bone. In the Chick, about the middle of the second week, the hinder parasphenoid ("basi-temporal") is composed of a right and left osseous centre (Phil. Trans., 1869; Plate 82, fig. 2, *b.t.*).

The great constancy of this bone as an azygous basi-cranial splint in the Ganoids, Teleostei, Dipnoi, Urodela, and in most of the Anura, makes one surprised to see it with its apex bitten off in *Rana pipiens* and *R. halecina*; it would not be more remarkable to find it modified by *median division*, also, and here this seems to be the case.

21. *Cystignathus typhonius.*—Adult female; 1¼ inch long. Porto Rico.

This small species has a much more ossified skull than the last large kind, just as the skull of the small *Cyclorhamphus marmoratus* (Plate 20, figs. 1–3) is much stronger than that of the large *C. culeus*, as Mr. GARMAN has shown.

On the whole this skull is very similar to the last, but its occipital condyles (Plate 16, figs. 6 and 7, *oc.c.*) are wider apart. The epiotic prominences (*p.s.c.*) are less, and the pterotic processes (*pt.o.*) are much shorter. The parotic wings are much less, for this skull is only *one-thirtieth* wider than long. The canals are well marked (*a.s.c., h.s.c., p.s.c.*), also the whole occipito-auditory region is ossified, and this bone runs up to the optic fenestra (II.). Between and a little below these spaces the floor of the skull is cartilaginous; but from thence up to the middle of the septum nasi, along the axis, the endocranium is all ossified, and the bone runs into the extremely large ethmo-palatine (*e.pa.*), leaving only the pre-palatine hook (*pr.pa.*) soft. Above, the bone is continued nearly to the front of the broad transverse snout, so that there are only a few places where cartilage remains. Behind, only the outside of the auditory floor is left below, and the pterotic angle (*pt.o.*) above, whilst behind these parts only the condyles retain their bark of cartilage. The three fontanelles (fig. 6) are quite normal, but they are rather small, as there is very much ossified overgrowth of cartilage (*tegmen cranii*).

The orbital region is almost oblong, rather narrower in front than behind, and bulging very little. The roof (*al.n.*) of the nasal region is well developed; the floor is of less extent than in *C. ocellatus*, but the pro-rhinals (*p.rh.*) are large and hooked. The huge bony ethmo-palatines (*e.pa.*) help the bones to enclose the large circular nostrils (*i.n.*), which are wide apart; the palatine bones (*pa.*) are well developed, falciform, and with an additional inferior sharp crest.

The pterygoids are feeble and sinuously inbent, with a much shorter part for the pedicle (*pd.*) than in the last; they have consumed nearly all the cartilage. The Eustachian passages (*eu.*), lying in their angle, is sub-pyriform in outline.

The quadrate is well ossified by the quadrato-jugal (*q.j.*), and the condyles (*q.c.*) are large well-made trochleæ. The annulus (*a.ty.*) is large and perfect; the stapes (figs. 6, 7, and 10, *st.*) is oblique, emarginate, and has a boss; the feeble columella has a large undivided emarginate unossified proximal part; the bony bar (*m.st.*) is feeble and curved. The extra-stapedial (*e.st.*) is a small irregular hook, and its ascending process is fibrous.

The mandible (fig. 8) is rather slender; it has a long condyle (*ar.c.*), a sharpish coronoid crest (*ar.*), a short dentary (*d.*), and a well-formed mento-Meckelian (*m.mk.*).

The stylo-hyal (fig. 7, *st.h.*) is confluent above; the hyo-branchial apparatus (fig. 9) is similar to the last, very elegant in form, has its front lateral lobe dentate in front, its basi-branchial ossicle (*b.br.*) divided in the middle, and there is a band of bony deposit along the edges of the basal plate; the thyro-hyals (*t.hy.*) are long, straight, and slender.

The nasal valve-cartilages (*n.l¹.n.l².*) are well developed.

The investing bones are also well developed; the fronto-parietals (*f.p.*) cover the temporal and part of the epiotic region, just roofing the lesser fontanelles, behind, where they form a straight margin; also in front they form a straight margin, and only just cover in the large fontanelle (fig. 6). The nasals (*n.*) are large, but less irregular, convex, and ribbed than in the last. The marginal bones (*pr., m.r., q.j.*) are all normal;

they are slight, but dense; there is a small septo-maxillary (fig. 6, *s.mx.*). The postorbital process of the squamosal (*sq.*) is straight, the supra-temporal tract small, the descending part normal. The parasphenoid (fig. 7, *pa.s.*) is not so attenuated in front, nor do the wings turn backwards as in the last kind; the tract between the wings is strongly ribbed in a cruciform manner, and the whole hinder part is largely anchylosed to the ossified endocranium.

The vomers (fig. 7, *v.*) are similar to those of the last, but they are broader, and with smaller notches in their margin. To the discrepancies between the last and the type, we must add (1) the highly osseous state of the skull, (2) the ossified ethmo-palatines and quadrates, (3) the anchylosed parasphenoid, (4) the fibrous supra-stapedial, and (5) the additional centres of the basal plate.

Fourth genus. *Pleurodema*.

22. *Pleurodema Bibronii.*—Adult female; 1½ inch long. Chili.

The skull of this species is very evenly semi-oval, and its greatest breadth is to its length as 8 to 7; it, therefore, is one of the short skulls (Plate 18, figs. 1, 2).

At first sight its main difference from the "norma" seems to consist in its open fontanelle (*fo.*) and the absence of the two secondary fontanelles, which open, like sky-lights, in the hind skull of more typical forms.

More attentive observation, however, brings out several aberrant characters.

The occipital condyles, the basi- and superoccipital tracts of cartilage, and the foramen magnum (figs. 1 and 2, *oc.c., f.m.*) are all normal. The occipito-auditory region is more extended, laterally, and more ossified than in the type.

Moreover, there is here a character which I have already described in another aberrant Neotropical Frog, namely, *Pseudis*; this is a tract of cartilage running over the crest of the posterior canal (*p.s.c., e.o.*), and which, in *Pseudis*, is due to segmentation of the originally continuous bony matter answering to both prootic and ex-occipital —that is an unmistakably *archaic* character. The basioccipital cartilage (fig. 2) is only half the width of the upper tract (fig. 1); this latter is a large wedge-shaped tract (*t.cr.*) which runs to, and forms the hind margin of, the great single fontanelle (*fo.*).

This is bounded by the ex-occipitals behind and by the prootics in front (*pr.o., e.o.*), and the latter run forwards just as far as the cartilage above, up to the optic fenestra (II.); they therefore take in the whole "alisphenoidal" wall as well as the "prootic" region. The "tegmen tympani" (*t.ty.*) is extended far outwards, and this *parotic* projection is largely cartilaginous; beyond the horizontal canals (*h.s.c.*) the auditory region loses one-third of its breadth, and this is bevelled off from the hind margin, mainly. In front the cartilage is nearly one-third the breadth of the whole auditory region; behind, it is nearly one-half, that is *above*. But below (fig. 2) the bony matter reaches to the edge of the sub-convex vestibular region (*vb.*), with its fenestra and cartilaginous plug or stapes (*st.*).

The interorbital region is rather pinched than swollen in the middle, and gently narrows, forwards; from the hind margin of the optic fenestra (II.) to the front of the true ethmoidal region, half behind, is cartilage, and half, in front, is bony; the latter tract is the "girdle-bone" (*eth*.). This bone has a convex fore, and a concave hind, margin; it nearly reaches the fontanelle above and spreads a little into the ethmoidal wings, right and left. There is no trace of a superorbital "cave."

The nasal region is normal as to its wide floor (*s.n.l.*), gradually lessening, backwards; it has large pro-rhinals (*p.rh.*); and it wants a pre-nasal rostrum; but the roof (fig. 1) is very narrow, and the thick wedge-shaped end of the septum nasi (*s.n.*) is ossified separately: this bone (*p.n.*) is a low triangle, with its apex backwards.

The large sub-oval orbital spaces, and the strongly bent bow of the palato-suspensorial are normal, but wider than usual; the pterygoid bone (*p.g.*), with the pedicle (*pd.*), are also normal; but the palatine bone (*pa.*) is composed of two sub-equal pieces. These are large in proportion to their cartilaginous model, and the outer piece is large and falcate. The quadrate hinge (*q.c.*) is normal; above it the quadrato-jugal (*q.j.*) has ossified a good tract of the cartilage. In the rounded angle formed by the forks of the pterygoid the Eustachian passage (*eu.*) is of medium size, is oblique, turning outwards and forwards.

The "annulus" (*a.ty.*) is broad, rather small, and imperfect above; the stapes (fig. 4, *st.*) is large, oval, sub-pedunculate, and obliquely emarginate on its antero-superior edge.

The columella (fig. 4) is pistol-shaped, there is no interstapedial cartilage separate from the medio-stapedial; this bar is gently arcuate (*m.st.*), and ends in a broadly spatulate extra-stapedial (*e.st.*) with a strong supra-stapedial (*s.st.*) confluent above.

The mandible (fig. 3, *d.*, *ar.*, *m.mk.*) is perfectly normal, and so is the hyo-branchial apparatus; the stylo-hyal (fig. 2, *st.h.*) is articulated, above.

The investing bones (figs. 1, 2) are on the whole similar to those of a Common Frog of the same size, but they are less solid. The parasphenoid (*pa.s.*) has a similar form, and so have the vomers (*v.*); these, however, are turned outwards much more. The pre-maxillaries, septo-maxillaries, maxillaries, quadrato-jugals, and squamosals have the typical form, but are all rather feeble.

But the fronto-parietals (*f.p.*) are very arrested; and although they reach from the ethmoidal wings to the junction of the anterior and posterior canals (*a.s.c.*, *p.s.c.*) they are each only two-thirds the width of the naked space between them.

The contrasts between this and the typical skull arise mainly from the following characters:—

1. There are no secondary fontanelles.
2. The fronto-parietals are abortively developed, leaving the fontanelle largely open.
3. The occipito-auditory ossifications, right and left, are generalised, and not properly divided into prootics and ex-occipitals.
4. There is an "anterior septal bone" in the nasal region.

5. The palatine ectostoses are in two pieces on each side.
6. The quadrate is partly ossified.
7. There is no distinct inter-stapedial.
8. The supra-stapedial is confluent, above.

Fifth genus. *Lymnodynastes*.

23. *Lymnodynastes tasmaniensis.*—Adult female ; 1¾ inch long. Tasmania.

This is a long skull (Plate 18, figs. 5-8), with short auditory, long nasal, and average orbital, regions. The length is the fortieth of an inch greater than the breadth, and the outline of the face is half a long oval. The sides are rather straight in the jugal region, for the broad and long nasal region makes the fore edge a regular semicircle. The whole ethmo-nasal territory is, as compared to the skull cavity, as 2 to 3 in length; the average fore and aft extent of that territory, as compared with that of the cranium proper, is as 1 to 2. The occipital condyles (*oc.c.*) are of moderate size, and of the usual distance from each other; they project but little, are more posterior than inferior, and the emargination of both floor and roof is crescentic.

The auditory capsules are of moderate size and become narrow at the tegmen (*t.ty.*); the cranial interspace between them is very wide, especially in front.

The ethmoidal region (*eth.*) is also wide, and the broad flat skull is almost like an hourglass, being much the smallest in the middle, and thus the suborbital fenestræ are almost oval in shape, the arcuate palato-pterygoids bounding them externally. The whole pre-cranial region is very flat and outspread.

The tegmen cranii (*t.cr.*) is more complete in this skull than in most of those known to me; it resembles in this respect the skull of the Skate, the roof growing along from the sides, so as to leave, when the roof-bones are removed, a space little more than a third as wide as the narrow interorbital region, and little more than a third the length of the cranial cavity.

Also those secondary spaces, so characteristic of the Anura, are absent, and the wide inter-auditory space, above, is completely chondrified. The tract, moreover, in front of the fontanelle (*eth.*) is large both ways, so that a very unusual amount of the girdle-bone is exposed. The ex-occipitals and prootics (*e.o.*, *pr.o.*) are confluent for some extent at their inner edge, above; below, there is a good space of cartilage dividing them. Above (fig. 5), the whole tegmen tympani (*t.ty.*) is soft, and a wedge-shaped tract of cartilage is seen between the bones. Behind this tract the ex-occipital is seen to reach outwards beyond the horizontal canal (*h.s.c.*) and inwards to such a distance from the middle that the superoccipital region (fig. 5) is left of unusual width. Below, the basioccipital synchondrosis (fig. 6) is only half that width, and is quite normal.

The prootics run forwards, above (fig. 5, *pr.o.*), over the optic passage, and inwards they each grow into a large falcate tract which nearly reaches the end of the fontanelle, and then come within a moderate distance of each other, near the middle. The roof of cartilage has a concave edge behind, as well as in front; and from the curved tracts of bone the cartilage runs forwards on each side to the end of the fontanelle, and exists, between the prootic and girdle-bone, as an obliquely four-sided space, which occupies about a fourth of the orbital edge of the skull.

Below (fig. 6), the prootics and ex-occipitals being of less extent, the cartilage is present as a large wedge on each side running into the floor of the skull; nearly the whole vestibular floor is unossified, and in front of the ear, the prootics, not completely surrounding the foramen ovale (V.), leave a tract of cartilage which surrounds the whole of the optic fenestra (II.).

The outer edge of the girdle-bone (*eth.*) margins half the orbital rim; it is only typically developed as to extent, a clear tract of cartilage separating it from the orbital fenestra; while in front it only just reaches the long septum nasi (*s.n.*).

This bone is large, because of the width and size of the matrix in which it is formed, and not because of overpassing its own proper territory, as in several kinds of Anura.

On each side the ethmoid bone does not quite cover the wings of the ethmoidal cartilage (fig. 5, *pa.*), from which the ethmo-palatine bars are extended; these transverse bars and all the hinge nasal region, are left unossified.

The roof of the nose (fig. 5, *al.sp.*) is very wide and enlarges from before, backwards; the floor of the nose (fig. 6, *s.n.l.*) is equally wide, and lessens from before, backwards. Here we see what has taken place since the larval stage—viz.: that the huge cornua trabeculae are now united together by a solid intertrabecular tract, ending, in front, in a rounded knob; whilst near this knob, on each side, a small ligular fork has been given off which passes between the laminae of the premaxillary : this small secondary "cornu" is the pro-rhinal (*p.rh.*). Above (fig. 5), the nasal roof throws the external nostrils (*e.n.*) far apart; each lamina has a rounded emargination behind the nostril, and then expands again until it passes into the substance of the ethmo-palatine band (fig. 5, *pa.*)

That band is twice the normal width, and spreads externally into a large fan-shaped plate (fig. 6).

The hind horn of this expansion (*pt.pa.*), as it passes into the pterygoid, becomes a mere thread, and the rest of the arch, although normal, is very feeble, and its two forks, the pedicle and quadrate region (*q.c.*), are short.

In this species it is not difficult to detach the palatine and pterygoid bones (*pa.*, *pg.*), for they graft themselves only slightly on the cartilage within. The palatines (fig. 6, *pa.*) are flat, lathy bones, rounded only at their inner end; they are sub-sigmoid, with sinuous edges, and an abrupt outer end; their inner ends pass largely under the face of the girdle-bone.

The pterygoids (figs. 5 and 6, *pg.*) are but little larger than the palatines; they

DEVELOPMENT OF THE SKULL IN THE BATRACHIA. 101

bend inwards in front of their fork, and the inner process covers the short cartilaginous "pedicle" very imperfectly. The quadrate region (*q.c.*) is short, but the condyle is large and reniform; it only reaches to the middle of the columella; this region is not ossified. The Eustachian tubes (*eu.*) are large and circular; the annulus (fig. 5) is small —two-thirds the typical size—and imperfect above. The stapes (figs. 6 and 8, *st.*) is large, oval, only straight-sided supero-anteriorly, not emarginate; for the columella is very small (fig. 8); it is pistol-shaped, has a bilobate unossified tract fitting inside the stapes, but no segment there. The medio-stapedial bone (*m.st.*) is dilated at the upper end, where it fails to ossify the bilobate tract; it then becomes a very slender rod, dilating gradually to its end, beyond which the unossified distal part is a small, narrow, cochleate extra-stapedial (*e.st.*), with no ascending process. The stylo-hyal end of the hyoid (fig. 8, *st.h.*) is narrow, and articulates with the auditory floor.

The mandibles (fig. 7) are long, and quite normal; the mento-Meckelian part (*m.mk.*) is unusually large, showing that the lower labial kept its size, after fusion with the mandible, more than is the case generally.

The hyo-branchial plate (fig. 7) is normal, the basal part is wide, but its processes are slender; the hypo-hyal lobe (*h.hy.*) is sharp and slightly perforate; the lateral lobes are smaller than usual, and the thyro-hyals (*t.hy.*) are very slender, diverging rods.

The investing bones (figs. 5 and 6) show a feeble skull; the fronto-parietals (fig. 5, *f.p.*) are thin curved shells of bone, hooked outwards, in front, where they bind on the outer angle of the girdle-bone, sinuous along their inner edge, where they form a waist to the fontanelle (*fo.*), and dilated behind into a large round lobe; these lobes, right and left, cover much of the unossified supraoccipital tract (*t.cr.*); externally, each bone just rises on to the swelling of the anterior ampulla (*a.sc.*).

The nasals (fig. 5, *n.*) are large, but leave a width of cartilage uncovered between them, equal to their own diameter. These bony shells are notched behind the outer nostril, in relation to it; their postero-external part, or handle, is short and sigmoid.

The parasphenoid (fig. 6, *pa.s.*) is typical, but very broad in the main shaft, and notched in front; the vomers (*v.*), as in *Pleurodema* (fig. 2), are placed almost transversely, and the inner nostrils (*i.n.*) open wide apart in the large rounded notch of each vomer. The main wing of the bone, in front of that notch, is, as usual, split into two sharp lobes. The dentigerous stem of each bone is a long thick rib, serrate with recurved teeth, subcrescentic in outline, and nearly reaching its fellow. Hence in opening the mouth of a Frog of this species the vomerine teeth are seen to run nearly straight across behind the inner nostrils, and to be scarcely separated at the middle; their real direction is inwards, and a little backwards, their ends lying under the fore margin of the girdle-bone.

The bones that fence the semi-oval face are normal, but delicate; the premaxillaries (*px.*) are wide, and the septo-maxillaries (*s.mx.*) small; they lie outside the second upper labial; both these cartilages (*u.l¹.u.l².*) are normal. The squamosals (*sq.*) are slight, and only just clamp the edge of the short tegmen tympani (*t.ty.*); the maxillaries (*mx.*) are thin, high shells in front, and have a slender jugal process; the quadrato-jugal (*q.j.*) only fastens on, does not ossify into, the quadrate.

As compared with the normal skull, this of *Lymnodynastes* has many curious points of difference; it would fit in among the skulls of Australian "Anura," generally, better than among those of the Palæarctic region, even those of its own "Family":—

1. It is a frail skull, the outer bones very feeble.
2. It is greatly depressed.
3. Its nasal region is one-third longer, and much broader than in the type.
4. The prootics are confluent with the ex-occipitals above.
5. The tegmen cranii is extremely developed, behind, in front, and also over the edges of the skull, so that there is only one small median fontanelle.
6. The fronto-parietals fail to cover the space.
7. The fore part of the palato-suspensorial cartilage is very broad and large, and the hind part unusually small.
8. The dentigenous stem of each vomer is very long and sub-transverse in position.
9. The annulus tympanicus is only two-thirds the normal size.
10. The columella is arrested and very small, without inter- or supra-stapedials.

Sixth genus. *Camariolius*.

24. (A) *Camariolius* (*Pterophrynus*) *tasmaniensis* (?)—Adult female; ¾ inch long. Australia.[a]

1. *Skull of the adult.*—This is one of the smallest of the Batrachia; mine had large ripe ovaries, and was only ¾ inch (9 lines) long; Dr. Günther (see note) gives 13 lines as the length of two, and 1 inch for the length of a third, species.

This is a *long* skull (Plate 19, figs. 1, 2), for its breadth is but little greater than its length; the contrast is much greater in a large number of "Anura."

Knowing this to be an adult, a satisfactory comparison can be made of its skull with that of the typical species.

This skull is the frailest I have yet seen in the group: and this not only as to the small quantity of *bone* that enters into its composition, but because of the economy as to *cartilage* also.

I shall describe, in their turn, other very minute skulls from the same region.

The occipital condyles (Plate 19, figs. 1, 2, *oc.c.*) are rounded in form, and largely inferior in position; the interspace is moderate, and gently emarginate.

The auditory capsules have the large ovoidal fulness of form seen in the young of the larger kinds, but, they are intensely ossified; their retention of the ovoidal form, with but little *tegmen*, is one cause of the relative narrowness of the skull.

---

* For a description of this genus see Günther, Proc. Zool. Soc., 1864, pp. 46–49. That author has examined this specimen for me, and considers it to be either *C. tasmaniensis*, or the one most akin to that species.

DEVELOPMENT OF THE SKULL IN THE BATRACHIA. 103

The wide temporal regions become concave in passing into the orbital, and then the skull narrows gently and bulges a little in front, near the ethmo-palatine "axils." The ethmoidal region (*eth.*) is covered by a very short "anterior tegmen;" the margin of this part has a large evenly rounded edge, and so also, in the other direction, has the "posterior tegmen," which is limited to the superoccipital region, and is barely twice the breadth of the narrow ledge in front. These two opposite margins of the great single fontanelle (*fo.*) are very far apart; the hinder is nearly as far back as the middle of the auditory capsules; the fore edge is nearly as far forwards as the arcuate ethmo-palatine (fig. 1, *pa.*).

The fontanelle, open over three-fourths of the cranial "barge," narrows in the temporal region, and only recovers three-fourths of its size behind; the roof-bones (*f.p.*) are at a considerable distance from it in every part.

The cartilage which bounds it laterally is narrow in front and widens in the temporal region to thrice its first breadth; this is the lateral remnant of the tegmen cranii, and it is, for the most part, *quite independent of the walls* (see fig. 2), which from a little behind the ethmoidal "axils" to the back of the optic *foramina* (II.) are entirely membranous. Thus the optic *fenestra* (the "foramen" is a hole through it) is excessively large; a good series of gradations in this respect are to be seen in the Australian Anura, and will be described in due time.

The floor of the skull is very narrow—only half the width of the roof—and is merely composed of the trabeculæ and an "intertrabecular" band of the same width. The nasal regions are quite normal, except that the subnasal angle is simpler than in the type. There is no "rostrum," and the pro-rhinals (*p.rh.*) turn inwards, and are not much dilated distally. The roof (fig. 1), widest behind, and the floor, widest in front, are normal; the septum (*s.n.*) is thick and clearly seen.

The ossification of this little semi-membranous skull is of great interest, for its very minuteness has enabled it to escape from the strict morphological bonds that keep the larger kinds in order.

Nearly all the ossification of the endocranium is behind the orbits; there are *more*, and there are *fewer*, osseous centres than in the typical kinds. The additional centres are the median bones that have been crushed out, so to speak, by the special law of Batrachian morphology; these have crept in again.

Fewer centres are seen in the lateral parts, for the occipito-otic centres (prootic and ex-occipital, *pr.o., e.o.*) are here, as in *Pseudis* and other generalised types, ossified fore and aft, without the transverse dividing band of cartilage, seen in those in which the bone begins only at the proper nerve outlets. And they are intensely ossified too; only a trifling edge of cartilage is left in the tegmen and in the floor (figs. 1 and 2), and in front the bony matter takes up half the "alisphenoidal" tract, giving a good bony margin to the "foramen ovale" (V.). The upper and lower median synchondroses are of moderate width; the lower is oblong, and is less than a third the width of the inter-condylar space, and contains the rudiment of a *basioccipital bone* (fig. 2) as a frail bony "cephalostyle."

The upper tract is of the same width at the foramen magnum, but widens out, wedge-like, to the fontanelle; it contains the rudiment of a *supraoccipital bone* as a slight endosteal deposit.

The antorbital (*axillary*) tracts of the ethmoid are ossified (*eth.*), not, however, as a "girdle-bone," but as in "Urodeles" as a pair of "sphenethmoids;" these occupy about three-fifths of the ethmoidal cincture—perfect here for a short distance. Below (fig. 2), the two bones are separated by a distance less than their own width; above (fig. 1), they come a little closer together, ending there as sigmoid "horns," turned forwards till they nearly touch the conchoidal nasal roof.

These bones, laterally, just run round the end of the long oval interorbital fenestra; in their interspace below there is between these two ecto-ethmoids an evident, *endosteal*, "mesethmoidal" tract like the azygous endosteal patches in the hind skull.

Here, once more, we come athwart the familiar "perpendicular ethmoid," and the basi- and supraoccipital bones.

The palato-suspensorial arches have thin bony plates (*pa., pg.*), and these, as well as the bones that surround the face, are normal, if compared with a young Common Frog of the first summer.

The vomers (fig. 2, *v.*) are small and toothless; the parasphenoid (*pa.s.*) is narrow, especially in the fore part, yet it is, essentially, typical. But the fronto-parietals (fig. 1, *f.p.*) are by far the slightest and most delicate I have yet seen; they reach to the conchoidal nasal roof in front, and partly cover the anterior canal (*a.s.c.*) where these needle-like bones become roughly pedate. These bones only cover the outer edge of the arrested lateral rudiments of the "tegmen cranii."

The nasals and the squamosals (*n., sq.*) are under-sized but quite normal; so also are the parts of the mandible (fig. 3), and the hyo-branchial plate with its processes (fig. 4).

The annulus (fig. 1, *a.ty.*) is open above and rather small. The stylo-hyal (*st.h.*) is articulated, above; the Eustachian passage (*eu.*) is of medium size and reniform, with the concavity looking outwards and backwards.

The stapes (fig. 5, *st.*) is between a lozenge and an oval in shape; it is large, and has a short stalk.

The cartilaginous enlargement on the oblique end of the columella is not segmented off as a distinct inter-stapedial. The medio-stapedial (*m.st.*) lessens rapidly from this cartilaginous process, and ends as bone, a little behind an oval, decurved, thick but small, *foot* of cartilage—the extra-stapedial (*e.st.*); there is no ascending process from this lobe. The labial cartilages (*u.l².u.l³.*) are normal. Many of the things that distinguish this from the typical skull depend upon arrest, and are such as can be seen in a young Common Frog, five or six months old.

DEVELOPMENT OF THE SKULL IN THE BATRACHIA.   105

24 (continued).—(B) *Camariolius tasmaniensis* (?).—Larva, ¾ inch (9 lines) long; legs, 1/10 inch. Same locality.

This Tadpole was the largest of several of this species examined by me, and its free hind legs showed that it has attained to its full larval condition; it was as long with its tail as its parent without.

In the possession of paired nasal sacs, each with a distinct roof, and of a well defined fenestra ovalis, this larva had gone beyond the Lamprey; but there was no bony deposit even below the skull (parasphenoid), and the hinder half of the long hypophysial space was still membranous; the rest of the space was composed of a very thin layer of young, half consistent cartilage.

This skull, therefore, was in the best possible state for comparison with that of *Petromyzon* and its allies.

Leaving out the prenasal structures this skull (Plate 15, figs. 6, 7) is nearly square; it is unusually short, and although the head was no larger than that of a *Blow-fly*, the chondrocranium had become a very solid structure; and cartilage was forming even in Tadpoles one-third the length of this—that is, in newly-hatched specimens corresponding to my earliest stage in *Bufo vulgaris* (Phil. Trans. 1876, Plate 55, fig. 1). The first-formed cartilage can now be well seen, as it is much more massive, and richer with proliferating cells, than the newer tracts.

The cartilage, which at first only enclosed the apex of the notochord as the ends of the trabeculæ, has now spread along the whole hind floor and a very definite tract crosses in front of the notochord. That rod is expanding to its full (spinal) size just where it emerges from between the parachordal bands (*ic.*); they pass outwards and backwards, and end in a free rounded point behind the ganglion of the 9th and 10th nerves (IX., X.), which lies in a notch in the outer border of each band. These bands are distinct until they reach the apex of the notochord, and then, as just mentioned, unite in front of it; there they form the hinder boundary of the pituitary space. Directly in front of the notch for the ganglion each parachordal dilates suddenly into a broad crescentic wing, which forms a concave floor for the antero-internal half of the auditory capsule (fig. 7) right and left. There is a shallow crescentic notch in the middle of the hinder margin of each wing; this forms the inner boundary of the fenestra ovalis (*f.o.*). In front, there is a deep round notch between the wing and the trabeculæ; *in* and *over* this space lie the more or less united ganglia of the 5th and 7th nerves (V., VII.).

The sub-orbital fenestra, and the band of cartilage which encloses it are, together, scarcely of larger extent than the auditory capsules behind. These have lost their simple oval form, for the three large canals and their globular ampullæ (*a.s.c.*, *h.s.c.*, *p.s.c.*) above, and the sacculus (*rb.*) below, have wrought the sac into their own form.

The alate basal floor is thin, and composed of rather young cartilage, but the auditory sacs are very solid, for they become cartilaginous directly after hatching, coevally with the trabeculæ and suspensoria.

They are now perfectly distinct from that floor, and a little pressure serves to shell them out under the cover-glass; this causes no tearing of the tissue. The large fenestra ovalis (*f.o.*) is as wide as one of the canals, and as long as the diameter of an ampulla; it lies along the middle, obliquely parallel with the horizontal canal; its antero-internal edge is *double*, being composed of both *floor* and *capsule*; the vestibule swells out on its inner side in a crescentic manner. From the inner edge of each capsule, above (fig. 6), there is a little tract of new roof-cartilage, which will finish the occipital arch. The basis cranii in the orbital region widens a little forward; the trabeculæ (*tr.*) occupy nearly a third of its breadth; the trabeculæ have met in the ethmoidal region, and are becoming fused together; the line of fusion is nearly equal to the intertrabecular space, behind (*i.tr.*), which is now half filled in with new cartilage. In front, the trabeculæ are free, diverge at more than a right angle, and curve outwards; these are the horns of the trabeculæ (*c.tr.*). Newer cartilage has appeared in several places; laterally, the trabeculæ have grown into orbital walls, and these walls are growing over the roof a little, especially in the ethmoidal region (fig. 6). Outside, in front of these rudiments of the "tegmen cranii," the "ethmoidal wings" have appeared, embracing the nasal sacs (*n.r.*).

Between these sacs the trabeculæ are very thick, where they are narrowest, and then suddenly expand into the "cornua." These latter, or their inner edge, are developing a thin, crenate expansion, ready to finish the *nasal floor*. The trabeculæ are most compressed where they form, below, the semi-circular inner boundary of the internal nostrils (*i.n.*). Over the sac, right and left, there is a small cartilage, like the valve of a small *Entomostracan*, attached to the ethmoidal wing: this is the nasal roof-cartilage (first "paraneural," *n.r.*); here it is more distinct than usual—more than I have ever seen it in this group.

The huge suspensorial bands reach by their distal condyles (*q.c.*) in front of the nasal sacs, whilst, behind, they curl themselves up against the intruded auditory capsules (*au.*), and bend suddenly upwards as they pass inwards, to form the narrowed (but really broad) pedicle (*pd.*).[*]

At this part the cartilage rises so high that it would seem to form the counterpart of the "ascending process" of the Urodeles, but it drops again before it passes into the trabecula, and although riding over the vidian nerve (VII¹.) it passes under the 5th (figs. 6, 7, V.). This thick high part will be further outwards, afterwards, and will form the otic process (*ot.p.*). In front, the quadrate passes inwards as a broad lamina from the saddle-shaped condyle, and grows into a hook—the pre-palatine rudiment (fig. 7) bounding the nasal passage, below, as the ethmoidal wing does, above.

The pterygo-palatine conjugational band is short, wide, and out-turned; on it, as yet, I see no definite rudiment of the post-palatine process. Over it, from the large projecting sulcate condyle for the hyoid (*hy.f.*), the orbitar process (*or.p.*) grows;

---

[*] The pedicle lies on a higher plane than the pterygo-palatine band; it is, like the ribbed edge, and orbitar process, a temporary tract of the suspensorium.

it is a very long, narrow leaf of cartilage, and lies in a recess of the ethmoid, nearly touching the wing; its fore edge is ribbed.

The mandible (*mk.*) is a thick, short, chubby cartilage—like a little short *ulna*; the lower labial (*l.l.*) meeting its fellow by a broad face, is thick and strong; and the *temporary* upper labials (*a.l.*) are thin crescentic leaves of cartilage.

The hyoid bar (fig. 7, *c.hy.*) is normal; above, there is the sinuous condyle, and the rudimentary styloid process; and ventrally it is dilated largely—most behind—and then suddenly less, it becomes the basi-hyal; this is composed of simple cartilage.

This chondrocranium is essentially like that of the youngest *Pseudis*, another member of the Cystignathidæ (see Plate 2); that is the *largest*, and this the *smallest* larval Frog I know. The larval skulls are, however, very uniform throughout the "Opisthoglossa;" yet they have non-essential differences that are of great interest. The skull of the adult *Cumariolius* is like that of a young typical Frog, several months after transformation.

This is seen in the general lightness of the investing bones—the moderate extension, backwards, of the quadrate hinges, and the divided condition of the ethmoid. But there are many things that cannot be put down to mere arrest; these are:—

1. The continuity of the membranous space, above; only one very large fontanelle.
2. The extension of the orbital (or optic) fenestra over three-fourths of the wall, in that region.
3. The intensely ossified and generalised condition of the occipito-auditory region, right and left.
4. The very slight and arrested, *tripartite* state of the elements of the girdle-bone; a "meso-ethmoidal" rudiment appearing.
5. An upper and a lower median rudimentary osseous centre in the occipital arch.
6. No septo-maxillaries.
7. Extremely rudimentary fronto parietals
8. Quadrate region partly ossified.
9. Columella rudimentary; no "inter-" or "supra-stapedial."
10. Vomers rudimentary and toothless.

Seventh genus.—*Cyclorhamphus*.

25. *Cyclorhamphus marmoratus.*—Adult female; 1¾ inch long. Vinco Caya, Peruvian Andes; height, 16,000 feet.

This is a stout, evenly semi-oval, short skull (Plate 20, figs. 1–6); the length is to the breadth as 6¾ to 8. The quadrate condyles only reach opposite to the middle of the stapedial plates, the epiotic eminences are almost flush with the occipital condyles, the roof is imperfect, and the ossification is intense, and attended with much anchylosis behind.

The ear is imperfect and the nasal roofs short, the vomers small, and the inner nostrils much nearer together—against rule—than the outer. This is one of the lowest and most generalised of the Ranidæ.

The occipital condyles (Plate 20, figs. 1, 2, *oc.c.*) are large, oval, and postero-inferior; they are separated by a concave notch of their own width. The whole hind skull has a remarkable form, for the auditory capsules are very large and thoroughly ossified, yet the canals project very much both before and behind, pushing the epiotic eminence almost as far back as the occipital condyles, and the anterior (prootic) swelling, caused by the anterior canals, into the orbital region. The parotic projections suddenly become one-half the size of the main capsule and are only one-third at the tegmen (fig. 1). The extreme edge of this part (inside *sq.*) is unossified, and there is a small semi-osseous tract in the superoccipital region; all the rest of the endocranium is ossified up to the middle of the nasal septum (fig. 2, *eth.*) below, and still further forwards above (fig. 1, *s.n.*).

Moreover, the parasphenoid behind, and the fronto-parietals for fully two-thirds of their extent, are anchylosed to the bone within. The temporal fossæ are deep rounded hollows; the orbital region of the skull, from thence, swells gently, and then remains of the average width up to the antorbital bars; then there is a slight superorbital projection, but it also is ossified.

The tegmen of this hard endocranium is extensive before, but more so behind; for that tract reaches so far as to lie over the front edge of the optic fenestra (II.). The edges of the open part are wide, so that the single fontanelle (*fo.*) is only one-fourth the length of the cranial cavity, and is narrow, and pinched in the middle. Here the roof is almost covered in, as in some *old* Skates' skulls.

The girdle-bone, as in some Salamanders, is not marked off behind; in front it runs far forward into the true nasal region, and right and left, ossifies the ethmoidal wings, but stops at the ethmo-palatines (fig. 1, *e.pa.*) exactly where the segment is in the genus *Bufo*. The nasal region (*n.r.*) is very remarkable, being, like that of *Bombinator* (Plate 25, fig. 1), merely a transverse double pouch ending in a broad sub-arcuate snout, which is scarcely covered by the nasals, which lie mainly on the proper ethmoidal tract. Hence the outer nostrils (*e.n.*) are very wide apart; but the floor being very narrow where the trabeculæ originally came first into contact in the internasal region the inner nostrils (fig. 2, *i.n.*), although large and round, are very near together for a Batrachian; the relative distances are exactly reversed in this case.

The *roof* (*n.r.*) is narrow from the ethmoid onwards, but the nasal wall (fig. 1) forms a pouch behind the outer nostril. The inner labial (*u.l*^{1}.) is small, and lies against the front of the broad snout; the outer piece (*u.l*^{2}.) is in front of the nostril. Below (fig. 2), the trabeculæ in becoming the floor have scarcely changed their form at all; but they have budded out into a pair of large falciform, secondary cornua at their inner angle; these are the pro-rhinals (*p.rh.*).

The palato-suspensorials are normal in the palatine region, and the bone (*pa.*) is of the usual *f*-shape; it forms but little union with the cartilage.

The pterygoid (*pg.*), on the other hand, has used up most of the cartilage to which it clings, and growing beyond the pith of the pedicle (*pd.*) makes this part pointed. On the right side, that independent ossification of the outer and inner elements which is so common in the "Anura," shows itself in a remarkable manner, for the very apex of the cartilage becomes detached, as a "meniscus," and then thoroughly ossified, as a small, free, diamond-shaped "metapterygoid" (figs. 5 and 5A, *pd., pd.m.*). In *Rana temporaria*, and others of that genus, the metapterygoid, which appears during metamorphosis, is formed in the outer fibrous layer: here it is endoskeletal. The quadrate region is neither long, nor far retreated (fig. 2, *q.c.*), it is all ossified by the quadrato-jugal, except the apex of the otic process, which is continuous with the slight remaining pith of the pterygoid, and the edge of the tegmen tympani (fig. 5, *sp., q., pg.c., pr.o.*). The condyle (*q c.*) which ends opposite the middle of the stapes (*st.*) is a large, well-formed trochlea, strongly clamped on its inner side by the pterygoid. That process forms a very obtuse angle with the pedicle, and here a very small crescentic slit can be seen just through the skin of the mouth, its convex side lying against the bone: this is the *blind* Eustachian opening (*eu.*). Outside the suspensorium a very small annulus (figs. 3 and 6, *a.ty., m.ty.*) is seen; it forms three-fourths of a circle, its diameter, and the breadth of the band, is about one-third that of a typical Frog.

The mandible (fig. 3) is very strong, especially in its hinder part; it is, however, quite normal.

The stapes (fig. 6, *st.*) is of the average size: it is a very regular oval, but has a rounded process behind, and an oval "boss" outside.

The columella is extremely delicate, and is not segmented. The medio stapedial bone (*m.st.*) occupies the dilated stapedial end, and runs along the thread of cartilage so as to be one-third the length of the whole rod, which ends as an arcuate extra-stapedial (*e.st.*); this is spatulate, and without a fork.

The stylo-hyal (fig. 2, *s.th.*) is confluent with a trace of cartilage on the outside of the floor of the vestibule; it enlarges in its descent (fig. 4, *c.hy.*), and becomes partly ossified. There is a small, sharp hypo-hyal lobe (*h.hy.*) in front, and one on the side of the short retral part; between these bars the "notch" is a very large crescent.

For the basal plate is very wide and short, and is of the same kind as we see in *Calyptocephalus* (Plate 21, fig. 4)—a near relative. There is an adze-shaped fore, and a styliform hind, lobe on each side; the thyro-hyals (*t.hy.*) are large, dilated, and separated by a space which is two-thirds of a regular oval. The proximal part of these bars take up half the basal plate, and they almost meet at the mid-line. The rest, all but the front edge and part of the fore-lobes, is semi-osseous.

Like some other sub-typical Frogs—*Pleurodema*, *Lymnodynastes*—the fronto-parietals (*f.p.*) do not finish the roof but are scanted in their front half, and end as rounded bars, each one-third the breadth of the roof. Behind, they are strong, smooth, sinuous, falling in at the temples, and swelling over the anterior canals, and are thoroughly anchylosed to each other, and to the subjacent bone. The premaxillaries and maxillaries (*px.*, *mx.*) are strong, smooth, high, and typical; and there is a large septo-maxillary (*s.mx.*) on each side. The falciform quadrato-jugal (figs. 1, 2, 3, and 5, *q.j.*) is perfectly continuous with the bony quadrate (*q.*); the squamosals (*sq.*) are like a hammer with a wide handle and a very short head; this latter part lies on the tegmen tympani but little (fig. 1, *sq.*).

The nasals (figs. 1 and 3, *n.*) scarcely hide the fore part of the girdle-bone, and leave the pouched short nasal roofs naked; they are narrow, convex, curved shells of bone, with a descending narrow process, outside.

The vomers (*v.*) are small and have a post-narial spur, but none in front of the passage; the fore part is narrow and bifid; the dentigerous lobe is oval and less than normal. The parasphenoid (fig. 2, *pa.s.*) is large, normal, rather blunt and ragged in its processes, and is somewhat confluent with the superjacent bone, behind.

Next to the skull of *Bombinator* and *Pelobates* this Frog shows most what is low and generalised in its skull, it differs from the "norma" in:

1. The large relative size of the auditory capsules.
2. The shortness of the nasal capsules.
3. The unfinished roof.
4. The intense and almost universal ossification of the endocranium, and the anchylosis of the outer roof and floor bones.
5. The differentiation of an endoskeletal metapterygoid on the *right* side.
6. The smallness of the upper part of the squamosal and of the applied "annulus."
7. The arrested state of the columella.
8. The almost entire closing of the first cleft.
9. The short, wide form of the basal plate, and its semi-osseous condition.
10. The narrowness of the internasal region, and small size of the vomers.

26. *Cyclorhamphus culeus*[*].--Larva; length, 3¼ inches; tail, 2 inches; hind legs, 7 lines. Puno, Lake Titicaca, Peru.

I am very fortunate in being able to give the structure of the larval skull in a *geographical neighbour* of the great *Calyptocephalus*; this is one among many of the things I owe to my friends AGASSIZ and GARMAN.

Moreover, among the various kinds of larval skulls worked out by me, this comes the nearest to that of that helmeted Frog; and as we know in the case of *Bombinator* and *Pelobates* how weak one skull, and how strong another, may be, in congeneric types, there is no difficulty on that head as to the relationship of these two Neotropical forms. *Cyclorhamphus marmoratus* (DUM. and BIB.) evidently comes between these two forms in respect of its adult skull (AGASSIZ and GARMAN, p. 277).

---

[*] For an excellent figure of the adult of this fine species, see AGASSIZ and GARMAN, "Bulletin of the Museum," No. 11, plate 1 (Cambridge, Mass., November 26, 1875).

This larval skull will be best described by comparing it with the next (Plate 22, figs. 6–9 with figs. 2–5); all these objects are shown as magnified five diameters; the Tadpole of *Calyptocephalus* was fully one-third larger than that of *Cyclorhamphus*, and in the latter the hind legs were four times (*relatively*) as long as in the former, and therefore it must have been somewhat more advanced in development, generally.

Indeed, this will be seen by comparison of the figures; yet this is too slight to effect the general form and relations of the parts.

The interauditory region, in this, is altogether narrower; the interorbital much more uniform, not narrowing, forwards, half as much; and the palato-suspensorial bars converge more. On the whole, however, these two skulls might easily be taken for those of two species of the same genus; part for part, and process for process, there is a very close resemblance between the two.

The notochord (*nc.*) is still found between the halves of the basal plate; the occipital condyles (*oc.c.*) have the same form; the ex-occipitals (*e.o.*) are more developed, answering to the longer legs, and are seen beyond the twin nerve-passages (IX., X.). The hinder "tegmen" only runs half as far forwards, and is not fenestrate, whilst the fore "tegmen" is finished and the single fontanelle (*fo.*) is a large evenly oval space, instead of being a smaller space, shaped like an oval leaf, with the stalk in front. The auditory capsules are not thrust so far out; they are naked, and show the canals (*a.s.c., h.s.c., p.s.c.*) through their diaphanous walls; the spurs growing from the "tegmen tympani" are very similar (*t.ty.**, *sp.c.*); the stapes (*st.*) is like that of the next.

The almost oblong interorbital region (fig. 7) is shorter, and scarcely bulges at all; its walls are perfect and run an edge over the roof, on which the bony "wall plates" lie. The front tegmen runs forward as a wedge-like mass, for it has coalesced with the more developed intertrabecular crest, and the "lamina perpendicularis" (*p.e.*) is now formed. The lateral ethmoidal wings (*al.e.*), and the upgrowths of the trabeculæ, outside the emerging olfactory nerves (I.), have conspired with the median crest and the roof to finish the plaster model for the future "girdle-bone." The first rudiment of the septum nasi exists, now, merely as the foremost part of the intertrabecula; beyond this frail commissure of the paired bands the cornua trabeculæ (*c.tr.*) diverge, arching both outwards and downwards; they are longer and narrower, and more diverged than in the next, making the lozenge-shaped interspace left by them and the upper labials (*u.l'.u.l².*) much wider.

The palato-suspensorial arch (*pd., p.py., q.*) shows its likeness and its unlikeness to that of the other species. The pedicle (*pd.*) is longer and the otic process is not distinct from the rounded "elbow" of the cartilage, which is bent backwards more. This, with the gradual convergence of the main bars, makes the orbital space wider behind; it is, also, shorter. The orbitar processes (*or.p.*) do not overlap so much as in the next, but keep outside the post-palatine rudiment (*pt.pa.*); this crest fails in its

* In fig. 6 this is wrongly lettered *t.cr.*

hinder lobe, which is merely a blunt projection. The trabecular and pre-palatine spikes ($c.tr.$, $pr.pa.$) help to surround the inner nostril ($i.n.$) in the same manner, and the quadrate region ($q.$) is alike in both, but most turned inwards in this.

The passages, only, of the 9th and 10th nerves are shown, but the common root of the upper and lower maxillary branches of the fifth ($V^{2,3}$.), the orbito-nasal (V^1.), the facial (VII.), and the long, undivided palatine or vidian are shown; the latter (VII^1.) is seen passing forwards outside of, and a little below, the orbito-nasal; it pierces the ethmoidal wing, then runs close to the mid-line, and supplies the upper lip. The optic (II.) and olfactory (I.) are seen *in situ*.[*]

The spiracular cartilage ($sp.c.$) is shorter; the upper labials ($a.l^a.a.l^b.$) are quite cut through into four pieces; the lower labials (fig. 8, $l.l.$) and the mandibles ($mk.$) are similar to, but still more solid than, those of the larger kind. The hyoid cornu (fig. 9, $c.hy.$) is extended out into an unciform stylo-hyal ($st.h.$) and the condyle ($hy.c.$) is narrow and oblong.

The extra-branchial pouches ($ex.br.$) are similar, but the small cerato-branchials ($c.br.$) are smaller still in this, and the second is confluent with the hypo-branchial plate ($h.br.$), as well as with its corresponding external bar.

All things, however, taken together, the want of conformity between these two in the chondrocranium is much less than between many other kinds that are congeners; the eye sees at once that they must belong to types very near akin to each other.

This is shown, also, in the parasphenoid (fig. 7, $pa.s.$), but it has a long point at its fore end, and its basi-temporal wings are pointed, but the two bones are very much alike: the great difference is in the want of any trace of granulation in this.

The fronto-parietals (fig. 6, $f.p.$) are also quite devoid of granulation; and their size is in extreme contrast with those of the larval *Calyptocephalus*; they belong to a somewhat riper stage, and yet, instead of being large expanded plates, are mere styles of bone, lying like wall-plates on the edge of the skull, dilated a little on the inside where they will be more or less segmented into two bones, and on the outside, where they bind on the auditory sac.

Third Family. "DISCOGLOSSIDÆ."

First genus. *Discoglossus*.

27. *Discoglossus pictus.*—Adult male; 2¼ inches long. South Europe.

This is a true Frog, it comes next the Ranine, and above the Cystignathine species, but is modified by having a discoid tongue, dilated sacral apophyses, and opisthocœlian vertebræ. The other Frogs that possess the first of these marked characteristics are

---

[*] The sudden curve, upwards, of the pedicle (fig. 6, $pd.$) gives it the appearance of lying over the 5th nerve at its exit; this is not the case, however, the actual end passes into the skull much lower down.

much less typical, and some of them are very aberrant and generalised. Hence it comes to pass that this is rather a motley group, and the question arises as to whether it should not be broken up and redistributed; if this were done the boundaries of the groups that should receive their own relations back again, would have to be made more elastic. This type would have to go to the Ranidæ; *Calyptocephalus* to the Cystignathidæ; *Pelodytes* would ask for admittance, either among the Bombinatoridæ, or the Alytidæ; whilst *Xenophrys*, and the kinds agreeing with it, would probably have to be made into a new family; that species is exceedingly generalised, and shows affinity with families far removed from the Ranidæ, far more clearly than with those typical Frogs. My mention of these things is as a protest against the family group as it now stands : the Zoologists must re-arrange it as they see fit.

As far as the skull is concerned, this type might have been left in the genus *Rana*, or still better, put with the species of Pyxicephalus ; with that of *P. rufescens* it agrees very closely. This is a typical semi-elliptical skull, whose breadth is to the length as 15 to 14 ; it is more frail, or less ossified, than in the average Frog's skull ; and its roof is imperfectly covered.

The extent of the three regions of the skull is normal, the orbital being the largest, the nasal next, and the auditory the shortest. The occipital condyles (Plate 20, figs. 7, 8, *oc.c.*) are large, semi-oval, almost directly posterior, and separated by a narrow notch.

The epiotic projections (*p.s.c.*) are gently convex, and come short, behind, of the condyles. The quadrate hinges are opposite the large twin nerve-passage (*q.c.*, IX. X.). The small ex-occipitals (*e.o.*) are wide apart both above and below, and are at a considerable distance from the equally arrested prootics (*pr.o.*); these latter bones enclose the foramina ovalia (V.), run out to the facet for the pedicle (*pd.*), and up over the anterior canal (*a.s.c.*). Beyond these bones the tegminal region (fig. 7) is only half the width of the canal region, and runs only a moderate distance beyond the horizontal canal (*h.s.c.*). The skull is well roofed, for the main fontanelle (*fo.*) is a rather small, long, heart-shaped space, and the small secondary fontanelles are wide apart on the large hinder part of the tegmen cranii.

The temporal region is wide, and from thence the orbital part of the skull narrows up to the rudimentary superorbital projections (*s.ob.*) ; these increase the width a little in the ethmoidal region, but, as in *Pyxicephalus rufescens* (Plate 14, figs. 1, 2), the cranial bout is very constricted in front. As in some small Ranidæ from Australia (*e.g.*, *Camariolius*), the girdle-bone (*eth.*) is imperfect ; it occupies only about a fourth of the interorbital region at its widest ; its halves are united by a narrow isthmus below, and scarcely meet above.

Most of the ethmoidal region, with its wings, and all the wide, well-developed, normal nasal region, are cartilaginous. Both roof and floor (figs. 7, 8) are wide, the septum (*s.n.*) is high and thick, the snout gently convex, with a slight prenasal bud ; and the pro-rhinals (*p.rh.*) and subnasal outer angles are large and well developed. So also is the annular growth of the nasal wall (*n.w.*), and the appendages (*n.l*[1].*n.l*[2].) that finish the nostrils (*e.n.*).

The inner nostrils (fig. 8, *i.n.*) are very large, short oval, and almost transverse; they turn a little forwards, within.

The palato-suspensorials are well developed, the fore part wide, with a slight prepalatine projection; the foot-like pedicle (*pd.*) projects well inwards, and the quadrate condyle projects well outwards, reaches back behind the stapes (*st.*), and is large and reniform, with a post-condylar lobe; there is no ossification beyond the setting on of the quadrato-jugal (*q.j.*).

The palatines and pterygoids (*pa., pg.*) are quite normal, but remain *ungrafted*. The Eustachian passages (*eu.*) are only half the size of the inner nostrils, and are reniform. The mandible (fig. 9) is perfectly normal; it is rather high. The annulus (*a.ty.*) is large and perfect. The stapes (fig. 11, *st.*) is oval, emarginate in front, and has a boss. The interstapedial (*i.st.*) is nearly as large, is gently notched below, and its fore third is ossified. The medio-stapedial (*m.st.*) is a strong phalangiform bar, and ends in a perfectly normal spatulate extra-stapedial, with a cartilaginous suprastapedial, confluent above (*e.st., s.st.*). The stylo-hyal is also confluent above, and the hyo-branchial apparatus (fig. 10) is perfectly Ranine. The investing bones are also quite Ranine, but the fronto-parietals (*f.p.*) are very narrow in front, and fail to cover the fontanelle perfectly; they are like those of *Pyxicephalus rufescens*, but do not unite, behind. The nasals (*n.*) are normal, but wide apart; all the marginal bones are normal, but the squamosal (*sq.*) has an exceedingly long postorbital process. There are no septo-maxillaries; the parasphenoid (*pa.s.*) is exactly like that of a *Rana*, but the vomers (*v.*) are not; they are sub-quadrate, with short snags. Each bone touches the septum nasi, the spike in front of the inner nasal opening is suppressed, the front part ends in a spike, and the post-narial spike is short. The dentigerous elevations are large, oval, and oblique.

This Frog differs from its narrow-backed relations, with forked tongues, in having:—

1. The whole skull feebly ossified.
2. The interorbital region very narrow in front.
3. The main fontanelle left partly uncovered by the roof-bones.
4. The four bones of the hind skull small, and the girdle-bone imperfect above.
5. No septo-maxillaries.
6. The squamosal very long in its post-orbital region.
7. The stylo-hyals and supra-stapedials confluent, above.

Second genus. *Pelodytes*.

28. *Pelodytes punctatus.*—Adult male; 1½ inch long. Europe.

The skull of this species is short, the breadth being to the length as 11 to 10; yet it looks shorter than it is, on account of the great breadth of the nasal region. In this respect it is intermediate between that of *Pleurodema* and *Lymnodynastes*, types to which it can claim no near relationship.*

But there are two European Frogs to which this type has some claims of relationship; these, *Bombinator* and *Alytes*, however, are very exceptional forms, having "opisthocœlian" vertebræ, whilst *Pelodytes* has them normal or "procœlian;" *they* have ribs, but this type has not (see Mivart, P. Z. S., 1869, pp. 290, 291, and 294). Nevertheless, in spite of all that, I would rather put this near them—the natives of the same geographical territory—than place it with any from a far country, having a great belief in the faculty of the *Anura* for modifying their internal structure, *during secular periods*.

The occipital condyles (Plate 23, figs. 1, 2, oc.c.) are large, oval, and posterior; the occipito-auditory region is wide proximally, and narrow distally. The interorbital region is rather wide, lessens gently forwards, and bulges a little; the nasal region is very broad, a little longer also than usual, and then the snout is very transverse. In the endocranium the fontanelle (*fo.*) is single and large, the roof-bones more than half hide it, and leave it of an hourglass shape.

Half the interspace between the occipital condyles is unossified, and the superoccipital cartilage (above *f.m.*) is of the same breadth; this is normal. But there is no distinction between the prootic and ex-occipital (*pr.o., e.o.*) right and left; this I take to be a primary confusion of parts, and not due to coalescence; if they have been separate at all, it has been secondary and temporary.

The bony matter half encloses the foramen ovale (V.); half the interorbital region is unossified; in this part the moderately large optic fenestra (II.) is seen. The other half is taken up by the girdle-bone (*eth.*), which simply ossifies its own *ethmoidal* territory in front, and only slightly passes into the wings; its axillæ are shallow, and its upper part loses half its extent at the middle, in bordering the great fontanelle.

The nasal roof and floor (figs. 1 and 2, *s.n., s.n.l.*—put by mistake on fig. 1) are well developed, the former widest behind, and the latter in front. The "pro-rhinals" (*p.rh.*) are small, and the angles of the floor large; there is a small "prenasal rostrum" running forwards from the thick septum nasi (*s.n.*).

The palato-suspensorial arches are quite normal, but the fore part is very broad, and the hinder part rather feeble; as in *Dactylethra* and *Bombinator* there is no palatine bone, and the pterygoid (*pg.*) is feeble like that of a young Common Frog; it scarcely affects the cartilage to which it is applied, and a considerable pad of cartilage forms the facet of the movable "pedicle" (*pd.*). The re-entering angle of the pterygoid and suspensorial cartilage is a right angle, and in it the Eustachian passage (*eu.*) is seen—it is large, oval, and transverse.

* Dr. Günther, who gave me my specimen, predicted that I should find it a very generalised type; this is quite true, for I am completely puzzled as to where it ought to be placed. The "Family" in which systematists put it, namely, the "Discoglossidæ," is merely a "Cave of Adullam," to which all sorts of irregular, lawless, and aberrant types are relegated. *Pelodytes* and *Calyptocephalus* have procœlian vertebræ, whilst in *Xenophrys* and *Discoglossus* they are opisthocœlian.

The annulus (*a.ty.*) is of the typical size, broad, and its horns are not united. The stapes (figs. 2 and 4, *st.*) is large and oval, with its fore and upper edge but little bevelled. The medio-stapedial (*m.st.*) is a strong and somewhat sigmoid rod, with but little cartilage at its proximal end, where it wedges in—without any interstapedial segment—between the stapes and the vestibule.

Beyond the bone the cartilage soon expands suddenly into a sub-peltate extrastapedial, with a ligulate supra-stapedial confluent above (*e.st., s.st.*).

The mandible (fig. 3) is normal, but very long and slender.

The stylo-hyal end of the hyoid (fig. 2, *st.h.*) is sub-acute, and loosely attached to the vestibular floor; it does not bind strongly round the hind margin of the Eustachian opening. Two-fifths of the band (fig. 3, *c.hy.*) at its lower end is reduced to a mere fibrous tract, and the hypo-hyals (*h.hy.*) are quite loosed from their proper stem and are confluent with the basal plate (fig. 3, *b.h.br.*), into which they run by a ligulate stalk. This distal part is a very large ear-shaped emarginate leaf of cartilage, which, by its oblique, external angle overlaps the front lobe of the basal plate, a somewhat smaller, more regular leafy growth, which runs almost transversely into the base by a shorter stalk. Behind this, on each side, there is the normal finger-shaped hinder lobe, a much smaller outgrowth of cartilage. At its root the fore end of the phalangiform thyro-hyal bone (*t.hy.*) is set into the cartilage; this is terminated by an unossified lobe; and the right thyro-hyal is considerably larger than the left. To the lower face of the proximal part of the right bone there is a curious *splint* applied: it is V-shaped, with all its points sharp; the short stem behind, and the forks, which are crooked, run, the right forwards and outwards, and the left obliquely outwards in the other direction, under the root of the left thyro-hyal.

This, which exists as a splint, I take to be an abortive attempt to produce an *ectosteal* basi-branchial bone; it will soon be described again in *Alytes*, and I shall have to refer to it when describing a more normal ("endosteal") basi-branchial in *Bombinator, Diplopelma, Callula, Engystoma, Rappia,* and *Pelodryas*.*

The fronto-parietals (*f.p.*) are similar to what we find in many sub-typical Anura, they are wedge-shaped shells of bone, with sharp fore and dilated hind ends. By these bones the great oblong fontanelle is reduced to an hourglass-shaped space, only half as large. The rest of the investing bones are normal; the nasals come within a moderate distance of each other, and are not very broad. With the upper labials—which are normal—there is, I believe, a small seed-shaped septo-maxillary. The premaxillaries (*px.*) are widely transverse; the maxillaries (*mx.*) thin and shell-like in front, and sharply styloid behind; and the styloid quadrato-jugal (*q.j.*) is not evidently grafted on to the quadrate.

The squamosals (*sq.*) are slight but normal; they fail to cover the short unossified "tegmen tympani" (*t.ty.*). The parasphenoid (fig. 2, *pa.s.*) is normal, and large in

---

* This is one among many instances in which the metamorphosis to which these *ichthyic* types have been subjected has only partially obliterated the form and structure of the Fish.

the fore part; its basi-temporal wings are narrow. The vomers (*v.*) are also normal, their dentigerous lobe is small and transverse; they strongly curl round the inner nostrils (*i.n.*), and have a moderate front lobe.

I shall have to refer to this skull once and again; it seems to me to belong to an archaic type of the "Anura," and to be a form very difficult to place, zoologically.

The main departures *from*, or failures in attaining *to*, what is typical, are as follows:—

1. There is only one fontanelle.
2. This is not covered by the roof-bones.
3. The ossification of the occipital arch and ear-capsules is generalised and continuous.
4. The nasal region is very dilated.
5. There are no palatine bones.
6. There is no inter-stapedial, the supra-stapedial is confluent above, and the extra-stapedial is sub-peltate.
7. The hyoid arch is very feeble, partly absorbed, and has its dislocated distal ends extremely dilated.
8. The front lobes of the basal plate are similarly dilated.
9. The thyro-hyals are non-symmetrical.
10. There is an irregular forked superficial basi-branchial on the under surface of the basal plate.

Third genus. *Xenophrys.*

29. *Xenophrys monticola.*—Adult male; 3 inches long; length of hind leg, 4⅜ inches. Darjeeling.

This fine specimen, the gift of Dr. GÜNTHER, is twice as large as the one described by him in his 'Reptiles of British India' (plate 26, fig. 11, 11', p. 414). *That* specimen only measured 19 lines from snout to vent, with a hind limb 31 lines long; *this* gives nearly double these lengths.

Moreover, *this* appears to be much better developed; it has a rudiment of the "interdigital membrane," and the fingers and toes have a discoid end *one-third wider* than the contiguous part of the digits.

I am very doubtful as to the position of this large Frog, and of its equally fine relative, *Megalophrys montana* (ibid., p. 412).

The figures of the skull of *Xenophrys* (Plate 23, figs. 5–10) have been purposely put on the same plate with those of *Pelodytes* (figs. 1–4) *for contrast*. The Family "Discoglossidæ" (GÜNTHER, "Batr. Sal.," p. 34; and MIVART, Proc. Zool. Soc., 1869, p. 294) must be one in which the members "agree to differ" to a very great extent, especially now Professor MIVART has added the "Asterophrydidæ" of Dr. GÜNTHER, and Professor COPE's Neotropical genus *Grypiscus*, which has *mandibular teeth*, and heads a "Sub-family" of its own. This type has opisthocœlian vertebræ.

I shall look for its true relations among the "Oriental" Batrachia, and these, when found, will most probably differ from it in one or other unimportant character, and yet correspond with it in non-essentials.

This is a short skull (Plate 23, figs. 5 and 6); the breadth is as to the length as 5 to 4, but the length is increased externally by the great retreat backwards of the quadrate condyles (*q.c.*), which make it, if measured by them, *one-tenth* more.

It is also an extremely flat skull, and has a decurved snout; it is very much unlike the skull of any of the "Ranidæ" known to me, and the crania that answer to it best are not yet described. I must therefore give an account of it as it is, and then refer to it afterwards, when its image appears again in other types. The occipital condyles (*oc.c.*) have large, oval, posterior faces, but they project very little and are separated by a gentle emargination.

With many deficiencies that make it lie some depth below the "norma," this skull is *ultra-Ranine* in its general form, for it is extremely depressed, the otic regions are wide wings, the jugal arch strongly bowed, and the hinge of the jaw is carried far behind the occipital articulation. The skull is of full width in the temporal region; then it does not become narrower, but much broader, towards the antorbital region. The margin above has a concave, and not a convex, outline; the walls do not bulge, but are scooped (fig. 6).

The dilated ethmoidal region is made still wider by superorbital expansions (*so.b.*), and in front the whole nasal territory is but little narrower than the ethmoidal. Thus the fore half of the cranium is in remarkable contrast with the hind half, and its copious cartilaginous matrix is very similar to that which is seen in Skates and even in the Chimæra; to make it still more archaic, the prenasal (*p.n.*) is a thick decurved beak, such as is seen in an early stage of the embryos of many Vertebrata.

The ossification of the occipito-auditory region is generalised, being continuous on the same side (*pr.o., e.o.*); the synchondroses above and below (*f.m.*) are rather wide and sub-equal. The fontanelle is quite covered; it is single and lanceolate, with the narrow end forwards; the tegmen, behind, is short; in front, it rather appears, than is, extensive, for the ethmoid (*eth.*) is of great extent. The occipito-auditory bones (*pr.o., e.o.*) nearly reach the fontanelle, under the roof-bones, above; below, they touch, over the parasphenoid, but leave the vestibular floor unossified. Above (fig. 5), the bony matter leaves merely the tegmen tympani soft; it is very narrow fore and aft.

In front, below, the bony matter almost encloses the foramen ovale (V.), and from thence half the interorbital region is cartilaginous in this tract; behind the middle of it is the large oval optic fenestra (II.).

The girdle-bone (*eth.*) occupies more than a third of the endocranium; it borders nearly a third of the fontanelle, and runs somewhat into the proper nasal territory. Below (fig. 6), it affects the wings a little; above (fig. 5). its edge is rounder; it runs quite clear of the superorbital "eave." A cross-shaped expansion, below, marks off the shallow axils of this depressed bony mass; these fossæ are large, but shallow. The *cross* is made more apparent by the extension forwards of the bony matter into the hind part of the septum nasi (*s.n.*), and the hiding of the outer part of the bone by the vomers (see figs. 6 and 9). Contrary to rule, the upper and lower outlines of the nasal region are very similar.

Above, a large and shallow, and, below, a narrow and deep, notch on each side, shows where the ethmoidal territory ends and the nasal begins, and these emarginations show that the nasal region is short; the thickness of the convex roof hides the form of the dividing septum nasi. Below (figs. 6 and 9), the girdle-bone ends in a sinuously-rounded margin, a short distance behind the notches in the sides. The bone is sinuous on its surface, first scooped in the middle, then swollen on each side, a groove sub-transversely dividing these swellings from the "wings." In front of the notches, the subnasal laminæ (fore part of trabecular cornua, *s.n.l.*) are ossified, each bone taking up one-third of the floor. A bone half the size of these has been formed in the cartilage of the median tract; it reaches from the nerve-passages (*n.n.*) to the decurved rostrum (*p.n.*). These three bones, like the ethmoid or "girdle," being ossifications of the three basi-cranial bars—like those found in so many types in the region of the brain cavity—may be entitled to the name of "serial homologues" of such segment-forming bony centres.[*]

The lateral subnasal ossifications are rugged and prickly at their margins; their middle is somewhat elevated, transversely; the "pro-rhinal process" of the cartilaginous matrix (*p.rh.*) is small.

The upper surface, formed mainly (*i.e.*, except at the middle line) by the nasal roof-cartilages (fig. 7, *al.n.*) is broad, gently convex, and passes insensibly into the prenasal rostrum (figs. 5–7, *p.n.*).

The suborbital space (*f.p.* to *pg.*) is large and sub-oval; its outer fence is the palato-suspensorial arch. The ethmo-palatine bar, like the rest of the cartilaginous pith of this arch, is slender, and runs outwards and forwards, like a continuation of the wing of the ethmoid; it is bi-aculeate in front (fig. 9, *e.pa.*). The post-palatine portion (behind *e.pa.*) can be seen to be but little affected by the palatine bone (fig. 6), which is a normal sub-falcate blade, separated by almost its own length from its fellow, and not reaching the pre-palatine spur in front. But the pterygoid (*pg.*) either ossifies or conceals much of the hinder part, yet the inner fork or pedicle (fig. 5, *pd.*), and the tract leading to it in the edge of the folded pterygoid (fig. 7, *pg., sp.*) shows that the axis is nowhere quite lost. But on the under surface (fig. 6, *pd.*) the pedicle is seen to be tied down by the pterygoid bone so that all motion is lost, and this inner tongue of the bone forms a strong squamous suture with the prootic and parasphenoid; this is a *Bufonine* character, as we shall see.

But the outer fork, or quadrate region, becomes much larger, and the part above the condyle is largely ossified by the quadrato-jugal (figs. 6, 7, *q. q.c., q.j.*).

---

[*] In Birds, ossifications of this kind in the precranial region of the base are very common; but in their compressed prognathous head the fore part of the intertrabecula is early absorbed; the paired trabeculæ end behind the intertrabecula and the "subnasal laminæ" are only exceptionally developed.

The condyle for the mandible (*q.c.*) is large, oblique, and reniform, and has the inner trochlea twice the size of the outer; the hinder and outer fork of the pterygoid bone runs vertically, splint-like, inside this far-retreated part of the suspensorium. The large Eustachian opening (fig. 6, *eu.*) is, by this, made to turn backwards, as well as outwards; its hind margin is a thick fibrous ligament.

The mandible (fig. 7) is quite normal and of great length, answering to the extent of the gape; the mento-Meckelian part (*m.mk.*) is large, and so is the cylindroidal condyle (*ar.c.*); the cartilaginous bar (*mk.*) is but little affected by the trough-shaped "articulare" (*ar.*); the dentary (*d.*) is a little more than half the length of the mandible.

The "annulus" is very remarkable; it is large (Plate 23, fig. 7, *a.ty.*), thick at the edges, oblong in shape, and whilst one horn is attached to the fore part of the portico formed by the squamosal (*sq.*) over the tympanic cavity, the other horn passes behind and under the hind part; its position is oblique, being carried backwards, below, by the suspensorium and its splint (*q., sq.*).

The stapes (fig. 10, *st.*) is thick and of a short-oval shape, with the fore margin emarginate. The columella fits by its thick, bilobate, scooped apex, within and around the fore part of the stapes. The inter-stapedial (*i.st.*) is represented by the larger, unossified lobe; but it is doubtfully segmented. The medio-stapedial (*m.st.*) has also some cartilage on the lesser lobe, and the thick bony end carrying the cartilage is almost discoidal. The narrow shaft is bent almost at a right angle on the dilated proximal part, and runs more than half way to the distal end. That end, gradually thickening to its middle, is the extra-stapedial (*e.st.*); at first it is merely a continuation in the same line as the shaft, but its distal two-thirds is bent down at a little more than a right angle, is thick below, and cochleate above. There is no supra-stapedial. This generalised, but large, columella is seen (fig. 7, *e.st.*) to emerge from the cleft, and then to pass, downwards and outwards, inside the membrana tympani. So far are these parts carried outwards, backwards, and downwards, that in the upper view (fig. 5, *a.ty.*) they are scarcely seen when the eye is focussed to the upper face of the skull; hence the *apparent minuteness* of the annulus in the figure showing that aspect.

I could find no cartilage in the hyoid arch from the Eustachian opening downwards until I reached the hypo-hyal region. There (fig. 8, *h.hy.*), there is a falcate hypo-hyal lobe which passes backwards and inwards to a basal plate of normal size (*b.h.br.*), and with the usual small posterior wings, and bony, divergent thyro-hyals (*t.hy.*); but there are no anterior wings to the basal plate. Here the absorption of the hyoid band is equal to what is seen in *Pipa*.

The investing bones, like those of the endocranium, are very unlike what we see in *Rana*, and in the "Ranidæ" generally.

The fronto-parietals (*f.p.*) just overlap the fontanelle and the side-walls of the cranial cavity; they are almost square over the hind part of that cavity, and then expand forwards to their end. Their hind margin has a pair of shallow notches, and their fore margin is cut away, but more irregularly, into four shallow emarginations. The ribbed orbital edges are higher than the middle, so that the top of the skull is gently concave; behind the superorbital edge there is a temporal notch. In front, these bones scarcely cover in the fontanelle; behind, they almost reach the foramen magnum.

The nasals (figs. 5 and 7, *n.*) are curious stalked shells of bone; the stalk is thick and is pointed below, the dilated part fits on to the convex cartilage, does not come near its fellow, and has a strongly crenate margin; much of the middle and all the fore part of the cartilaginous roof is exposed.

In this very "Phryniscine" skull (see Plate 41) the præmaxillaries are thrown beneath the prenasal rostrum (Plate 23, figs. 5–7, *p.x.*); they extend outwards a good way, but are narrow, have a small palatine process, and an average nasal process. As with the dentary below, so with the premaxillary, above, the bone is grafted on to a labial cartilage, which cartilage it largely ossifies; in this case it is the "inner upper labial" (*u.l¹.*). All the cartilages of the outer nostril are large and solid—we shall see the like again in the "Pelodryadidæ"—and whilst above the opening (*e.n.*) the nasal roof (*al.n.*) is very thick and crescentic, the front of the nostril is guarded by the large "outer upper labial" (fig. 7, *u.l².*), which is pedunculate and tridentate.

The maxillaries (*mx.*) are long, narrow bones; the jugal process reaches far back; and the quadrato-jugals (*q.j.*), which are half as high as the maxillaries, are rather long: they are extensively fused with the quadrate (*q.*). The real height of the oblique backwardly-turned squamosal (*sq.*) is equal to half the length of the skull. Its supratemporal part (fig. 5) is extensively developed over the auditory capsule, as well as over the marginal "tegmen;" the bone is cleft, as it were, to bind fore and aft upon the parotic mass.

The postorbital process (figs. 5 and 7) is falcate; the edge of the temporal part forms a crescentic porch to the ear, and the descending stem is a narrowish lath of bone binding on the retreating suspensorium.

The parasphenoid (fig. 6, *pa.s.*) is large, well-developed, and *Ranine*; its short "handle" is split, its basi-temporal "guard" is strongly sutured to the pterygoid (where it lies over the pedicle), and the pointed "blade" reaches as far as to the wings of the ethmoid; the trough of the bone is deep (fig. 7, *pa.s.*). The vomers (fig. 6, *v.*) are remarkable; they are thick, and bent like "knee-timbers," are wide apart, have a solid boss for the teeth, an out-turned front lobe which reaches the pre-maxillary angle, and a round notch for the internal nostril (*i.n.*), which passage is guarded by a long spike of bone on the outside.

As compared with the skull of the typical species, this is very abnormal; its most important modifications are:—

1. The extreme flatness of the skull.
2. The great extension outwards of the parotic wings.
3. The extremely backward position of the quadrate articular condyles.
4. The fusion above of the occipito-auditory bones.

5. The absence of the secondary fontanelles.
6. The increasing width, forwards, of the cranium, involving the great breadth of the girdle-bone.
7. Superorbital projection of the ethmoid cartilage.
8. The width, convexity, and solid rostral end of the nasal region.
9. The three anterior subnasal ossifications.
10. Slightness of palato-suspensorial arch, with its fixed pedicle.
11. Bony quadrate.
12. Oblong, oblique, open annulus tympanicus.
13. Generalised columella, without supra-stapedial process or infra-stapedial segment.
14. Absence of hyoid band and of antero-lateral lobes of basal plate.
15. Inferior position of pre-maxillaries.
16. Reversed shape of fronto-parietals.
17. Extensive double growth of suprateinporal region of squamosal.
18. Solidity and arcuate form of vomers.
19. Massiveness of upper labials, and union of the inner pair with pre-maxillaries.
20. Absence of septo-maxillaries.

These twenty marked characters of difference from the type show how far removed this Frog is from the Ranidæ; its generalised nature is so great that I shall have to refer to it for comparison when I come to the Polypedatidæ, the Hylidæ, the Phryniscidæ, the Engystomidæ, and the Hylaplesidæ.

Fourth genus. *Calyptocephalus.*

30. (A) *Calyptocephalus Gayi.*—Adult female; 5½ inches long. Chili.

(A) *The adult.*—This large, massive, and almost ganoid skull (Plates 21, 22) is, on the whole, very Ranine, but *it has retained* a copious overgrowth of granular subcutaneous bone, which makes it like a strong rough helmet.

Its general form is half a short oval; it is similar in outline to rather more than the narrow end of a *short* Hen's egg; for its outline narrows in again behind. Compared with its length (from snout to occipital condyles) the breadth is as 4⅓ to 3, and the quadrate hinges project 2 lines beyond the occipital hinges, making the skull *one-eleventh* longer at the outside. The actual measurements are—

1. Greatest breadth, 2 inches 5 lines.
2. Length from snout to occipital condyles, 1 inch 9 lines.
3. Extent of quadrate hinges behind them, 2 lines.

This skull carries the peculiar Batrachian type to its greatest degree, viz.: the great width as compared with the length.

So great is the development of the investing bones that the endocranium is almost entirely roofed over by them; and the orbital spaces, so large generally, have only one-third the extent above that they possess below.

The endocranium is not intensely ossified, and there is no increase of the number of bony centres either in it or in the external plates such as is seen in *Rana pipiens;* but it agrees with that large Frog in the moderate ossification of the inner skull, and in the extension backwards of the condyles for the lower jaw.

But on the whole, leaving out of view the dense overgrowth of the subcutaneous bones, this skull differs less from that of *Rana tigrina* and *R. hexadactyla* (Plates 6 and 7) than it does from the skull of *R. pipiens.*

It agrees with that of the Oriental Bull-frogs in the form of the endocranium; in the relative size of the cranial trough (much wider and larger than in *R. pipiens*), and in the severely typical paucity of bony centres, whether *external or internal.*

On the whole, there is little in this skull of divergence from the "norma," except the exuberant growth of its roofing plates; and in this genus, and also in *Ceratophrys,* it is in this overgrowth, and not in indefinitely numerous bony patches, that we have the mark of *ancientness:* the great size of the species, however, is in itself suspicious in a group whose members are for the most part small, and even very small.

The occipital condyles (Plate 21, figs. 1, 2; and Plate 22, fig. 1, *oc.c.*) are large, oval, postero-inferior in aspect, and separated by the basioccipital cartilage, which is gently emarginate, and one-third of the width of one condyle.

The superoccipital arch (Plate 22, fig. 1) is excavated above, and less than a line in front of the basal outline (Plate 21, fig. 1, *f.m.*); it is not finished with bone, for there is a considerable wedge-shaped tract of superoccipital cartilage.

Both the auditory and ethmoidal regions stand out at right angles to the axis, and the former is twice the extent of the latter. The interorbital region between these is of the average length, and twice as broad as in *Rana pipiens* (Plate 8); it narrows gently up to the ethmoidal region, and is almost straight-sided, not bulging, as in many types. The depth (fig. 3) is proportional to the width, and the *sectional form* is boat-like, gently widening upwards. The foramen ovale is partly rimmed in front by the prootic (fig. 2, V., *pr.o.*); the optic fenestra (II.) is of the medium size and is midway in a tract of cartilage that occupies half the interorbital region. The girdle-bone (*eth.*) takes up the rest; it spreads some way into the ethmoidal wings, but does not reach quite to the front of its own proper territory, nor ossify the limited superorbital lamina (*s.ob.*).

Behind, the ex-occipitals and prootics (fig. 2, *e.o., pr.o.*) are separated by a triangular tract of cartilage, and do not extend outwards as far as the parasphenoidal wings; thus the lesser vestibular and the larger tegminal tracts (fig. 2) are left soft.

The tegmen tympani (*t.ty.*) is largely extended outwards, beneath the over-arching squamosal (*sq.*), as a flat, sub-quadrate, pterotic tract; the vestibular floor is gently convex, and is confluent on its outside with the dilated stylo-hyal end of the hyoid bar (*st.h.*).

In the nasal region there is no rostrum to the septum (*s.n.*) in front; the prorhinals (fig. 2, *p.rh.*) are broad, short, pedate, and they fill the space between the nasal and palatine processes of the pre-maxillaries (*px.*). The outer angles of the subnasal laminae (*s.n.l.*) enter a similar space in the fore end of each maxillary (*mx.*); the laminae themselves bulge downwards, and the septum ends below on a higher plane (fig. 2, *s.n.*). Both these inferior (trabecular) plates and the nasal roof (*al.n.*, *al.sp.*) are moderately developed, and the latter can be seen to project beyond the fore end of the nasals (figs. 1 and 3, *al.n.*, *n.*). The small round outer nostrils (figs. 1 and 3, *e.n.*) are unusually near together, are formed in a scooping of the nasal roof—or rather that roof runs round them, and they are protected by the large, solid inner and outer "upper labials" (*u.l'.u.l².*).

The inner nares (fig. 2, *i.n.*) are twice as wide apart as the outer, twice as large, oval, and directed cross-wise; they are quite included in the space between the middle process of the vomer and the ethmo-palatine cartilage and bone (*pa.*).

That cartilage (*pr.pa.*) is of the usual form, and that bone is not more than usually developed; it is straight in direction, but sigmoid in form, with a dilated (external) blade. The post-palatine region of cartilage is unossified for a considerable extent, and as it passes into the pterygoid region is at first only covered on its inner side by a delicate stylo of bone, for the pterygoid (*py.*) in its fore part, whilst it still clings to the cheek, is aborted considerably by the huge maxillary (fig. 2, *mx.*).

Becoming free, before it forks, this bone suddenly enlarges; then in front of the Eustachian opening (*py.*, *eu.*) it sends, straight inwards, a large tongue of bone having an ectosteal relation to the pedicle, which reaches almost to the foramen ovale (V.). By this process, as in *Xenophrys* and *Bufo*, this partially absorbed dorsal end of the pier of the mandible is hidden and *fixed*; we miss here the gliding joint seen in *Rana*, and many other types.

The outer and hinder fork of the pterygoid passes backwards, and then outwards, in a geniculate form, to clamp the inner face of the quadrate region of the suspensorium (*q.*).

This region, ossified below by the quadrato-jugal (*q.j.*), ends in the large "trochlea" for the lower jaw (*q.c.*); it is a very elegant pulley with a large inner, and a smaller outer longitudinal convexity, and a deepish concavity between.

This part, with the condyle, the practical suspensorium, but only made out of the distal end of the larval cartilage (see Plate 22, figs. 2, 3, *q.*), is directed almost equally outwards, downwards, and backwards (Plate 21, fig. 1–3, *q.*).

The large, oval, Eustachian passages (*eu.*) are turned a little backwards, and not directly outwards; the "annulus" (fig. 3, *a.ty.*) is of medium size; is open above, and is a rather broad band. The mandible (fig. 3) is quite normal; it is a strong bar, with a large cylindroidal condyle, with a moderate coronoid crest to the articulare (*ar.*), an unusually long and high dentary (*d.*), and a thick, short mento-Meckelian bone. The stapes and columella (figs. 2–6), partake of the general stoutness of the skull; the former (*st.*) is a solid, sub-pedunculate mass, sub-oval in form, but scooped to receive the columella. This latter part has no separate inter-stapedial segment; but there is a large thick unossified lobe to the end of the medio-stapedial (*m.st.*) which represents it. Between this and the main stem there is a deep notch, faced fore and aft with cartilage, then the bony bar gradually lessens, is arched, and ends in a terete rod of cartilage which dilates into the thick, oval, bi-convex extra-stapedial (*e.st.*). From the inner face of this lobe there arises, at a sharp angle, the terete, stout supra-stapedial (*s.st.*), which is confluent with the "tegmen" above.

The stylo-hyal end of the cerato-hyal (figs. 2 and 4, *st.h.*, *c.hy.*) seems like a flabelli-form continuation of the vestibule; it is wide at first, then narrows as it passes directly outward, margining the Eustachian opening; is then wider again as it creeps along the inside of the quadrate, and after this becomes a little less again, as it passes, sigmoid in form, to the hypo-hyal region (*h.hy.*). There is no enlargement, but it forms a round loop, and then gently dilates as it passes into the basal plate (fig. 4, *b.h.br.*).

The antero-posterior extent of the basal plate is small, as in the "Hylidæ;" its lobes are large, the foremost flabelliform, the hindmost—*here* quite as large—is stalked and emarginate. There is no bone, save in the "thyro-hyals" (*t.hy.*); they are of moderate length, divaricate well, are stout, and soft-footed.

This skull, with its strong roof and sides, is constructed, externally, of seven pairs of bones, and an odd one; in the young the main roof-bones were divided across, which gives seventeen "parostoses" originally investing the proper endocranium; the plates applied to the cartilage, and *grafted upon it*, are not counted, but the two "dentaries" in the lower jaw bring the sum up to nineteen. The other bony tracts in skull and face amount to seven pairs and an odd one—that, however, the girdle-bone, was double once, so that there are sixteen bony centres that are properly "endoskeletal;" in all, there *were* thirty-five; there *are* thirty-two bones in this skull.

But the "quadrato-jugal" is grafted on the quadrate and becomes partly endo-skeletal, whilst, on the other hand, the bony centres that were formed in the chondro-cranium (ex-occipital, prootic, &c.), are largely recruited, in their growth, from the perichondrial layers of membrane, thus they link on to the palatine and pterygoids that begin in membrane, and get their "endosteal" additions afterwards. But the truly *parosteal* "sub-species" of bony centre concerns us now; I shall describe this group, and then show what was the form of the original "chondrocranium," *on* and *in* which this strong architecture is perfected.

Behind, the roof-bones come close to the edge of the foramen magnum (*f.m.*), and are built across from side to side, without interruption, over the condyles of the quadrate region; the hind margin of this wide tract is emarginate, crescentically.

The nasal, frontal, and saggital sutures form one line of division from snout to occiput, and each *region* of this long suture is about equal. The two pairs of main roof-bones, viz.: the parietals and frontals, are now only one pair (*f.p.*), the coronal sutures having early disappeared.

This double fronto-parietal tract is twice as wide over the occiput as over the orbits; its hind margin, like its front margin, is almost straight, but the edge on each side uniting it with the squamosal (*sq.*) converges forwards, and is a roughly dentate suture—this answers to the squamous suture in Man.

Each squamosal (*sq.*) is a bone with a descending and retreating handle, and a roughly pentagonal blade; these are the temporal and pre-opercular regions of the bone.

The pointed fore end of the temporal plate—its postorbital projection—is united by suture with the upper edge of the maxillary (fig. 3, *mx.*); this is a sinuous suture, and behind it the edge of the squamosal rises and falls twice: first outside the pterygoid (*pg.*), and then over the "annulus tympanicus," forming the ledge under and to which it is attached (fig. 3).

The preorbital rim is formed by the nasals (*n.*), these are very large, about half the size of the fronto-parietals.

In front of their concave preorbital edge they form a large descending pedate process (fig. 3), which rests upon, and is united by suture with, the fore end of the maxillary, on each side.

The median spur of each nasal is blunt, and the fore edge has a large round notch, exposing the nasal roof (*al.n.*), which is itself notched for the nostril (*e.n.*).

The main roof-bones dip into the orbit (fig. 3, *f.p.*), forming an orbital plate.

The arcuate premaxillaries (*px.*) form a divided but strong key-stone to the great arch of the face; their palatine and nasal processes are well developed and normal.

The maxillaries (*mx.*) are notched in front, where they articulate—without the intervention of a septo-maxillary—with the premaxillary.

The lower dentary edge of the bone is gently arched, and the jugal process reaches as far back as the tegmen tympani (*t.ty.*). Half the depth of the side of the skull is formed by the maxillary, whose edge is cut away below the eye, and is again deeply notched behind the postorbital suture; it then becomes the rapidly lessening jugal process.

The quadrato-jugals (*q.j.*) are one-third the length of the maxillaries, but are only the size of the jugal processes; they strongly bind on the lower part of the squamosal (fig. 3), and are grafted largely on the quadrate, and there pass under the squamosal; this is best seen in the palatal view (fig. 2).

That view also shows how large the palatal plate of the maxillary is, especially where it is locked to the pterygo-palatine arch (*pg.*).

The parasphenoid (figs. 2 and 3, *pa.s.*) is a large bone, and reaches to within a short distance of the foramen magnum, behind, and to the beginning of the unossified septum nasi in front.

It projects by a very obtuse angle behind, its "basi-temporal wings" are broader than the main bar, and end by a ragged oblique edge, which runs to a point in front, reaching nearly to the Eustachian opening (*eu.*).

The middle of this hind part is raised into a lozenge-shaped convexity, but the main bar, which is three-fifths the width of the trough of the cranium, is itself a deep trough.

Between the foramina ovalia (V.) it is compressed a little, is widest between the optic nerves (II), and then scarcely lessens to its rounded ragged fore end, where it reaches the thick bosses of the vomer (*v.*) that carry the teeth.

These bones (fig. 2, *v.*) are very thick in their dentigerous portion, where they form a lobe that runs outwards and a little forwards; the thinner part forms a strong spike in front of the internal nostrils (*i.n.*), and a broad diverging lobe in front of the spike, which reaches the maxillary under the angle of the "subnasal lamina" (*s.n.l.*).

This cranial building is thus finished below and above; its roof, floor, side-walls, partitions, chambers, and outworks are all essentially such as we see, on a small scale, in the Common Frog. But the great size of this skull, and the extension under the skin of thick dense bone, ornate with a honeycomb pattern of hollows and papulate ridges passing into pearly grains, makes it in appearance very unlike its simple "norma." These modifications may be classed under one head as modification No. 1.

2. The upper fontanelle appears to be single.

3. The long inner process of the pterygoid aborts the *joint of the pedicle*.

4. There are distinct superorbital ledges.

5. The quadrate is largely ossified by the quadrato-jugal.

6. There is no inter-stapedial segment.

7. The extra-stapedial is suborbicular, and its supra-stapedial process is confluent above.

8. The stylo-hyal is confluent above.

9. The basal plate is very short, the hypo-hyals have no lobe, and the hind lobe is very large and emarginate.

These are the few, and for the most part gentle, differences between this skull and that of the norma.

30 (continued).—(B) *C. Gayi.*—Larva; 4¾ inches long; tail, 2¼ inches; hind legs, ⅛ inch. Chili.

In the large Tadpole of this species the chondrocranium (Plate 22, figs. 2-5) is seen to differ in many points from that of the larvæ of the American Bull-frogs (Plates 3 and 4), and of *Pseudis*.

It may be compared with the same (my first) stage in *Rana pipiens* (Plate 3, figs. 1-3), *Rana clamata* (Plate 2, figs. 5-7), *Pseudis paradoxa* (Plate 2, figs. 1-4), and with a somewhat more advanced stage of *Cyclorhamphus calcus* (Plate 22, figs. 6-9), and with the larval skulls of *Cystignathus*, *Hyla*, *Bufo*, &c., in other Plates.

Its true relationship will be at once seen by a comparison of the figures on this

Plate (22) with the others. *Cyclorhamphus* is its nearest ally, by far, and in this comparison we have dug deeper for evidence of kinship than in any comparison that could be made of the skull in the adult.

Small hind legs have budded out in this larva, and as a correlate of these growths three investing and two intrinsic bones have appeared in the skull; these lift it out of the sphere of simple "chondrocrania."

The notochord (*nc.*) is still of considerable size, but the copious development of hyaline cartilage has already obliterated several morphological landmarks.

The occipital condyles (*oc.c.*) are now well formed, and the arch itself is very wide, rather low, and has a huge doorway (*f.m.*). The large oval ear-sacs are completely chondrified, and are confluent with the basis and tegmen cranii; they abort its walls largely, growing into the sides of the skull; the old lines of junction can, however, be traced.

The cranium proper is very wide, especially behind, and the tegmen is developed, already, nearly as much as in Sharks—more than in most Skates; it has two lesser fontanelles in it (*fo'.*). The interorbital part is wide, gently bulging, and gradually narrowing to the ethmoidal region, which is now in the act of closing in upon the fore part of the brain capsule. Above (fig. 2) the principal fontanelle (*fo.*) is only a quarter the size of the general tegmen of cartilage, and in front it is unenclosed—the lateral halves of the *primary ethmoidal region* of the skull-wall have not met over the brain cavity. Also the unfinished state of the skull is seen still further in this, namely, that the "perpendicular ethmoid" (*p.e.*) is still a mere oval upgrowth of cartilage—a tuberous ascending development of the intertrabecular tract which is still visibly distinct from the trabeculæ, all the way from the pituitary floor to where the trabeculæ become again free in front as the "cornua" (*c.tr.*).

In front of that little swelling of cartilage (fig. 2, *p.e.*) the foundation of the septum nasi is seen in the narrow foremost part of the cartilage which conjugates the two trabeculæ.

We get here additional light upon the vegetative growths of cartilage that finish, by closing in, the cranial capsule, in front. Where the walls have most converged in front they there suddenly grow out as the "ethmoidal wings" (*al.e.*), running by continuous growth of cartilage into the pterygo-palatine bands (*p.py.*).

The olfactory nerves (1) escape here from inside the thickening wall which gives off the wings; in the middle the mesethmoidal "tuber" (*p.e.*) becomes a flatter structure, and grows into a vertical partition wall between the nerve-outlets, the foremost of the three *sense-capsule fenestræ*. The right and left roofs—growths of the ethmoidal wall—unite with each other and with the middle wall; and besides this, in front of the cranial cavity, each trabecula at its outside develops a process of cartilage, similar to, but less than, the median intertrabecular tuber. These three, growing round the olfactory nerves, finish the round passages (fenestræ), and, with the lateral ethmoidal walls (growing into the roof), close in the skull-building.

DEVELOPMENT OF THE SKULL IN THE BATRACHIA. 129

Directly in front of these ethmoidal crests the trabeculae are at their narrowest, are closest together, their scooped edge forming the inner margin to the internal nostrils (*i.n.*); the complete hind rim is formed by the crescentic fore edge of the pterygopalatine band (*p.pg.*). In front, the quadrate part of the suspensorium (*q.*) and the trabeculae (*c.tr.*) of the same side each send out a bud (*pr.pn.*), and these approaching growths are united by a fibrous band which finishes the nasal opening. The cornua trabeculae maintain their average width up to the frontal wall, but are narrower and are longer than usual, diverge very gently, and end in a moderately pedate process, which is turned outwards.

The nasal roofs are not yet chondrified, the eyeballs have been removed with the other soft structures, but the auditory capsules are shown, for they are built into the walls of the cranium as side chambers. Below (fig. 2), there is no columella at present, but there is an open space, and into this space—the fenestra ovalis—the oval "stapes" (fig. 2, *eb., st.*) fits.

The roof of the future tympanic chamber (*t.ty.*) is very largely developed; outside the large horizontal canal (*h.s.c.*) it sends out a small hinder, and a large front, spur; the latter is the confluent "spiracular cartilage" (*sp.c.*)—the "annulus" that is to be; it just touches and overlaps the otic process (a free elbow now) of the suspensorium (*ot.p.*).

The anterior canal (*a.s.c.*) is covered with bone (*f.p.*), the posterior (*p.s.c.*) is naked, and bulges gently above and behind. A small circular plate of bone, the ex-occipital (*e.o.*), has grown inside the double nerve-passage (IX., X.) and encloses and subdivides it externally.

The roof is covered to a great extent by the fronto-parietals (*f.p.*); these are many times larger than those of any other larva at this stage dissected by me (see Plates 2, 3, 4; also Plate 22, figs. 6, 7).

The two bones are very granular and rugose, quite unlike what is seen in the other Tadpoles, they, together, have an emarginate bow-shaped hind edge spread over the large ear-capsules up to the horizontal canal (*h.s.c.*), which they partly cover, and then occupy two-thirds of the interorbital region of the roof. Each bone is deeply notched on the inside, half-way; this is the beginning of the segmentation into two bones; the space thus left bare is roofed by the front part of the hinder tegmen. Behind the notch the bones come closer together, ready to form a *sagittal suture*, in front they diverge to their pointed end and have a rounded inner edge.

But the parasphenoid (fig. 3, *pa.s.*) is also rugose, pitted, and shows signs of a "ganoid" nature; this is lost in the adult, as if the larva retained more of the nature of some ancient armed type than the adult. The bone is similar in form to that of *Cyclorhamphus* (fig. 7), and not much larger, relatively; it is very large in both kinds; it is broader altogether, and notched at the tip; and the basi-temporal wings are, like the tip, less acuminate than in the other Tadpole (fig. 7).

This condition of the roof and floor-bones in a larval Batrachian just putting forth its hind legs is of extreme interest; and the size, massiveness, and sculpturing of these bones is surprising when compared with what is seen in the large Tadpoles of cognate "Ranidæ."

The nerves figured as issuing from the skull in this view are the 1st (I.), the optic (II.), the trigeminal (V^1.), $V^{2,3}$.), the facial (VII.), with its vidian branch (VII^1.) which runs forwards to the upper labials; the 8th is out of sight; the 4th and 6th were not drawn, and the 9th and 10th (IX., X.) are indicated by their twin-passage.

The "palato-quadrate" arch in this Tadpole is full of interest (figs. 2 and 3, *p.pg.*, *pd.*, *q.*); it is very massive, and its hinge is opposite the beginning of the terminal third of the cornua trabeculæ (*c.tr.*).

Altogether these bars are twice as wide as, and much thicker than the trabeculæ, and every thing about them is large and solid. They are almost parallel with the skull, converging very little; the pedicle (*pd.*) is short, for the elbowed and forwardly turned bar is of great width and takes up the room; its "otic process" (*ot.p.*) is a rounded bud overlapped by the spiracular cartilage (*sp.c.*). The main bar becomes half its former width near the pterygo-palatine conjugation (*p.pg.*); there it sends up over the temporal muscle that large, outwardly rimmed scooped and sessile leaf, the "orbitar process" (*or.p.*), which at its base has the infero-external crescentic hollow for the condyle of the hyoid (*hy.f.*).

As yet the pterygoid band is not free from the ethmo-palatine, but the palatine region of this great arch has all its three *sub-regions* well marked out. The first of these is that which runs into the trabecula, namely, the ethmo-palatine (*p.pg.*), it is half as broad as the trabecula and runs outwards and forwards behind the inner nostril (*i.n.*). Running inwards in front of that passage is the pre-palatine spur (*pr.pa.*); and from the wide space between the outer and the conjugating bar, in the front part of the long subreniform orbital space, a falcate lobe of cartilage grows backwards and inwards, taking up about one-fifth of the large membranous tract: this is the "post-palatine" (*pt.pa.*), whose axis will afterwards be coincident with that of the pterygoid band with which it also will be confluent.

The distal quadrate region (*q.*) is more than twice as wide and thick as the trabecular cornu; it is sinuously flat above and sub-carinate below; its condyle is saddle-shaped, like that of a human "humerus," and looks a little inwards.

The free mandible (Plate 22, fig. 4, *mk.*) is a stout phalangiform cartilage, hooked externally like the "ulna," whose condyloid notch it imitates; its distal part is solid, broad, and excavated at the end for the lower labial (*l.l.*).

This latter segment now runs across to its fellow with but little downward bend; the two meet by a straight fibrous joint; each piece is oblique and bulging, where it fits into the end of the mandible.

The upper labials (Plate 22, figs. 2, 3, *u.l.*) are nearly divided from behind, forwards, into two flat, oblique plates, the outer of which has a long, blunt, angular process; here the subdivision of the labial tract, right and left, takes place very early.

The hyoid cornu (Plate 22, fig. 5, *c.hy.*) is very massive, with a semi-globular condyle and a flabelliform body; its basal piece (*b.hy.*) is composed of simple cartilage.

The basi-branchial (*b.br*¹.) is a solid cartilaginous disk, with the rudiment of a second (*b.br*².) behind it.

The hypo-branchial (*h.br.*) is a triangular plate, which passes outside the postero-internal angle of the hyoid, and has articulating with it the four normal "extra-branchials" (*e.c.br.*). They give rise near the lower end to processes that run into the septa on the inner and upper surface of the pouches; these (*c.br.*) are (all but the last) small rays, with a pedate free end: they are "cerato-branchials."

Fourth Family. "ALYTIDÆ."

Genus *Alytes*.

This Family, as enlarged by Professor MIVART (P. Z. S., 1869, p. 291), suggests this remark, namely, that that assemblage of characters by which the Alytina, Scaphiopodina, and Uperoliina are characterised, after all only serves to faggot them together; and the possession of neck-glands, dilated sacrum, maxillary teeth, and sharp toes by *Alytes* in Europe, by *Scaphiopus* in North America, and by *Hyperolius* (*Uperoleia*) in Australia, proves nothing as to the genesis of these three genera.

Before all things, in classification, natural geographical grouping has to be looked to, and then afterwards the modification into Families, Genera, and Species.

All the Australian Anura I have dissected, as yet, look to me rather like branches of one Australian stock than scattered scions of Families and genera from distant geographical regions; and whilst appraising at their full value every character that can be discovered in any type whatever, I shall keep as far as possible the distribution of the types before my eyes. The Dog-faced Opossum is far nearer akin to the Kangaroo, in spite of his canine features and form, than he is to the true Dog of other regions.

31. *Alytes obstetricans.*—Adult female; 1 inch 10 lines long. Europe.[*]

The skull of this type is extremely like that of the delicate Australian Tree-frogs, *e.g., Hyla Ewingii* and *H. phyllochroa* (see Plate 31); whilst in the next instance, an Australian Frog, the skull might be taken for that of one of the "Ranidæ," and a not very aberrant form of the European or the Oriental species of *Rana*. Yet a careful examination of this skull will show that the skull of *Alytes* has some important points of coincidence with that of *Pelodytes* and *Bombinator*; and the Obstetric Frog is more likely to be one branch of the same stock as these other European kinds than a migrated relative of an Australian species, or the Australian species a migrated relative of this.

[*] My specimens of this Frog are the gift of Professor TROESCHEL, of Bonn.

This skull (Plate 24, figs. 1, 2) is short, flat, and a very delicate semi-osseous structure altogether; the length is to the breadth as 5 to 6, and the quadrate condyles (*q.c.*) reach as far back as the fore edge of the stapes. The occipital condyles (*oc.c.*) are large, separated by a very gentle emargination of the basal plate, and *directly posterior* in aspect.

The auditory regions are greatly winged outwards, becoming narrow externally; the common occipito-auditory bones are single on each side, reach as far out as to the wings of the parasphenoid below, and touch the narrow squamosal above; thence the outside of the vestibule (*vb.*) and the tegmen tympani (*t.ty.*) are left soft.

So also is a rather wide tract above and below at the middle; whilst in front the prootic region of the bone reaches the foramen ovale (fig. 2, V.), which it half encloses. The optic fenestra (II.) is large, and is well surrounded by cartilage, and this cartilage is continued beyond the middle of the orbital territory; two-fifths of this territory only is taken up by the girdle-bone (*eth.*) This bone just reaches the "wings" and the fore end of its own proper territory, and above (fig. 1, *eth.*) it is imperfect; the right and left half, which are united, but leave emarginations below (fig. 2), are not fused above: their horns just meet in the fore margin of the anterior fontanelle (*fo.*). The "tegmen cranii" is very short in the ethmoidal region; it forms a flat and widish walltop, runs a band across the roof in the postorbital region, and becomes superoccipital in front of the middle of the auditory regions, right and left. The main space thus spanned is divided into two fontanelles (*fo.*, *fo'.*); the front space is one-third larger than the hinder, and is emarginate behind; the hinder space is transversely oval. This is a rare modification; behind, there are, as a rule, two secondary spaces, and in several types none.

The tegmen grows out into a small superorbital "eave" in front, and then there is a wide stretch of cartilage—ethmoid and ethmo-palatine. Outside the small superorbital projection of the "wing" there is a very large superorbital cartilage (*s.ob'.*); it covers nearly a third of the orbital space; is roughly oval, its large end foremost, and it is placed obliquely, with its hollow face downwards. This is also a rare character, it occurs in *Phyllomedusa bicolor*; but in that type it is much smaller. The common septum of the ethmoidal and nasal regions (fig. 1, *s.n.*) is large, and is continued in front as a prenasal rostrum (*p.n.*). As in *Bombinator* the alae of the nasal roof (fig. 1) are very little developed along the sides of the septum, the cartilaginous pouches being nearly all in front, lying on, and coalesced with, the "subnasal laminae." These plates (fig. 2, *s.n.l.*) are normally wide, and have long narrow pro-rhinals (*p.rh.*).

The muzzle being very obtuse, the nostrils are wide apart, they are protected by the usual inner and outer labials (*a.l¹.n.l².*), which are large and well developed.

The palato-suspensorial arches are wide apart, and the orbital spaces are oval, the pre-palatine spike (*pr.pa.*) is large, the stem (*e.pa.*) wide, and the post-palatine and pterygoid tracts are but little affected by the palatine and pterygoid bones (*pa., pg.*), which are quite typical. This cartilage is seen in both forks—the pedicle (*pd.*) and the quadrate (*q.*)—the latter is partly ossified near the trochlea or condyle (*q.c.*) by the quadrato-jugal (*q.j.*).

The Eustachian openings (*eu.*) are large, reniform, and transverse; the annulus (*a.ty.*) is large and its horns are confluent, above.

The stapes (fig. 5, *st.*) is large, truncate in front, and has a large knob for muscular attachment.

The inter-stapedial (*i.st.*) is well developed as a "sesamoid" cartilage; the medio-stapedial (*m.st.*) is roughly pistol-shaped; the extra-stapedial (*e.st.*) is a large spatula, giving off a ligulate supra-stapedial (*s.st.*), which is confluent, above.

The mandible (fig. 3) is normal, but the dentary (*d.*) is much longer than usual, and the articular portion has an endosteal "articulare" (*ar¹.*) as in *Bombinator*.

The basi hyo-branchial plate (fig. 4) has the notch very deep; the cerato-hyals (*c.hy.*) expand to twice their upper size, in the middle, and have a small hypo-hyal lobe; they are free above. The lateral lobes, fore and hinder, are well developed, but the thyro-hyals (*t.hy.*) are near together in front, and are short; they have a pedate unossified free end, and each has, on its outside, a small oblong nucleus of cartilage sticking to the bony shaft.*

Here again we see, as in *Pelodytes*, the V-shaped perichondrial bone, with its angle backwards, lying on the under face of the basal plate near the thyro-hyal; it is an attempt to form a basi-branchial bone.

The investing bones correspond very closely with those seen in the skulls of *Pelodytes* and *Bombinator*, and they are also very similar to those of the more delicate Australian *Hylæ*. The fronto-parietals (*f.p.*) are thin laths of bone, lying on the skull-wall, and curved so as to bind round the two fontanelles and the front of the ear-capsule, just at its inner edge. But they fail to cover the superoccipital region, and only partly hide the girdle-bone.

The nasals (*n.*) have the usual form, and in their fullness of size and shape cover in the imperfect nasal roof.

The premaxillaries (*px.*) are extended widely; the maxillaries, quadrato-jugals, and squamosals (*mx., q.j., sq.*) are normal but feeble; the quadrate is partly ossified by the second of these; the third only binds on the edge of the tegmen tympani.

The parasphenoid (fig. 2, *pa.s.*) is large in its fore part, but its wings are narrow externally, and the hind part is but little produced. The vomers (*v.*) are very large, but in this wide-muzzled Frog they only reach the septum nasi inwardly, and outwardly do not touch the maxillaries. The dentigerous part is oval and transversely placed; the interspaces between these tracts is almost equal to the tracts themselves. There are no septo-maxillaries.

It is impossible to compare this skull with that of *Bombinator* and *Pelodytes* without seeing that these three have many things in common, and that in so far as they agree

* Mr. Howes has shown me, in his exquisite dissections of the Sturgeon at the South Kensington Museum, that the visceral arches of that fish have many small nuclei of cartilage inside the periosteum.

with each other, they differ from the more typical Anura of the European sub-division of the Palæarctic region. They differ greatly from the normal *Bufo*, as well as from the normal *Rana*, and they are isomorphic, more or less, with types in the "Notogæa," to which they can claim only the remotest Batrachian relationship.

The main characters by which Alytes differ from the common typical Frog are:—
1. The flattened, very feeble skull, as a whole.
2. The open state of the fontanelles, through the deficient growth of the roof-bones.
3. The fusion of the normal pair of secondary fontanelles into one.
4. The presence of a clear superorbital edge to the skull, and the articulation with it of a large, distinct superorbital cartilage.
5. The nasal roof imperfect, and the "rostrum" well developed.
6. No septo-maxillaries.
7. Occipito-otic bones continuous.
8. Quadrate region partly ossified.
9. Supra-stapedial confluent.
10. An endosteal "articulare."
11. Cerato-hyal wide, and basal "notch" very large.
12. A V-shaped ectosteal basi-branchial, and *periosteal* cartilages on thyro-hyals.

Fifth Family. HYPEROLIIDÆ.

Genus *Hyperolius*.

32. *Hyperolius* (*Uperoleia*) *marmoratus*.—Adult female ; $1\frac{1}{2}$ inch long. Parramatta, Australia.

This type represents the "Uperoliidæ" of GÜNTHER ("Batr. Sal.," p. 39), but Professor MIVART (P. Z. S., 1869, p. 291) melts this lesser group into the "Alytidæ;" the correspondence in certain characters, however, gives them no real title of near relationship. This more extended group of "Alytidæ" must be taken as a convenient *temporary* zoological bundle; a very small spark will devour the "tow" that binds these alien types together.

This skull (Plate 24, figs. 6, 7) is evenly semi-elliptical in outline; its breadth is a little greater than its length, and the quadrate condyle (*q.c.*) is only opposite the foramen ovale (V.).

This skull is at once seen to be both generalised and arrested; yet it belongs to a minority among its Australian congeners, in having its fontanelle covered. Altogether it is more like the skull of a *young Common*, than of an *old Obstetric*, Frog.

It is, indeed, more like the typical skull than that of the "Cystignathidæ" of Australia—*Lymnodynastes, Camariolius* (Plates 18 and 19)—whose skulls approach those of the *Tree-frogs* of the same region. The occipital-condyles are almost hemispherical; they are posterior, and are separated by a straight-edged interspace larger than themselves. Below, there is a narrow, above, a wider and widening tract of cartilage, dividing the bony masses of the ear and occiput, which are moderately extended, wholly ossified, and furnished with a large, wide tegmen tympani (*t.ty.*), which overhangs the vestibule (fig. 7) to an unusual extent. These right and left bony regions reach to the fore edge of the foramina ovalia, below, which they almost encircle.

The skull is wide behind, and narrows evenly, with scarcely any bulging up to the axils of the ethmoidal region. The covered fontanelle (fig. 6) is very large; it reaches to within a short distance of the foramen magnum, behind, and leaves very little tegmen cranii in front. The edges formed by the walls of the skull are straight, and those walls have scarcely any "coping." The converging and slightly bulging interorbital walls are unossified from the foramina ovalia to the ethmoidal axils; in the hinder third of these cartilaginous tracts, the optic fenestra (II.) is seen to be of moderate size. The girdle-bone (*eth.*) scarcely covers its own territory in front, and behind, only ossifies the axillary or pinched part, reaching some way back below; above (fig. 6) it is imperfect, its halves scarcely meeting at the mid-line.

The nasal region is entirely unossified, there is no prenasal rostrum, and the prorhinals (fig. 7, *p.rh.*) are well developed; the subnasal lamina (*s.n.l.*) is very broad, and the roof (fig. 6) is of normal width. The external nostrils (*e.n.*) are only half as wide apart as the internal (*i.n.*). The palato-suspensorial arch is quite normal, both in its cartilage and in its bones (*pa., pg.*); but both the pedicle (*pd.*) and the quadrate (*q.c.*) are short; the latter is scarcely affected by the quadrato-jugal (*q.j.*). The Eustachian openings (fig. 7) are rather small, and directed obliquely outwards and forwards. The annulus (fig. 8, *a.ty.*) is a very small crescent, one-third the average diameter. The stapes (fig. 10, *st.*) is large, ear-shaped, and ossified, except at the edges; it has a strong apophysis.

The columella has no inter-stapedial segment, the medio-stapedial (*m.st.*) is pistol-shaped, the large, proximal pedate end not ossified, and the terminal part is a thick, short spatulate extra-stapedial (*e.st.*), with no ascending process.

The mandible (fig. 8, *m.mk., d., ar.*) and the hyo-branchial apparatus (fig. 9, *c.hy., b.h.br., t.hy.*) are perfectly normal; the stylo-hyal end (fig. 7, *st.h.*) is confluent, above.

The labials (*u.l¹.u.l².*) and the investing bones are, on the whole, quite normal—that is if compared with those of a *half-grown* Common Frog. The fronto-parietals (*f.p.*) are very wide behind, deficient inside, in front, and on the *left side* (reversed in fig. 6) behind, are unusually developed over the arch of the anterior and posterior canals.

The parasphenoid (fig. 7, *pa.s.*) is large and normal, but the vomers (*v.*) are mere films of bone, bordering the inside of the inner nostrils; they are toothless.

There are no septo-maxillaries; the maxillaries (*mx.*) send up a distinct process to articulate with the "manubrium" of the nasal (*n.*), which has the normal conchoidal shape, and comes near its fellow, above.

The pre-maxillary (*px.*) has united, by its nasal process, with the inner upper labial, so that this cartilage, like the lower labial, is partly ossified.

The wide contrast between this skull and that of *Alytes* is seen at a glance by a comparison of the figures; it differs from the "norma" in the following particulars:—

1. The skull, as a whole, corresponds with that of a young specimen of the typical kind.
2. The occipito-auditory region is quite ossified, showing no distinction of front and hinder centres.
3. The skull, proper, is very broad behind, and largely unossified in the interorbital region, the girdle-bone being imperfect above.
4. The annulus and Eustachian tubes are small.
5. The stapes is very large, projecting, and bony.
6. The columella is arrested; there is no inter-stapedial segment or ascending process.
7. The stylo-hyals are confluent above.
8. The premaxillaries are grafted on to the 1st upper labials.
9. The vomers are very small and toothless.

Note—That the highly ossified hind part of the skull is in great contrast with the feebly ossified fore part; the large size of the stalked stapes is in great contrast with the arrested columella and the feeble annulus.

Sixth Family. "BOMBINATORIDÆ."

Genus *Bombinator*.

33. *Bombinator igneus.*—Adult female; 1¾ inch long. Europe.

The semi-oval skull is of the average form; the length is as to the breadth as 7 to 8, and the quadrate condyles reach to a point opposite the fore edge of the stapes; in *Alytes* they were opposite its fore edge, but the breadth of the skull was much greater.

This and the next type have many things in common, but their skulls are in strong contrast, on account of the feeble ossification of the one, and the unusual degree of hardness attained by the other; the contrasted figures will show how they differ (Plate 25, figs. 1–4, and 5–11).

The main difference in form of outline between the skull of *Bombinator* and that of the Common Frog, is that in the former the snout is much wider; the interorbital region is much alike in both, it is rather narrow, lessens from before, backwards, and bulges gently in the middle.

The auditory region is also wider, and the whole skull more archaic, and also much arrested.

The occipital condyles (*oc.c.*) are separated by an interval their own breadth, and but little emarginate; they are *supero-posterior* in position as in *Pipa*, which is a rare modification. The roof over the foramen magnum is also gently emarginate, and both in roof and floor the bones (*e.o.*) are divided by a moderate wedge of cartilage.

The prootics and ex-occipitals (e.o. to V.) are one undivided mass of bone, right and left, and the tracts are margined by cartilage, outside and in.

The outer tract, above, is the widish oblong tegmen tympani (t.ty.), beyond the horizontal canal (h.s.c.); below, the vestibular region (vb.) is largely unossified.

In front, the bone reaches to the foramen ovale (V.). Thence the cranial trough is unossified for three-fifths of the region of the orbits; in the hinder half of this tract the optic fenestra (o.s., II.) is very large.

The girdle-bone (eth.) does not ossify all the ethmoidal region in front; it reaches farther below than above, and there it projects somewhat into the "wings," and reaches forwards most in the middle.

The fore part of the great, single, open fontanelle (fo.) cuts away a semi-circle from the girdle-bone; three-fourths of it is uncovered by the arrested roof-bones. The fore part of the ethmoid and all the nasal territory is unossified; the septum nasi (s.n.) is large and clearly marked, and the subnasal laminæ (fig. 2, s.n.l.) are very large, even at their narrowest part, mesiad of the internal nostrils (i.n.); their notched outer edges nearly reach these passages. The "intertrabecula" only forms as a knob in the prenasal region; the angles of the subnasal laminæ are triangular, and spread into the maxillaries at their fore end.

The secondary cornua, or pro-rhinals (p.rh.), are long, slender, and out-turned.

The nasal roof-cartilages form a mere rim to the septum nasi (fig. 1, al.n., s.n.), but they swell out into a curious bag of cartilage in front. This bag lies on the fore part of the subnasal lamina, and is confluent with it; the premaxillary encloses part of this bag, as well as the pro-rhinal band.

The thick upper and fore part of the nasal roof stops abruptly, the cartilage dipping down in front; after forming the bag, it grows up again, behind the large outer nostril (e.n.), which is protected, in front, by the three-toothed outer and upper labial (u.l².); this is attached to the inner labial (u.l¹.), a large oval segment, inside the nasal process of the pre-maxillary.

These nasal pouches look like those of *Dactylethra*, but in that type these bagpipe-shaped pouches are formed out of the large upper labial; the true nasal roof being a small band over the nasal sac, on each side (see Phil. Trans. 1876, Plate 59, figs. 1, 3, 5, and 6, u.l.).

The ethmo-palatines (e.pa.) are wide, diverging, cultrate bands, emarginate in front, with a wide pre-palatine blade (pr.pa.), and having the post-palatine bar (pt.pa.) continuous with the pterygoid as one arcuate cartilage, but little affected behind by the normal pterygoid bone (py.). There is no *palatine bone*, but the pterygoid runs far forwards; behind, it bifurcates as usual, to invest the large pedate pedicle (pd.), and to clamp the inside of the quadrate stem (fig. 2, sp.) This region is only a little affected by the quadrato-jugal (q.j.); its condyle (q.c.) is small, but normal; it reaches only a little behind the very small Eustachian pouch (eu.).

There is neither annulus tympanicus, or columella; the stapes (figs. 1 and 4, st.) is therefore an uninjured oval, having no bevelling or emargination in front; it forms a fit lid for the vessel of the vestibule (*eb.*). The stylo-hyal end of the hyoid bar (figs. 2 and 4, *st.h.*) is small, pedate, and articulates loosely with the ear-mass behind the almost closed first cleft.

The mandibles (fig. 3) form an elegant arch; the mento-Meckelian (*m.mk.*) is not much developed; the dentary (*d.*) is small, and the articulare (*ar.*) is large, has a high coronoid crest, and within its trough behind the articular portion of MECKEL's cartilage has a large endosteal nucleus (fig. 3, *ar'*.), as in *Chelone viridis*. The hyo-branchial structures show the same *semi-transformed* condition as the other parts of the skull, and some very generalised characters, besides.

At first the stylo-ceratohyal band is narrow (fig. 3, *c.hy.*); it then widens largely, narrows again, has a knee-like dilatation at its fore part, and then goes straight backwards as a narrow hypo-hyal (*h.hy.*).

The notch between the lateral parts of these structures is three times the extent of the solid basal plate (*b.h.br.*); and this basal plate has a bony "basi-branchial" (*b.br'.*) in its centre. The thyro-hyals (*t.hy.*) are normal, but the posterior lateral lobes are twice as large as those hinder forks, and instead of being a slight cartilaginous snag, they are large, and largely ossified.

The anterior "lateral lobes" are very large; they, also, have retained much of the larval cartilage, and each flabelliform outgrowth is notched deeply and sinuously in front.

The investing bones are very similar to those of newly metamorphosed Frogs and Toads of the more typical sorts. The frontal portion of the two roof-bones (fig. 1, *f.p.*) is merely a narrow lath, lying like a wall-plate on the flat top of the skull-wall; this ends at the second third of the interorbital region. The parietal portion widens, covers the hinder part of the large single fontanelle, spreads a little over the temporal region, runs sinuously over the hind skull, and ends some distance in front of the foramen magnum. The right bone overlaps the left, and the suture is very irregular.

The nasals (*n.*) are large thin shells of bone, coming near each other by their round backs, and having two shallow emarginations outside. The premaxillaries (*px.*) are of great transverse extent, but their processes are feeble. There are no septomaxillaries; the maxillaries (*mx.*) are long thin bones, with a rather high shell-shaped fore end.

The quadrato-jugals (*q.j.*) are but little united to the quadrate; they and the squamosals (*sq.*) are feeble. The parasphenoid (fig. 2, *pa.s.*) is a short, wide, thin bone, normal in its processes, but ending far behind the antorbital region in a lathy ragged manner, like that of a young Common Frog. The vomers (fig. 2, *v.*) are very large, have a massive dentigerous lobe that nearly reaches the septum nasi, are coiled round three-fourths of the inner nostril (*i.n.*), and grow out in front to the suture between the premaxillary and maxillary.

Here we have a combination of characters that betokens an *old* type and an *arrested* form; and these things are not the same, although they often co-exist in the same species.

The great difference between the skull of this kind and that of *Pelobates* partly arises from the *Bufonine* stoutness of the latter, and the *Hyline* delicacy of the former; and this in general form as well as in the internal *histological* differences.

The skull of *Pelobates* is as much overwrought, in comparison of the "norma," as that of *Bombinator* is underwrought; moreover, the former has a rudimentary columella, whilst *Bombinator* has none.

Pelobates is as remarkable in showing *ichthyic* bony patches in the cerato-hyals as *Bombinator* in the basi- and hypo-branchials. *Bombinator*, in its skull, looks a little towards *Dactylethra*—that is, in the condition of the nasal capsules; and it is, for the most part, like a young typical Frog.

Its divergence from the last-mentioned species, or the type, is seen in the following particulars:—

1. The fontanelle is wide open, through the arrest of the roof-bones.
2. The ossification of the occipito-auditory regions is continuous.
3. The occipital condyles are *supero*-posterior.
4. The nasal capsules are almost entirely deficient in their roof-cartilage along the septal region.
5. There are no septo-maxillaries.
6. The investing bones are all feeble, except the vomers.
7. There is no annulus tympanus.
8. The Eustachian passage is a very small diverticulum.
9. There is no columella.
10. There is no palatine bone.
11. The coronoid process of the "articulare" in the mandible is unusually distinct, and there is an "endosteal" articulare besides the outer plate.
12. The cerato-hyal is greatly dilated distally, and the notch in front of the basal plate is very large.
13. The limited basal plate has a bony basi-branchial centre.
14. The posterior lobes of the basal plate are very large, and ossified as a second pair of hypo-branchials.

Seventh Family. "PELOBATIDÆ."

Genus *Pelobates*.

34. *Pelobates fuscus.*—Adult male; 2 inches 5 lines long. Europe.

In these types extremes meet under an apparent external harmony. Beneath similar Taxonomic characters there are hidden the greatest contrasts in these two European representatives of two of the lowest Families of Frogs.

In the architecture of their skull they are as nearly much contrasted as any two species that can be found in the whole group of the Anura; we shall see how much of this is due to the feeble condition of one, and to the stout, stony strength of the other.

In the skull of *Pelobates* (Plate 25, figs. 5–11) the length of the skull as compared to its greatest breadth is only as 7 to $8\frac{3}{4}$; the occipital condyles project so far beyond those of the quadrate that the median length is one fourteenth greater than the extent of the skull at the sides.

Here we miss the excessively wide gape of *Xenophrys* and the American Bull-frogs; *there*, all was for elasticity; *here*, all is for compactness.

The occipital region is moderately wide; the great width of the skull is mainly due to the suspensorial structures (*sp., q.*).

The endocranium, best seen from below, and the end (figs. 6 and 8) is wide, but not flat as in *Xenophrys* (Plate 23, figs. 5–7); yet it increases in width from behind, forwards, as in that type; it has also a very broad but short nasal region, which is, as in *Xenophrys*, a well-fused convex mass, with a thick, short, decurved "prenasal rostrum" overhanging the front; but the premaxillaries under the skull are more normal. Strong as are the outworks (ectocranial elements) of the skull, the equally strong endocranial territories can be made out very clearly, on the whole, without unroofing.

The occipital condyles (figs. 6 and 8, *oc.c.*) are large, sub-reniform, postero-inferior in position, and separated by a space two-thirds their width, which is gently emarginate. The whole occipito-auditory region, right and left, is completely ossified up to the foramen ovale (V.); the super- and basioccipital tracts of cartilage are quite obliterated. There is cartilage from the front of the foramen ovale to the front of the optic fenestra (II.), and then the girdle-bone (*eth.*) occupies all but small supero-lateral tracts in the front; the sides of the nasal region, only, are unossified. The girdle-bone thus occupies the fore part of the orbito-sphenoidal region, all its own territory, and all the middle part of the nasal.

Yet the terminal part was probably formed by a separate centre (or centres), as in *Xenophrys*, anchylosis taking place afterwards.

From whatever aspect the endocranium is examined (figs. 5–8) great strength is seen, the end view (fig. 8) especially shows how strong the occipital arch is, terminated by the large condyles, and flanked by the massive auditory capsules. These show the large arches of the canals (*a.s.c., h.s.c., p.s.c.*); on the outside right and left, the two "tegmina" (fig. 5) are clamped by the large squamosals (*sq.*) as in a vice.

The outspread girdle-bone (*eth.*) has shallow "axils," and sharp angular wings; these articulate with the palatines (*pa.*). The ali-nasal cartilages (figs. 5, 7, *al.n, n.w.*) are crescents of cartilage outside the double bony roof, nasal and prenasal; and the external nostrils (*e.n.*) are half embraced by their concave edge; these openings are well finished in front and at the side by the normal inner and outer upper labials (*u.l¹.u.l².*). Below (fig. 6), we see that the bifurcate cornua trabeculæ—pro-rhinals and angles of subnasal laminæ (*p.rh.*, *s.n.l.*)—are not ossified. The pro-rhinals are triangular, pointed and in-turned, as in the genus *Bufo*; inside them the passages (*n.n.*) for the terminal branches of the "orbito-nasal" nerves are large.

The thick roof hides the fontanelle, but I suspect it to be single, as in the thin skull of *Bombinator* (fig. 1).

That I should claim for the pterygoid and palatine bones (*pg.*, *pa.*) the title of *endoskeletal* and "ectosteal" will not seem very strange if the condition of these bones in *Pelobates* is considered. All, or nearly all, the cartilaginous palato-suspensorial arch has been eaten by these strong bones. Here the palatine (fig. 6, *pa.*) has the form of a hatchet blade, for it has taken up the prenasal spike; it is deeply notched, and half the long-oval, transverse inner nostril (*i.n.*) is in this notch, and the hole is finished inside by the vomer and wing of the ethmoid (*v.*, *eth.*). Curving elegantly round in the hollow of the maxillary (*mx.*) the palatine meets the pterygoid (*pg.*); seen from below, these two bones appear of equal length : a thing contrary to rule.

The pterygoid (fig. 6) has the strength of a "flying buttress," jammed in between the dentary and palatine laminæ of the maxillary (*mx.*), and then springing obliquely across to clamp the front of the hind skull over the reduced and useless "pedicle," it fixes and binds the cheek to the cranium. That inner fork, like a strong foot, stands stoutly on the prootic, and is sutured to the parasphenoid (fig. 6, *pr.o.*, *pa.s.*). The outer fork passes back outside the minute Eustachian pouch (*eu.*), and clamps the inside of the quadrate (*q.*), the distal part of this huge "pier." By grafting of the quadrato-jugal (*q.j.*) on the cartilage, this part also is well ossified, all but the large convexo-concave reniform condyle (*q.c.*).

Answering to its pier, the mandibular arch (fig. 10) is unusually stout; yet it does keep the Meckelian rod—its pith—unossified. The mento-Meckelian (*m.mk.*) is well developed; the dentary (*d.*) is continued from it as a strong splint along the front two-thirds of the arch; and the "articulare" (*ar.*) forms a strong inwardly crested trough for the primary cartilaginous rod.

I can find no "annulus tympanicus;" the spiracular cartilage is probably combined with the edge of the "tegmen." The stapes (figs. 8 and 11, *st.*) is rather large and has an "umbo;" its form is three-fourths of a circle, with the hind margin cut away to fit against the sub-convex bony margin of the fenestra ovalis, behind. The columella (*co.*) is a little phalangiform bone, as long as half across the stapes, thick behind, thin in the middle, and knobbed at its free end; it fits against the inferior margin of the stapes, just below the facial nerve (VII.). The stylo-hyal (*st.h.*) has a pedate end which *articulates* with the auditory mass close behind the minute Eustachian pouch (*eu.*). The continuation of the hyoid bar gently increases in size until it comes to the hypohyal region (fig. 9, *e.hy.*, *h.hy.*); that part is a large rounded ear of cartilage, which, behind, passes by a broad stem into the basal plate (*b.h.br.*). The cerato-hyal has two small osseous centres in it; a rare modification.

As in the Hylidæ, the pre-basal notch is of great depth, and the basal plate of small extent, axially; three-fifths of this is taken up by the huge, very diverging, widely podate, thyro-hyals (*t.hy.*).

The posterior lateral lobes are normal; the anterior are small, irregular lobes.

In this skull the roofing and outer walls are in harmony with the inner parts of the building; the bones are very solid, and their outer surface is almost Ganoid, being richly sculptured with honeycombings and clear granules, like the surface of a large thick-walled *Globigerina bulloides*.

The roof in this old male is one bone, its parietal region projects as a thick triangular process over the archway of the occiput, and the nasal end (*n.*) reaches to the verge of the bony prenasal (figs. 5 and 7, *n.px.*, *eth.*); there it ends in two broad, rounded lobes, divided by a sharp notch—all that remains of the nasal suture. This bilobate end of the roof is formed by the nasals, which were very large whilst distinct; they form a right and left wing over and down the antorbital region; this part is of great breadth, quite unlike the usual spike that articulates with the maxillary. There is a regularly crescentic orbital margin to the frontal region, right and left; then a short, stout, postorbital process; then a rounded notch in the temporal region (belonging to the parietals); and then an oblique, crenate margin up to the terminal spike.

This large, completely anchylosed, convex slab of sculptured bone (fig. 5) overhangs the flat orbital edge of the endocranium, its width, in the orbital region, increasing from before backwards, whilst the inner skull, in that part, increases from behind, forwards (fig. 6).

The only part of the slab which articulates with other investing bones is the nasal wing—right and left; much of the extended ear-mass divides the parietal edge from the squamosal (*sq.*), and the nasal lobes do not reach the nasal processes of the premaxillaries (figs. 5 and 7).

The parasphenoid (fig. 6, *pa.s.*) is large and typical; the blade is sub-carinate and trough-like; the guards, on each side, are also hollow above, and receive the ear-masses into their hollows; they bend downwards, right and left, in accordance with the dipping of the hind floor of the skull, towards the suspensoria (fig. 8). The handle of the parasphenoid is short and bifid; it does not reach to the edge of the base, behind.

The vomers (fig. 6, *v.*) are curious foot-shaped bones, and are extremely solid, especially the dentigerous "toe"—an oblique, oval mass, looking straight towards its fellow, from which it is separated by a space half the width of the lobe. The "heel" of this bone lies over the internal nostril (*i.n.*), from thence it becomes a thick wedge reaching to the junction of the premaxillary with the maxillary.

The solid dentigerous lobe of each vomer, and the internal nostrils, form together a transverse series of long ovals, broken by the exposed base of the girdle-bone, and bounded by the bony palatines.

The premaxillaries (*px.*) are strong, wide, have a small flat palatine, and a large bowed sub-vertical nasal process (fig. 7, *n.px.*), which overlaps the inner and upper labial

($u.l.^1$) hardly as large as usual. Attached to this cartilage is the second or outer labial ($u.l.^2$); it, also, is crescentic and solid; its concave face lies towards the nostril.

Below this, and wedged in between the premaxillary and maxillary, there is a granular septo-maxillary (fig. 7, *s.mx.*). The fore margin of the high, long, and thick maxillary (*mx.*) is notched where it receives the small bone; its greatest height is where the nasal wing rests obliquely upon, and is sutured to, it. The upper surface is cut away in crescentic manner, and this margin and the orbital edge of the frontal region, together, form most of the oval orbital space; it is only deficient where the post-orbital process fails to meet the antero-superior angle of the squamosal (*sq.*) in the temporal region; the temporal muscle is thus uncovered at its " origin."

Behind the suborbital margin the maxillary rises to form a suture with the post-orbital process of the squamosal, and then lessens to its end, which is a blunt point overlapping the quadrato-jugal (*q.j.*). This latter bone is high, thick, and quite one with the ossified quadrate region (*q.*).

The squamosal (figs. 5, 7, 8, *sq.*) answers well to the rest of the outer garniture of this strongly-built skull; its facial granular plate is large, oblique, has two large rounded emarginations above, and a lesser round notch below. Its stem (fig. 7) is strong, sinuous, and notched below; its supra-temporal tract runs over the tegmen tympani, binding hard upon the prootic, up to the horizontal canal (*h.s.c.*), to the form of which it is adapted. The end view (fig. 8, *sq.*) shows how solid and convex this bone is, and what a stout knee its " pre-opercular," or descending process makes, where it binds upon and clamps the suspensorium, and articulates with the pterygoid by its hinder edge.

I shall compare this skull now with that of the type, as I have done with that of *Bombinator* which has just been described. The most striking modifications here seen are as follows:—

1. The form of the skull, which is very short, high, convex, and of a regularly semi-oval contour.

2. The intense ossification of all the parts, outer and inner, and the sub-ganoid condition of the exposed parts of the investing bones.

3. The anchylosis of all the *six* roof-bones.

4. The complete union of all the *four* occipito-auditory centres.

5. The great extent of the girdle-bone, from the optic fenestra to the fore edge of the skull.

6. The inferior position of the premaxillaries.

7. The obliteration of the joint of the pedicle.

8. The lateral junction of the squamosal and maxillary.

9. The complete ossification by the quadrato-jugal of the quadrate.

10. The minute size and imperforate condition outwards of the Eustachian passage as the tympanic cavity.

11. No annulus tympanicus.

12. Columella extremely minute, generalised, and completely ossified.
13. The two cerato-hyal centres, the short basal plate, and the minute size of the antero-lateral lobes.

I might, perhaps, add
14. A single fontanelle. But this is doubtful; there is only one in *Bombinator*.

I. A. *b*.—PLATYDACTYLA.

First sub-division.—Tree-frogs with narrow sacral apophyses and no neck-glands.

First Family. POLYPEDATIDÆ.

First genus. *Polypedates.*

35. *Polypedates chloronotus.*—Adult male; 2 inches long. India.

The difference between this skull and the next (Plate 26, figs. 1, 2; and Plate 27, figs. 1, 2) lies mainly in its general form; its shape is half a long ellipse, and the length is to the breadth as $8\frac{1}{2}$ to $7\frac{3}{4}$—an almost exact reversal of the proportions seen in the other.

The mid and hind skull are greatly affected in their shape by this, the parotic projections being much less and the orbital region much longer; it is also much narrower. The nasal region has lost little of its breadth, but the ethmo-palatines are shorter and lie within less divergent maxillaries; its three fontanelles are normal. The occipital condyles (Plate 26, figs. 1, 2, *oc.c.*) are very similar in both; they are oval, posterior, and separated by a concave emargination, which is narrower in this species than in the other. Here the ossification is more intense in the hind skull; the bony masses of the two sides (*pr.o., e.o.*) are confluent both above and below, and there is scarcely any cartilage left—it is only seen at the outermost part where the pedicles articulate—and the bone reaches half-way between the foramen ovale and the foramen opticum (V., II.). But the girdle-bone (*eth.*) is somewhat more restricted, and only reaches to the middle of the orbits; likewise in front it does not run along the septum nasi, but the whole bone stops short a little behind the margin of the proper ethmoidal region.

The sinuosity of the bounding lines of the mid-skull is similar in both, but the curves are gentler in this narrower skull. The actual breadth of the nasal roof and floor is but little less in this kind, and the snout is more directly transverse as in *Rhacophorus maximus* (Plate 26, figs. 5, 6). The pro-rhinals (fig. 2, *p.rh.*) are large, and the septum nasi (*s.n.*) is very thick and solid, as is seen both from the upper and lower surfaces. The double palato-suspensorial bow (with its outworks) is rather tightly drawn; and the quadrate hinges (*q.c.*) retreat a little less than in the next kind.

Here the whole structures is very *Ranine*, but unusually narrow from side to side, yet the form of the palatines and pterygoids (*pa.*, *pg.*) and the amount of unossified cartilage is quite normal; as also is the pedicle (*pd.*), but the quadrate (*q.*) is partly ossified by the quadrato-jugal (*q.j.*).

The Eustachian openings (*eu.*) are quite as large as the inner nostrils; they are oval, and are turned outwards and forwards.

The annulus (*a.ty*) is large and complete, and the stapes and columella (figs. 1 and 2, *st.*, *c.st.*) are like those of *P. maculatus*; the stylo-hyal (*st.h.*) is confluent above. The mandible and hyo-branchial apparatus (figs. 3, 4) are normal, but the articulare (*ar.*) is fastening upon the ossifying cartilage to make a more solid lower jaw than that of most species; also the hypo-hyal (*h.hy.*) has a free lobe in front, as in the "norma," which is not present in *P. maculatus*.

That species approaches *Rhacophorus maximus*; this comes very near *Hylarana temporalis* (Plate 29, figs. 1–5).

The investing bones only differ from the type in being fitted to a skull with a wider cavity and narrower cheeks.

As regards conformity to the Batrachian type, this kind is one of the highest:—

1. The whole cranio-facial structure is longer, and with a broader skull-cavity.
2. The occipito-auditory bones are all confluent, right and left.
3. The columella has an imperfect supra-stapedial.
4. The cartilage both above and below the mandibular joint is partly ossified.
5. There are no septo-maxillaries.

On the whole, this is perhaps the least modified from the typical Ranine skull of any of the "Platydactyla."

36. *Polypedates maculatus.*—Adult male; 2 inches 1 line long.   India.

There are two noticeable varieties of the Oriental Polypedatidæ, namely, those with short and flat, and those with long and narrow skulls; this (Plate 27, figs. 1–4) belongs to the first variety, which culminates in *Rhacophorus maximus* (Plate 26, figs. 5–9). The outline is evenly semi-elliptical; the length is to the breadth as 8 to 9, and the condyles of the quadrate come short of those of the occiput by one-thirtieth of the whole length of the skull.

The term *Hylarana* given to one of the genera in this Family perfectly expresses the relationship of this group, whose members are almost exactly intermediate between the species of *Rana* and the species of *Hyla*.

In this skull the antero-posterior extent of the nasal and orbital regions is equal, but that of auditory, only one-half; this latter region, however, is of great transverse extent.

The chondrocranium, as a whole, has had a fair half of its bulk converted into bone, and the skull in all respects has a medium strength; it is much depressed, and the orbital region is at least one-third wider than in most of the species of *Rana*.

The small, oval, directly posterior occipital condyles are separated by a concave space more than the width of both together, and more than half of this basal intercondylar tract is cartilaginous.

Also a wide "cross" of cartilage intervenes between the four main hind-skull ossifications, below; above, the superoccipital synchondrosis is narrow, and the prootic and ex-occipital (fig. 1, *pr.o.*, *e.o.*) of the same side are entirely confluent, the bony substance reaches to the squamosal (*sq.*) under which the "tegminal edge" is soft. Even below, the trigeminal nerve (fig. 2, V.) passes through the bone, which is not far from the girdle-bone, over the optic nerve (II.). Some cartilage and three small membranous fontanelles lie across the hind skull, behind the large heart-shaped principal space. The breadth of the skull at the top lessens along the girdle-bone, in which, however, it spreads out again gently in front. This bone (*eth.*) takes up three-fifths of the orbital region, it passes, in front, nearly to the boundary of its own (ethmoidal) region, and above, runs a little along the septum nasi (*s.n.*). Over its wings this common ethmoid has a small superorbital "eave" of cartilage.

The nasal region is very large both ways; it ends abruptly in front, has large crescentic roofs (fig. 1), a long and moderately deep septum (*s.n.*), and a wide floor (fig. 2). The pro-rhinals (fig. 2, *p.rh.*) are slender, but well formed and uncinate. The superadded cartilaginous valves of the external nostrils (*u.l¹.n.l²., e.n.*) are large, and those passages are wide apart.

Very wide apart indeed are the internal nostrils (*i.n.*); they are large oval holes converging forwards. The palato-suspensorial arch is large and well developed; all its parts are wide, for *width* characterises the whole skull and all its parts. Yet, in essentials, both the cartilage and its grafted bones (*pa., py.*), with the condyles of the quadrate and the pedicle (*q.c., pd.*), are quite normal.

The Eustachian openings (*eu.*) are less than the inner nostrils, but are large; they are oval, with the long axis directed outwards and a little forwards. The annulus (*a.ty.*) is large, broad, and perfect; the stapes (fig. 4, *st.*) is large, oval, and flattish: it is broader in front than behind. The inter-stapedial (*i.st.*) is a thickish subquadrate segment, nearly half the size of the stapes; the medio-stapedial is long, straight, and its imperfectly ossified, wide, proximal end is scooped out; the extra-stapedial (*e.st.*) is a reversed spatula—it is narrower distally than proximally, and has only a fibrous supra-stapedial band.

The narrow stylo-hyal (fig. 2, *st.h.*) is confluent above; it increases only gently in size downwards (fig. 3, *c.hy.*), has no hypo-hyal lobe (*h.hy.*), and soon reaches the basal plate as it turns backwards. That plate (*b.h.br.*) is short, wide in front, narrow behind, has small fore-turned anterior, and long, slender, posterior lobes; the thyro-hyals (*t.hy.*) are long and slender; they diverge so as to be at a right angle with each other.

The mandibles (fig. 3) are quite normal, but are long and slender; the dentary (*d.*) is half the length of the ramus.

The investing bones are normal but partake of the general flatness and breadth of

DEVELOPMENT OF THE SKULL IN THE BATRACHIA.

all things else in this skull. The roof-bones (*f.p.*) run short in front; the parasphenoid (*pa.s.*) has pointed arms under the ear-capsules; the vomers (*v.*) are wide apart and have only a smallish crop of teeth; there are no septo-maxillaries; the premaxillaries (*px.*) are wide but feeble, so also the maxillaries (*mx.*) have a wide facial plate, but it is thin; the quadrato-jugals (*q.j.*) are not grafted on the quadrate; the squamosals are but little developed over the tegmen (fig. 1, *sq.*).

Here, in a few things, the "norma" is not reached; besides the general breadth of the parts—

1. The auditory occipital masses of bone are confluent above.
2. The lesser fontanelles are very small and have a still smaller space between them.
3. There are small superorbital projections.
4. There are no septo-maxillaries.
5. The supra-stapedial is membranous.
6. The hyo-branchials are feeble.

This is a poor list of indictments against this species as coming short of the *norma*; the flatness of the general shape, and of the individual parts, are also no great modifications in the morphology of this very *Ranine* skull.

Second genus. *Rhacophorus*.

37. *Rhacophorus maximus.*—Adult male; 3⅛ inches long. North India.

This large species bears the same relation to the Tree-frogs of India that *Pelodryas* does to those of Australia, and *Phyllomedusa* to the Neotropical kinds.

That which characterises the Oriental Polypedatidæ, generally, is here carried to excess, namely, the great dilatation of the endocranium, especially in the nasal region. The length is seven-eighths as great as the breadth; the quadrate condyles (Plate 26, fig. 5, 6, *q.c.*) are opposite the passage for the vagus nerve (fig. 6, X.).

The cranial "barge," from the foramen magnum to the great transverse snout, is twice the breadth of an average skull of a species of *Rana* of the same size as this kind. The fore skull, leaving out the massive premaxillaries, is as long as the mid skull, and is wider: it is extremely wide in the ethmoidal region. The mid skull lessens gently to the temporal region, where it is pinched in; in front it is three times as wide as it is deep (figs. 5, 6, 7).

The antero-posterior extent of the hind skull is half that of either of the other two regions, yet the breadth of the cranial cavity is not lessened there, but the parotic wings extend far out, as the great breadth would indicate, beyond the horizontal canals (fig. 5, *h.s.c.*); those wings are rather narrow. The occipital condyles (*oc.c.*) are rather small for so strong a skull, they are posterior, and are separated by an arched (emarginate) line of their own breadth.

The occipito-auditory ossifications are all continuous, as in *Polypedates chloronotus*, and just a little cartilage is left on the outside of the wings; the bony matter reaches half way between the foramen opticum and the foramen ovale (II., V.).

The tract of unossified skull-wall round the optic passage is only one-fourth the extent of the mid skull, for the girdle-bone (*eth.*) is very large; in front, above, it occupies half the nasal roof (fig. 5, *al.n.*), and below, it runs up to the dentigerous lobes of the vomers (*v.*).

Under the roof-bones (fig. 5, *f.p.*) the three normal fontanelles can be seen, and as those bones only reach, in front, halfway from the foramen magnum to the transverse end of the snout, the girdle-bone is but little roofed in above. Below, the bone passes into the ethmoidal wings, but above, there is a considerable selvedge unossified; this is the apiculate superorbital (*s.ob.*).

The pre-cranial mass, which is the fore half of the ethmoidal, and all the nasal region, is square, and its breadth is equal to the widest part of the roof in the orbital region. Its fore edge is almost directly transverse, being only gently rounded.

The depth of the nasal region is only moderate (fig. 6, *s.n.*) is very thick; the floor and roof (figs. 5 and 6) are equal in width; the pro-rhinals (*p.rh.*) are small and wide apart outside the nerve outlets, which also are wide apart.

The external nostrils (figs. 5 and 7, *e.n.*) are very wide apart, and the nasal roof (*al.n.*) is very thick over them; so, also, the second upper labial (*u.l².*) is a thick crescentic valve. The 1st upper labial (*u.l¹.*) is rather large, and the nasal process of the premaxillary (*p.x.*) is grafted upon it, ossifying much of its substance. The internal nostrils (*i.n.*) are very wide apart. The palato-suspensorial arches are not large in proportion to the skull proper; and they have only the average ossification, the two bones (*pa., pg.*) leaving a considerable tract unaffected. The palatines (fig. 6, *pa.*) are thin, straight, splint-like bones becoming twice as wide at their outer end as at their inner. The curved and bifurcated pterygoids (*pg.*) are moderately strong; the inner fork on the pedicle is half as long as the outer, which runs inside the quadrate (*q.c.*). Both the pedicle and the quadrate are formed of a solid mass of cartilage, especially the former (*pd.*), which forms, here, the most massive condyle I am acquainted with, at this part. This sinuous facet glides on a definite tract of basal cartilage, which is confluent, externally, with the stylo-hyal (*st.h.*). The quadrate, above the reniform condyle (*q.c.*), is not ossified to any noticeable extent.

The Eustachian openings (*eu.*) are large, oval, and have their long axis directed forwards. The annulus (fig. 7, *a.ty.*) is large, broad, and complete; its edge has a strong rim.

The mandible (fig. 7) is normal; it is rather slender, and has a high coronoid process to the articulare (*ar.*).

The stapes (fig. 9, *st.*) is oval, and not raised externally to any appreciable extent. The inter-stapedial (*i.st.*) is massive and semi-ossified; it runs some distance on the shaft. The medio-stapedial (*m.st.*) is a rather slender, terete rod; the extra-stapedial (*e.st.*) is large and spatulate; the (*s.st.*) supra-stapedial is cartilaginous and confluent, above.

DEVELOPMENT OF THE SKULL IN THE BATRACHIA.   149

The hyo-branchial plate (fig. 8, *b.h.br.*), with its outgrowths, falls short of the type in having no hypo-hyal lobes, and in the small size of the front lateral lobes; the thyro-hyals (*t.hy.*) are rather slender, and only diverge gently.

The investing bones are very similar to those of the type, but modified so as to fit to a more outspread cranium. The roof-bones (*f.p.*) are strong, very wide, and as in most of this Family are very scant in front, especially near the mid-line. The nasals (*n.*) make no pretence of covering the skull in their region, for they are very narrow crescentic shells of bone. The premaxillaries (*p.x.*) are, perhaps, the largest in the Order, bounding, as they do, so wide a nasal fore wall.

There are no septo-maxillaries; the maxillaries (*m.x.*), the quadrato-jugals (*q.j.*), and the squamosals (*sq.*) are all strong, well developed, and normal. So also is the parasphenoid (fig. 6, *pa.s.*); it is very sharp, both behind and before. The vomers (*v.*) stretch far across the nasal floor, but fail to meet; they have the usual four-snagged form; the large thick tooth-bearing lobes are arcuate, and their concave margin is behind.

The relation of this type of skull to the others in this Family is evident; this is the fullest expression of the *Polypedatine* type, and thus the greatest divergence from the *form* of the *Ranine* skull is obtained. We have to remark upon—

1. The short auditory and occipital region, the wide orbital, and the wide *and long* nasal region.
2. The fusion of the bones of the hind skull.
3. Unusual solidity and strength of the pedicle.
4. The inter-stapedial ossifying as part of the *shaft*, apparently without absolute segmentation of the cartilage.
5. The confluence above of the supra-stapedial and stylo-hyal bands.
6. The feebly expressed cartilages of the hypo-branchial plate.
7. No septo-maxillaries.
8. A distinct superorbital cave.
9. The grafting of the premaxillary on the " 1st upper labial."

All these modifications are gentle and non-essential.

Third genus. *Ixalus*.

38. *Ixalus variabilis.*—Adult female; 1 inch 1 line long.   Ceylon.

This skull differs very little indeed from that of *Polypedates maculatus*; it would differ still less from that of a young individual of that species of its own size.[a]

It is a short skull, and the length bears to the breadth the proportion of 7 to 8; it is therefore a trifle narrower than that of *Polypedates maculatus*. The outline is extremely regular, and is half an ellipse, whose long diameter may be put as 13 and

---

[a] Dr. GÜNTHER is now inclined to consider this species as the young of *Polypedates maculatus*; but the well developed supra-stapedial band, and the absence of an inter-stapedial and of secondary fontanelles, makes this doubtful.

its short as 8, for the outline of the face stops a little short of the axial length; the mandibular hinges are slightly less retreated than in the large kind. The occipital condyles (Plate 27, figs. 5, 6, *oc.c.*) are similar; the cartilaginous tracts at the midline are larger, and the projection of the ear-mass less, and less ossified; the "tegmen" (fig. 5, *t.ty.*) is soft. The hind part of the orbital region is not so wide; the girdle-bone is of less extent, and the superorbital tract of cartilage is not present.

The nasal region is similarly broad, on the whole, but the outer nostrils (*e.n.*) are nearer together, and the inner nostrils (*i.n.*) are scarcely oval. What difference of length over breadth there is, is shown differently—the holes are transverse. There are no secondary fontanelles (fig. 5); the "annulus" (*a.ty.*) is smaller, but complete; the Eustachian openings (*eu.*) are similar in both kinds.

The columella has no separate inter-stapedial (Plate 29, fig. 6, *m.st.*); and the supra-stapedial (*s.st.*) is confluent above; the stapes (*st.*) has a boss. The investing bones differ in these two kinds, as those of the old and young of the same species differ; there are no vomerine teeth.

The mandibles and hyo-branchial apparatus (fig. 7) are very similar, but the former are more evenly arched, and the latter has a *young* appearance, for the front lateral lobes are not such elegant "ears," and they are not definitely marked off from the hinder projection (*b.h.br.*). The basal plate is longer, and the thyro-hyals (*t.hy.*) are slenderer, and are inbent.

This small *Ranine* Tree-frog differs from the type in the following particulars:—
1. In the general broad, short, outspread condition of the skull.
2. There are no secondary fontanelles.
3. There are no septo-maxillaries.
4. There are no vomerine teeth.
5. The tympanic annulus is less.
6. There is no inter-stapedial segment, and the supra-stapedial is confluent above.
7. There is no lobe to the hypo-hyal.

Fourth genus. *Hylarana*.

39. *Hylarana malabarica.*—Young; $\frac{3}{4}$ inch long. India.

The specimens dissected by me were only one-fourth the length of the full-grown adult, which attains to nearly 3 inches.

Nevertheless I found the skull in these young individuals (Plate 28, figs. 1, 2) to be much more perfectly ossified than in the adult of *H. temporalis* (Plate 29, figs. 1, 2), next to be described.

This is one of the *longest* skulls in the "Order;" its length is to the breadth as $7\frac{1}{2}$ to $6\frac{1}{2}$—a Frog's skull which is as long as it is broad, is long as compared to most "Anurous" skulls.

The proportionate length of the three regions of the skull is very different from what is seen in *H. temporalis*, where the muzzle is very broad: here the fore and hind regions are equal, and each is three-fourths the length of the mid skull, which is both long and broad.

The occipital condyles (Plate 28, figs. 1 and 2, *oc.c.*) are postero-inferior, and are moderately large. A considerable tract of cartilage remains in both the basi- and superoccipital regions; there is a triangular *endosteal superoccipital bone* (figs. 1 and 5, *s.o.*); and the bony tracts (*pr.o., e.o.*) of the same side are quite confluent. Below, the bone leaves an even crescent of cartilage at the margin of the floor of the vestibule (*au.*); above, the bone is snagged, and leaves an irregular cartilaginous tegmen (fig. 1).

I could find no secondary fontanelles, but a very large long main space with no lateral ingrowth for a considerable distance along the sides, quite unlike the next. Moreover, this deficiency of the "tegmen cranii" shows itself both behind and before: behind, the roof is devoid of cartilage up to the ear-capsules; and in front the roof extends a very short distance over the hemispheres. As this is a *young specimen*, the ear-masses and the nasal roofs show much of their primary form; they are not drawn into new shapes by the overgrowth of the peripheral bones and arches.

Hence comes the *narrowness* of the skull in the auditory region, for the tegmen tympani is merely a narrow selvedge to the ovoidal ear-sacs; this would be at least somewhat modified in an old individual. On the whole, the orbital region, or mid-skull, is very *Polypedatine*, but is broadest behind, narrows gently, and widens out again towards the ethmoidal region.

What is very remarkable in so young a skull is the trespassing of the *so-called* "prootic" round the front of the foramen ovale (V.), even so as to include the lesser cranial nerves between it and the foramen opticum (II.). Therefore, so early, the orbital region is half of it bony, the girdle-bone (*eth.*) being as long as the cartilage behind it. It runs into the septum nasi (*s.n.*) both above and below; then it takes in all its own cartilage, athwart the ethmoidal "wings;" above, it scants a little, and leaves the edge of the ethmoidal roof, at its junction with the nasal roofs (fig. 1), naked.

These roofs are very large and broad (fig. 1), larger than the floor (fig. 2); the intertrabecular wedge (or *wall*) between both grows into a distant "prenasal rostrum" (*p.n.*) in front, which is a new thing in an Oriental member of this family. Nevertheless, the snout is very broad, and the rostrum, which is drawn as *dissected out* in the figures, in reality turns downward, as in embryos of all kinds that possess this projecting cartilage. In this kind, as in its congeners, the nostrils (*e.n.*) are very wide apart, nearly as wide as the "choanæ" or inner openings (fig. 2, *i.n.*). The prorhinals (*p.rh.*) and the labials (*u.l¹,u.l².*) are well developed.

The arrest of the "parotic wings," in a skull with such long regions, causes the facial "bow" to be strongly bent; otherwise all is normal; the bones (*pa., py.*) are already of the normal size and strength, and leave a cartilaginous palato-suspensorial of the average size.

The forks of this arch differ greatly, for the pedicle (*pd.*) is very short, and the quadrate, with its condyle (*q.c.*) runs downwards and outwards to a considerable distance; *backwards*, it only reaches to the fore end of the stapes (*st.*); above the condyle the quadrato-jugal (*q.j.*) has spread into the cartilage, ossifying it.

The Eustachian openings (*eu.*) are large and oval; they are turned forwards; the annulus (*a.ty.*) is large and perfect; the stylo-hyal (*st.h.*) is confluent above. The stapes (fig. 4, *st.*) is large, convexo-concave, oval, and emarginate in front; the interstapedial (*i.st.*) is a thick mass, ossified largely, and joined by a suture to the mediostapedial (*m.st.*), which is a straight rod. The extra-stapedial (*e.st.*) is a wide spatula, and its ascending fork (*s.st.*) is cartilaginous and confluent above.

The mandible (fig. 3) is perfectly normal, and the mento-Meckelian (*m.mk.*) is well ossified. The coronoid crest of the articulare (*ar.*) is moderate, and the condyle is sub-reniform. The hyo-branchial apparatus is normal, but its actual shape is much like that of the next (*Rappia*, fig. 8); the hypo-hyal lobe is small, and the ceratohyal broad (*h.hy.*, *c.hy.*); the semi-oval notch in front is short, the front lateral lobe very large and separated by a small space from the hind lobe; the basal plate (*b.h.br.*) is both long and broad, as compared with that of other kinds, and the thyro-hyals (*t.hy.*) are long and slender.

Altogether this part is exactly intermediate between what we find in the ordinary Oriental Polypedatidæ, and in the next but one, a small Ethiopian type.

The investing bones are such as would be found in a Common Frog's skull of the same size and age. The fronto-parietals (*f.p.*) are more *Ranine* than in the last, being pointed above; they would be more truncated in an old specimen. The nasals (*n.*) are better developed, and cover more of the nasal roof than in its congeners. The parasphenoid (*pa.s.*) is very large, both in the median and transverse parts; and the vomers (*v.*) are well developed and dentigerous. I find no septo-maxillaries; the marginal splints (*px.*, *mx.*, *q.j.*, *sq.*) are feeble but normal. An intermediate place between *Hylarana temporalis* and the West African *Rappia* is asked for this species: we shall see this in the two next instances. Compared with the type, we have some curious discrepancies:—

1. The great size, both length and breadth, of the orbital region.
2. The intense ossification of the occipito-auditory region.
3. A well-marked endosteal *supraoccipital* bone.
4. No septo-maxillaries.
5. The confluent condition of the apices of the stylo-hyal and supra-stapedial.
6. The intense ossification of the inter-stapedial.
7. The large size and close contiguity of the lateral lobes of the basi-hyobranchial plate; none of these are *deep-seated* differences.

40. *Hylarana temporalis.*—Adult male; 2¼ inches long. Ceylon.

Dr. GÜNTHER ("Rept. of Brit. Ind.," p. 427) gives his reasons for believing this insular species, which at that date (1864) he had received from Ceylon only, to be distinct from the continental form—*H. malabarica*. I have just described that species, and when this has been compared with it then all doubt will vanish as to their distinctness; both kinds attain the length of *three inches*, or thereabouts; my specimens of *H. malabarica* are all young, and only a quarter the length of the adult; yet the ossification of their skull is much more perfect than in the *adult* of the Ceylonese kind; there are also other differences, as we shall see.

This is a long skull (Plate 29, figs. 1, 2), for the length is equal to the breadth; it is extremely like that of *Polypedates chloronotus* (Plate 26, figs. 1, 2), but is not so long; in that species the length is about a *twelfth* greater than the breadth; and that kind has its ossifications intenser, and has no evident superorbital "eave."

Although belonging to the narrow-skulled group of this Family, the cranium proper is in reality very broad, half as broad again as in an average species of *Rana*, and therefore settling this as a character of the Oriental flat-toed Frogs.

They are, indeed, a broad-muzzled, flat-skulled race, but when flattened most their skull is very *Ranine* as compared with that of a true *Hyla*.

The occipital condyles (Plate 29, figs. 1, 2, *oc.c.*) are large, posterior, and separated by a crescentic emargination half their breadth. Measured along the axis, the occipital region is half, and the nasal two-thirds, the extent of the orbital, which is very large. The "tegmen cranii" covers two-thirds of the skull, and grows in well from the sides, so that the larger cordiform fontanelle is small, and the secondary spaces very minute.

The parotic wings stand out well, and the canals (*a.s.c.*, *h.s.c.*, *p.s.c.*) are prominent. The ossification, right and left, is only equal that of the type, and the four normal centres (*pr.o.*, *e.o.*) are separated by extensive tracts of cartilage. Above, the prootic stretches out in front so as to underlie the squamosal a little. Below, the floor of the vestibule (*au.*) is unossified, for the ex-occipitals, which flank the posterior canal above, do not reach to the fenestra ovalis below. In front, the prootics just reach the optic fenestræ (II.). Of the remainder of the long orbital region the hinder half is unossified; the rest, or the region of the "girdle," is partly ossified, and partly *calcified*, the latter tract marking out the usual extent of the bone, which however is only fully developed in front into an irregular ∪-shaped bone, thick and rugged, but whose halves do not meet above (fig. 1, *eth.*). In front, this imperfect girdle scarcely covers the proper ethmoidal territory, and the whole nasal region, and likewise the extensive superorbital crescents (*s.ob.*), are unossified. Leaving these out, the form of the orbital region of the skull is oblong, modified by a slight bulging, and a gentle increase of breadth, forwards.

The nasal territory (*al.n.*) is large and four-square, with the roof spreading wider than the floor; the septum (*s.n.*) is thick, and both above and below we can see clearly how much has been added to the original "trabecular horns" by the solid intertrabecular wedge (or *wall*), and how much by the large shells of cartilage that form the roof. Their division also, into the floor (with its angles that bury themselves in the maxillaries), and the new "horns" or pro-rhinals (*p.rh.*) that have budded from their infero-medial face, is clearly seen.

The broad snout is gently emarginate in front, and the nostrils (*e.n.*) look outwards and forwards; they are wide apart, and are protected by the normal valves (*n.v.n E.*). The large circular inner nostrils (*i.n.*) are still wide apart, not much nearer together than in the more flat-headed "Polypedatidæ."

The palato-suspensorial arches are perfectly normal, but slender; their bones (*pa., py.*) partake of a like slenderness, and the well-retreated condyle of the quadrate (*q.c.*) is large. So also is the condyle of the pedicle (*pd.*); it is sub-convex; the other is deeply bilobate: above this hinge the suspensorium is ossified to some extent by the quadrato-jugal (*q., q.j.*). The Eustachian openings (*eu.*) are large, oval, and look outwards and forwards. The annulus (*a.ty.*) is very large, broad, and perfect.

The mandible (fig. 3) is long, slender, and normal; it has a rather high coronoid process to the articulare (*ar.*); the dentary (*d.*) is half the length of the ramus, and *rises into a crest* behind the mento-Meckelian (*mk.*), as in the Oriental *Ranæ* (Plates 6 and 7).

The stapes (fig. 5, *st.*) is large, sub-oval, and has a boss for muscular attachment; the inter-stapedial (*i.st.*) is a large, thick, sub-oval cartilage; the medio-stapedial (*m.st.*) is a short pistol-shaped bone, very large and expanded above; the extra-stapedial (*e.st.*) is large and spatulate; and the supra-stapedial (*s.st.*) is almost entirely fibrous.

The hyo-branchial apparatus (fig. 4) is normal, but all the "lobes" are small and narrow.

The investing bones are intermediate between those of a middle-sized *Rana* and a broad-headed member of its own Family; both the roof-bones (*f.p.*) and the para-sphenoid (*pa.s.*) reach further forwards than in its congeners. The nasals (*n.*) are angular, jagged shells, and leave a large cruciform tract of cartilage uncovered.

The outer bones of the face are all slender; there are no septo-maxillaries; and the quadrato-jugals (*q.j.*) are grafted upon the quadrate cartilage. The vomers (*v.*) are large, have a considerable dentigerous lobe, and the three other spikes are well developed. Compared with the — "norma," this skull is—

1. Remarkable for the breadth of the cranium proper; for the great length as well as breadth of the mid skull; and for the width of the fore skull, or nasal region.
2. There are no septo-maxillaries.
3. The superorbital "eaves" are large.
4. The girdle-bone is imperfect.
5. The stapes has a *boss*, and the stylo-hyal is confluent.
6. The supra-stapedial is imperfect.

These are all slight modifications, and on the whole this skull is extremely *Ranine*, and agrees especially with the Oriental *Ranæ* in having a crest on the fore end of the dentary, and a large perfect annulus.

Fifth genus. *Rappia*.

41. *Rappia* ——? sp.—Adult female; ⅝ inch long. Lagos.

This specimen was collected, with others, several years ago at Porto Novo, near Lagos, by R. B. WALKER, Esq., and was sent to me by Mr. T. J. MOORE, of Liverpool, on March 24, 1874.

Dr. GÜNTHER was doubtful of the species, although there is a *Rappia lagoensis* (see GÜNTHER, P. Z. S., June 23, 1868, pp. 478-490; a figure of that species is given in plate 40, fig. 2, and a description at p. 487).

Dr. GÜNTHER gives the length of *P. lagoensis* as 28 millims.; my specimen measured only 21 millims., although a female. In the paper just referred to, 15 species of this genus are given—all African. In the "Batrachia Salientia," p. 85, these are described under the generic term *Hyperolius*, a name now used instead of *Uperoleia* (ibid., p. 39) for a genus related to *Alytes*. The only Australian kind described in that work is *H. bicolor* (p. 89) from Port Essington. The author remarks of this, whose skull I shall describe next, that—" This species is very probably the type of a separate genus." This suggestion is a true one; the Australian type has an exceedingly different skull from that which I shall now describe from West Africa.

This skull (Plate 28, figs. 6, 7) greatly resembles the last but one (figs. 1, 2); and the thought that occurs is this—Are the African *Rappiæ* dwarfed "races" of various species of the "Polypedatidæ;" types which attain so large a size in the Oriental region?

This is a rather long skull; the length and width are equal, and it tends towards the triangular form. The condyles of the quadrate (*q.c.*) do not reach so far back as in the last but one, they end in front of the *fenestræ ovales*.

The occipital condyles (*oc.c.*) are similar—they are large and posterior; but the basal emargination is less in this species. In conformity with the greater breadth across the ears, the simply oval form of the capsules is lost, and there is a considerable tegmen tympani (fig. 6). The general form of the mid and fore skull is very much alike in both kinds, but they are both a little broader (relatively as well as really) in this small *adult*; and the nasal roofs are like those of a metamorphosing *Rana temporaria* (see Phil Trans., 1871, Plate 8, fig. 1). The four bony tracts of the hind skull (*pr.o., e.o.*) are like those of the same of Young Frogs; whilst the girdle-bone is like that of quite old individuals, and the skull, altogether, is a curious mixture of *old* and *young* characters.

The notochord (figs. 6, 7, *nc.*) is persistent; there is, as in the last but one, only one large fontanelle, the tegmen cranii being very short in front, short behind, and not present along the sides in most of the orbital region.

Below (fig. 7), the prootics and ex-occipitals slightly overlap the wings of the parasphenoid (*pa.s.*); but the cruciform synchondrosis is very large. Below, the ex-occipitals reach half way between the vagus-passage (X.) and the stapes (*st.*);

above, they climb up to the inside of the posterior canal (fig. 6, $p.s.c.$), which they flank; there they are wide apart.

The prootic ($pr.o.$) runs forward to the foramen ovale (V.), and outwards as far as to the tegmen tympani (fig. 6); they show mostly in front, and to a very moderate extent either above or below. Three-fifths of the orbital region is unossified, and the optic fenestra (II.) in this type is larger than in the other (fig. 2). The girdle-bone ($eth.$) is more developed than in it, and it has ossified exactly all the cartilage that belongs to the ethmoidal region, for it has carefully avoided the anterior sphenoidal, the ethmo-palatine, and the true nasal, regions. The nasal region is very instructive; the proper roofs (fig. 6) are large cartilaginous pouches that have coalesced with the ethmoid behind, and with the large intertrabecula (nasal septum), ($s.n.$) in the middle; they even exaggerate the large size of the nasal region in the Polypedatidæ, for they are very tumid, postero-laterally; a clean chink separates them from the ethmo-palatine bars ($e.pa.$).

In front, the septum grows out into a well-marked prenasal rostrum (figs. 6, 7, 9, $p.n.$). (In figs. 6 and 7, both this and the pro-rhinals ($p.rh.$) are drawn as dissected out, and shown beyond the outer bones; they were examined by transmitted light.)

The form of the larval cornua trabeculæ is clearly seen below (fig. 7), as they first bend inwards at their first coalescence, and then spread out in front; their secondary cornua—the pro-rhinals (figs. 6, 7, 9, $p.rh.$)—are long, slender, and pedate.

The secondary upper labials ($u.l^1.u.l^2.$) are well developed; and the outer segment is an almost perfect "annulus," with an outer lip over the external nostril: this passage is far from its fellow, as in this "Family" generally; three-fourths as far as the internal nostrils ($i.n.$), which, however, are very much larger than the outer. The facial "bow" is not so strongly bent as in the last, and here we have an arrested condition of the applied bony tracts ($pa., py.$), which are extremely feeble, and like those of very newly curtailed Common Frogs. The cartilaginous palato-suspensorial bar is well developed; the pedicle ($pd.$) is large, and the quadrate region ($q.c.$) is thick, and of the normal length. The condyles, however, are very peculiar; they are obliquely saddle-shaped, and have not the usual resemblance to a kidney.

But one of the rarest characters, for an *adult*, turns up here: this is the retained orbitar process ($or.p.$), a structure which is large in the Tadpole, in all the Anura, and is to be seen in *Polyodon* (see BRIDGE, Phil. Trans., 1878, Plate 57, $or.p.$); I find it also in the Sturgeon.

It remains here as a round leafy lobe ($or.p.$), at the point where the quadrato-jugal ($q.j.$) and squamosal ($sq.$) meet.

The Eustachian tubes are large and reniform; they turn outwards and a little forwards. The annulus ($a.ty.$) is rather small, and does not unite above; the stylo-hyals ($st.h.$), coalesce with the auditory capsule. The stapes (fig. 10, $st.$) is large and oval: the inter-stapedial ($i.st.$) is large and semi-osseous: the medio-stapedial ($m.st.$) is a pistol-shaped rod of bone, which ends in a tongue-shaped cartilage, the extra-stapedial ($e.st.$); it has no ascending process.

Here, again, in the large bony inter-stapedial and the suppressed supra-stapedial we have a curious mixture of what is arrested and over-developed.

The mandibles (fig. 8) are well developed and quite normal, except that the articular condyle is very deeply sulcate, so as to be the exact *reniform* counterpart of the quadrate condyle of other types; this is in conformity with the very deep scooping of the part on which it rolls (figs. 6 and 7, *q.c.*).

The cerato-hyal band (fig. 8, *c.hy.*) and the hypo-hyal lobe (*h.hy.*) are quite normal, but the large front notch, as in the other kind (fig. 3), is unusually short.

The whole plate is large, and here the lateral lobe is notched in front, is of large extent, and is only separated from the hinder small lobe by a gentle emargination, there is an approach to this in the other (fig. 3), but here (fig. 8) we have it well developed, and it is a rare condition.

The thyro-hyals (fig. 8, *t.hy.*) are rather large and only moderately divergent, they show age in a remarkable way; the inner edge of each is produced into a large wing of *periosteal bone*; moreover, the basal plate (*b.h.br.*) is all converted into *true bone*, except at its edges.

More or less calcification often occurs and is typical; sometimes *true bone* appears close to the thyro-hyals; here, however, the bony substance reaches to the front notch, and runs far into the lateral lobes.

The investing bones are exactly such as we see in newly transformed Frogs of the common kind. The fronto-parietals (*f.p.*) are quite similar to those of the young *Hylarana* (figs. 1 and 2), and the nasals (*n.*) are small, crescentic shells, covering very little of the cartilage. The marginal bones (*px., m.e., q.j., sq.*) are frail splints, like the palatine and pterygoid ectostoses. The parasphenoid has similar wings to the other, and its main part is short but very broad; here again we have the characters of a young Frog. The vomers (fig. 7, *v.*) are very minute *toothless* crescents, and there are no septo-maxillaries.

Many of the characters of this *dwarf* are those of a young typical *Rana*, but the intense ossification of certain regions show that it is an adult arrested at certain points. It differs from the "norma" in the following particulars:—

1. The cranial notochord is large and persistent.
2. The occipito-otic centres are arrested.
3. There is only one, and that a very large, fontanelle.
4. The mid and fore skull are very wide, and the latter has the roofs very tumid, as in young species of other kinds.
5. There is a prenasal rostrum.
6. The palato-suspensorial ectostoses are arrested.
7. The quadrate condyle is of an unusual form, being deeply scooped, and its counterpart in the mandible being reniform with a very definite "hilus."
8. The retention of the *larval* "orbitar process."
9. The intense ossification of the *inter-* and the absence of the *supra-*stapedial.

10. The unique ossification of the large basal plate, and the crested growths of the thyro-hyals.

In the next kind of dwarf "Polypedatid,"—a smaller species than the last and from another region,—the signs of arrest are more numerous, and some of these stoppages must have taken place early in the larval stage; moreover there is but little sign of compensating growth, in over-ossification and the like, such as is seen in the one just described.

42. *Rappia* (*Hyperolius*) *bicolor*.—Adult female; $\frac{3}{4}$ inch long. Dog-trap Road, Paramatta, Australia.

This is a rather long skull, for the length and greatest breadth are equal. The quadrate condyles (Plate 19, figs. 6, 7, *q.c.*) end opposite the middle of the mediostapedial (*m.st.*). The mid skull is one-third longer than the other two regions, which are equal to each other in this respect; they are both moderately broad, but the mid skull is very broad.

The interauditory region is nearly as wide as the interorbital, and this latter increases in width forwards. The "tegmen cranii" is very slight in front, largely covers the hind brain, and there is left a fontanelle (*fo.*) which is evenly egg-shaped, and has its wide end in front.

There is a very narrow band of cartilage bounding the fontanelle on each side, but this is not continued downwards into the *wall*, which is membranous, except for a short distance at each end.

There are two rather large secondary fontanelles in the extensive interauditory roof.

The occipital condyles (*oc.c.*) are small, sub-reniform, and postero-inferior; they are separated by a straight basal tract equal to both in width. The basal cartilage is equal to *one* condyle in width, and the upper or supraoccipital cartilage is as wide as the intercondyloid space.

The four bones (*pr.o., e.o.*) are larger than in the last; the ex-occipitals reach the fenestra ovalis below, and are wider above; the prootics cover the anterior canal (*a.s.c.*), nearly, floor its "ampulla" below, and go round, but not in front of, the foramen ovale (V.). There is a moderate tegmen tympani (fig. 6), and the canals and vestibule bulge out above and below (*h.s.c., a.s.c., p.s.c., eb.*).

The tegmen cranii becomes a narrow band directly it passes into the interorbital region, for the trabeculæ (*tr.*) have not enlarged, relatively, since the legs of the Tadpole began to bud out, only the intertrabecular tract of cartilage was added.

The optic *foramen* (II.) is large, but the optic *fenestra* (or fontanelle, *o.fo.*), which is large in several types of small Australian *Anura*, here occupies nearly all the orbital region; it is *one-third the length* of the skull. The bulging floor of the skull widens towards the ethmoidal region, and then, behind the true nasal region, there is a complete *girdle* for a short distance—not a complete *bony* girdle, however, yet there are three tracts of bone, as in the young of *Rana temporaria* (Phil. Trans. 1871, Plate 8, fig. 7, *eth.*).

The broad end of the egg-shaped fontanelle (*fo.*) is bounded by a very narrow tegminal tract; this tract is ossified, at first, evidently, as in the young of *Rana temporaria* of the first autumn, by ossification of the fibrous investment, and then by some degree of *endosteal* deposit. In *Dactylethra* (Phil. Trans. 1876, Plate 59, fig. 1, *s.eth.*) the ossified membranous tract is large and T-shaped, but it does not unite with the extensively ossified chondrocranium; but in the *phaneroglossal* Anura, any external bone there may be soon lost in the deeper ossification.

In this "dwarf" the superior marginal crescent of bone seems to have started in growth from symmetrical points, but of this I am not certain; it may have been single and median at first, and then have become partly sub-divided afterwards—a very common thing in this group. In this narrow ethmoidal girdle the "axils" are ossified (*eth.*), and their limited lateral tracts unite with the arch of bone above, inside the superorbital band. That band (*s.ob.*) is double, there is a narrow tract over the anterior angle of the orbital space, and a wider lobe projecting from the ethmoidal wing; these two are separated by a deep rounded notch.

The true nasal region is altogether unossified; it is of a good size both in length and breadth. The septum (*s.n.*) narrows a little, forwards, from the perpendicular ethmoid, and ends in front in a shortish, rounded rostrum (figs. 6, 7, 9); this rostrum is margined by the nasal roofs, and the front of this region is sinuous, and retires considerably, right and left. The pro-rhinals (figs. 7 and 9, *p.rh.*) are smallish and falcate. The broad subnasal laminæ (fig. 7, on each side of *s.n.*) grow out into ear-shaped angular processes. The external nostrils (*e.n.*) are as wide apart as in other "Polypedatidæ;" they are almost as far apart as the inner openings (*i.n.*); the valvular cartilages (*u.l².u.l².*) are normal.

The palato-suspensorial arches are normal, but feeble, so also are the bony tracts applied to them (*pa., pg.*); both the hinder forks (*pd., q.c.*) are short, and the quadrate is not ossified.

The Eustachian openings (*eu.*) are nearly circular; they are two-thirds the size of the inner nostrils; the annulus (figs. 6, 10, *a.ty.*) is rather small, and imperfect above.

The mandible (not figured) is normal, but feeble; the stapes (fig. 10, *st.*) is nearly semicircular, and has a long "boss;" the inter-stapedial is not cut off from the dilated and forked end of the stout arcuate medio-stapedial (*m.st.*), but is a massive, emarginate lobe of cartilage. The extra-stapedial (*e.st.*) is the short unossified end of this non-segmented columella; it is an inverted saddle-shaped process, and has no ascending ray. This small "key" is embedded in a thick cushion of fibrous and fatty tissue. The stylo-hyal end of the rest of the hyoid arch is not coalesced above; the descending bar (fig. 8, *c.hy.*) is of the normal width for the greater part of its length, but widens before it turns back into the hypo-hyal region (*h.hy.*); this part has no front lobe, but on its outer side there is a small oval "extra-hyal" (*ex.hy.*). The front notch of the basal plate (*b.h.br.*) is deep, the plate itself of normal length and width, but the front lateral lobe is absent; the hind lateral lobe and the ossified thyro-hyals (*t.hy.*) are normal.

The investing bones are, on the whole, like those of a very young Frog; but the fronto-parietals (fig. 6, *f.p.*) are small, falcate shells of bone; their ragged inner edges pass a very little way beyond the boundary band of cartilage, behind; they expand over the antero-medial part of the ear-masses, but are wide apart even there.

The nasals (*n.*) are normal, but cover only a third of the roof cartilage; there are no septo-maxillaries, and the marginal bones (*px., mx., q.j., sq.*) are feeble and thin, like those of young Common Frogs.

The parasphenoid (fig. 7, *pa.s.*) is relatively large and well developed; all its processes are pointed at their free ends, and the basi-temporal projections are uncinate. The vomers (*v.*) are small, arrested, toothless, sub-falcate, and rather ragged tracts of bone, like those of the young of any normal type.

The figures show, of themselves, how far short this small Frog comes of the development seen in its nearest *Ethiopian* and *Indian* relations; it falls short of the "norma" in the following particulars:—

1. The arrest, during an early larval condition, of the chondrocranium, both above (in front) and along the sides.
2. The feeble ossification, especially in the ethmoidal region, so that the girdle-bone is but little developed, broken up into three centres, and widely unfinished, below.
3. The distinct prenasal rostrum.
4. A very large, open, main fontanelle.
5. A bilobate superorbital (and preorbital) tract.
6. General feebleness of facial structures, both cartilaginous and bony.
7. Small Eustachian hole, and annulus tympanicus.
8. Absence of inter-stapedial segment, and supra-stapedial ray, with stunted extra-stapedial.
9. Modification of hyo-branchial apparatus. *a.* No hypo-hyal, or front lateral lobe; and *b.* A distinct extra-hyal, either a rudiment of the pectinate "inter-branchial" of the Selachians or a remnant of the first extra-branchial of the Tadpole.

Note—That the *open window* in the orbital space is like what I have described in another small Australian type, namely, *Camariolius* ("Ranidæ") (same Plate); it is seen in a less degree in some of the small Hylæ (Plate 31, fig. 7), and in the little *Bombinator* Toad (*Pseudophryne*) (Plate 42, fig. 2); all from the same region.

Second Family. "Hylodidæ."
First genus. *Hylodes.*

43. *Hylodes martinicensis.*—Adult female; 1¼ inch long. Martinique.

This skull differs in many respects from that of the Oriental "Polypedatidæ;" and the species of *Hylodes* evidently bear the same relation to the *Cystignathine* Frogs of the Nearctic and Neotropic regions, that the Indian "Polypedatidae" do to the *typical* Frogs of their own region.

In general development the skull is about equal to that of *Polypedates maculatus* and *chloronotus* (Plates 27 and 26), but the ear-organs are more arrested; altogether, however, it is a lighter and more elegant skull, and is almost exactly intermediate between the skull of a *Cystignathus* (of the same size), on one hand, and that of such a true *Hyla* as *H. rubra* (Neotropical), (Plate 33, figs. 6, 7) on the other. I shall compare it with that of *Cystignathus ocellatus* and *typhonius* (Plate 16).

The length is nine-tenths of the greatest breadth (Plate 29, figs. 7, 8), and the cheeks rapidly narrow in towards the broad, rounded snout.

The fore and middle regions are of equal axial length; the hind skull is only two-thirds their length, but it is of great breadth; the ossification, outer and inner, is typical; the whole structure is elegantly light, but of considerable strength. The occipital condyles (*oc.c.*) are small, project but little, are directly posterior, and are nearly twice their own breadth apart; they are separated by a gently emarginate line; the condyles for the mandibles reach as far back as the fore edge of the stapes. The upper and lower median cartilaginous tracts are rather wide, and the superoccipital edge (fig. 7, *f.m.*) is almost as far back as the basioccipital. The prootics and ex-occipitals (*pr.o., e.o.*) of the same side are continuous, and the bone reaches in front to the optic fenestra (II).

The bone, outside, reaches beyond the horizontal canal (*h.s.c.*) and then there is a large lozenge-shaped tract of cartilage which ends in the tegmen tympani (*t.ty.*), whose margin stretches outwards and forwards.

In front of the ampullae of the anterior and horizontal canals (*a.s.c., h.s.c.*) there is a blunt hook of bone, looking outwards; this process is also developed largely in *Siren lacertina* (HUXLEY, Art. "Amphibia," Encyc. Brit., 9th edit., p. 758; and WIEDERSHEIM, "Das Kopfskelet der Urodelen," plate 1, figs. 11, 12, V. F., and plate 2, fig. 18, V. F.). I have worked out this part in *Siren*, and in the corresponding stage in the larvae of *Triton cristatus*, and I find it to be a foregrowth of the plaster of cartilage developed beneath more or less of the auditory capsule—an outgrowth from the bevelled edge of the basal plate, or "investing mass."

The hinder roof (tegmen cranii) reaches to the optic fenestra; it has two lesser fontanelles in it. The main fontanelle is a long oval, half the extent of the roof.

The fontanelle is narrow because the tegminal edges are wide, equal to the scanty tegmen in front—the top of the girdle-bone (*eth.*); there is a small open space there, and a lozenge-shaped naked tract of the ethmoid. The temporal fossae are large and round, being margined outside by the prootic spurs; then the rather narrow skull steadily widens forwards, as in *Cystignathus ocellatus* (Plate 16, figs. 1, 2).

Infero-laterally, two-fifths of the mid skull is unossified, the large optic fenestra lying at the end of the soft tract, right and left.

In this flattish skull the girdle-bone barely reaches the true nasal region, and does not ossify its own alæ; its axillæ are shallow. Although elegantly arched in front the snout is very wide, and the oblique outer nostrils (*e.n.*) on each side are sub-tubular and very large; they have outside them well developed labials (*u.l*¹.*u.l*².).

The roof and floor (fig. 7, *al.n.*; fig. 8, *s.n.*) are very wide tracts of cartilage; there is no prenasal rostrum, and the angles of the floor and the pro-rhinals (fig. 8, *s.n.*, *p.rh.*) are large, and end in out-turned hooks; the septum nasi, above and below, is thick and clearly defined. Up to the point where the true palatine cartilage (fig. 8) begins, the ethmoidal wing, which is generally thick and somewhat pointed forwards behind the outer margin of the inner nostril, is here unusually developed, as in the genus *Cystignathus* (Plate 16) (indeed, in *C. typhonius* it is ossified continuously with the ethmoid, Plate 16, figs. 6 and 7). Here it is not ossified, but is of great breadth, and is seen outside the post-narial spike of the vomer (*n.f.*, *v.*)—a part which is also, here, four times its usual size. Beyond the notch which separates the ethmoidal from the palatine regions the cartilage is rather feeble and narrow, and so is the whole palato-suspensorial arch, except in its quadrate region behind.

In conformity with this, the two pairs of bones (*pa.*, *py.*), although quite normal, are slender and rather feebly developed.

The pedicle (*pd.*) is a small osseo-cartilaginous "foot," gliding on the sub-concave facet outside the prootic spur. The angle between the forks of the pterygoid is more than a right angle, and in it we see the large, oval, Eustachian opening (*eu.*)

The cartilage of this arch is nowhere obliterated except above the joint (*q.c.*) where the quadrato-jugal (*q.j.*) has freely grafted itself.

The condyles are large and normal, the inner convexity being much the larger of the two. Outside the suspensorium and its T-shaped splint (*sq.*) there is a smallish crescentic annulus (fig. 11, *a.ty.*); it is three-fourths of a circle, its width and concavity moderate, and its hinder horn two-fold. The stapes (fig. 11, *st.*) is oval, and has a boss; the columella is slender, the partly ossified proximal end is smaller than usual, and there is no inter-stapedial segment.

The bony rod (*m.st.*) is almost straight, slender, and partly unossified distally; at that part segmentation has taken place, cutting off the *symplectic* element, or extra-stapedial. This cartilage (*e.st.*) is like the bill of a *Spoonbill*, and has no supra-stapedial fork. The mandible (fig. 9) is quite normal, the condyle (*ar.c.*) is reniform, the coronoid crest (*ar.*) is definite, the dentary (*d.*) is nearly half the length of the ramus, and the mento-Meckelian (*m.mk.*) rather long.

The stylo-hyal (*st.h.*) margining the Eustachian opening is confluent with the auditory capsule; it passes into a slender cerato-hyal (fig. 10, *c.hy.*); outside its flexure there is a small extra-hyal, and from the arch there is a long slender hypo-hyal horn (*h.hy.*), which curves towards its fellow in front of the great pre-basal notch.

The basal plate (*b.h.br.*) is of the average width, and withal, very long also; its lateral lobes are narrow, the hinder longer than the front pair; these grow outwards and forwards, those outwards and backwards.

The thyro-hyals (*t.hy.*) are as long as the basal plate at its middle; they are slender, straight, moderately divergent, and well ossified.

The investing bones are thoroughly Cystignathine, and very elegant; the fronto-parietals (fig. 7, *f.p.*), are dilated in the temporal and ethmoidal regions, are hollowed out to form the temporal fossa, and spread well over three-fourths of the hind skull, where they end by a straight margin. They widen steadily in the orbital region, and taken together, their fore-edge is crescentically emarginate—so much so that the great fontanelle is not quite covered, and a large lozenge-shaped tract of the girdle-bone is left naked. The diverging outer angle of each lies on the corresponding nasal. The nasals (*n.*) have the normal shape; they are large, convex shells, and each touches the septum nasi by its inner edge; their bluntish fore ends leave the snout uncovered for some distance.

The marginal bones (*px., mx., q.j., sq.*) are well developed and perfectly normal; the squamosal, however, only lies on the edge of the tegmen tympani; its post-orbital process is bent outwards, and is curved and long; the descending part is normal.

There is a very small, sesamoid septo-maxillary (fig. 7, between *u.l².* and *al.n.*).

The parasphenoid (fig. 8, *pa.s.*) is long, well-developed, and normal; the bone is thickened at the cross, and the transverse processes are carinate, the thickening running along them as a ridge; the "handle" is unusually long and slender.

The vomers (fig. 8, *v.*) are very large triradiate plates; there is a *post*-narial, but not a *pre*-narial spur; and the front part is triangular, running under the edge of the dilated subnasal lamina (right and left of *s.n.*); these, right and left, are far apart.

The hinder half of each bone, on the contrary, converges towards its fellow, each almost reaching the point of the parasphenoid. From the hind margin a thick rib grows, covered with a rasp of retral teeth; these lobes reach outwards further than the inner edge of the inner nares (*i.n.*), their thick end is outside, and they are furthest forward there. This hind margin is slightly arcuate, so that they run across the fore palate as a crescentic rasp, just broken at the mid-line; they are quite Cystignathine.

Most of the divergencies from the "norma" in the skull in this species are very gentle, and yet quite appreciable; this is a true *Neotropical* Frog, with the toes dilated.

Its main points of difference are as follows:—

1. The nasal region is extremely large, both wide and long; and the orbital region steadily widens towards it.

2. The hind skull is unduly ossified.

3. The prootic region has a projecting spike, as in *Siren lacertina*, and the larva of *Triton cristatus*.

4. The palato-suspensorial arch is very slender.

5. The quadrate is largely ossified.

6. The annulus is small and widely open.

7. There is neither an *inter-* nor a *supra*-stapedial.

8. The processes of the basal plate are all very slender, and there is a pair of small extra-hyals.

9. The squamosals have a very narrow supra-temporal part, and the vomers are of great size, especially their dentigerous lobe.

Second genus. *Acris*.

44. (A) *Acris Pickeringii*.—Adult female; 10 lines ($\frac{3}{6}$ inch) long. Cambridge, Mass., U.S.

Although essentially a *Hylodes*, this minute Frog is worthy to be put into a subgenus; I therefore retain for it the term *Acris*. The skull of this species is very valuable as showing the effect of *dwarfing* in a remarkable arrest of the chondrocranium, such as is seen in the dwarfs of other genera, in different and distant regions. The small species of *Rappia* from Australia, *e.g. R. bicolor*, and similar minute *O.cydactyle* Frogs from the same territory, *e.g. Camariolius*, have the same modification of their skull—*almost all windows*, with but *narrow strips of wall*; and this is also seen, but to a less extent, in the small *Bombinator Toads* of Australia, viz.: *Pseudophryne*. Hence it is evident that this economy of cartilage is due to the same cause as the general arrest in size of these species. On the other hand, some of the smallest kinds are most ossified, notably a species close akin to this, and to *Rappia bicolor*, viz.: the small *Rappia* from Lagos; and some of the small edentulous Anura show the same thing, *e.g., Hylaplesia*—a *Platydactyle* type: the smallest of the typical Frogs, *Rana pygmæa*, has also a very bony skull.

This skull (Plate 30, figs. 1–5) has its breadth slightly greater (one-twentyfifth) than its length; it is semi-oval in its facial outline, has a broad but rounded muzzle, and its quadrate condyles (*q.c.*) end opposite the middle of the columella.

The occipital condyles (*oc.c.*) are large, reniform, and postero-inferior; they are separated by a space nearly equal to the width of both, and this wide notch is crescentic.

There is a wide basal or median tract of cartilage, in which the remnant of the notochord is seen to be enclosed in a feeble bony sheath—a cephalostyle—the rudiment of a basioccipital (*b.o.*). Above the wide foramen magnum (*f.m.*) the roof-cartilage comes well back, and is seen to be partly calcified, showing the rudiment of a supra-occipital bone. This roof is complete up to the same transverse line as that from which the optic nerves (II.) emerge; from thence the roof is membranous nearly up to the septum nasi (*s.n.*). This is the large ovoidal fontanelle (*fo.*); it is two-fifths the length of the skull, and three-fourths the width of the interorbital region. The ossification of the hind skull is feeble, and by the two pairs of normal centres (*pr.o., e.o.*).

The ex-occipitals (*e.o.*) flank the posterior canals (*p.s.c.*) above, and surround the twin nerve-passages (IX., X.) below; the prootics (*pr.o.*) climb over the ampullæ of the anterior and horizontal canals (*a.s.c., h.s.c.*) above, and half surround the foramina ovalia (V.) on their upper edge, reaching from thence to the pedicles (*pd.*). Beyond the horizontal canal the capsule lessens to one-third its breadth, to form the "tegmen" (fig. 1). The fore and hind skulls are equal in length ; the mid skull is one-half longer, and relatively broad ; its sides are sinuous, for it dilates a little both in the temporal and ethmoidal regions. The actual floor of the chondrocranium is only half the width of the roof, and the optic nerves (II.) pass out of the end of a long, oval fenestra, which reaches to within a short distance of the ethmoidal wings. Measured from the hinder margin of the actual optic foramen to its fore end, this fenestra takes up four-fifths of the orbital wall, and the girdle-bone (*eth.*) does not even reach to the fore edge of the fenestra. A narrow band of cartilage bounds the single, great fontanelle (*fo.*), and as this widens in front it becomes bony ; the fore margin of the fontanelle is bony, but the tegmen cranii thus ossified is a very narrow selvedge, and this band only reaches to the very definitely marked nasal roofs, whilst the bone runs very slightly into the septum (*s.n.*). Laterally, there projects a narrow superorbital eave beyond the bone, and the ossification projects slightly into the wings.

The nasal septum (*s.n.*) is thick behind and bulbous in front ; its junction with the roof-cartilages is shown in front, where the short decurved "rostrum" (*p.n.*) has a small projection on each side of it, the *right* being the largest ; these are the ends of the crescentic roof-tracts, which are large and wide.

Below (fig. 2) the floor is very wide, and ends in front in the primary angles of the cornua trabeculæ, and in the smaller, secondary, decurved pro-rhinals (*s.n., p.rh.*).

The narial openings are wide apart, the external nostrils (*e.n.*) being three-fourths the distance from each other of the inner holes (*i.n.*) ; the outer openings are protected by the normal valves (*n.l¹.n.l².*). The palato-suspensorials are slender, but quite normal ; from the ethmoidal wing, which thickens round the hinder margin of the inner nostril, the ethmo-palatine becomes a moderate tape of cartilage, expanding into the spiked pre-palatine band, which runs backwards as a post-palatine, with a rounded hinder lobe projecting into the maxillary, and then becoming pterygoid (*pr.pa., pg.*). The forks of the suspensorium are apparently nearly equal, as seen from above and below, but the quadrate region and its condyle (*q., q.c.*) pass further backwards, reaching as well some distance downwards and outwards. The palatine bone (*pa.*) is the normal falcate thin piece, and the pterygoid (*pg.*) has the usual shape, but, contrary to rule, it covers the cartilage most above and least below. The bulbous pedicle is most seen above ; below, it is confluent with the stylo-hyal (*pd., st.h.*)—a rare modification, and seen again in *Phyllomedusa bicolor* (Plate 34, fig. 8). Above, in front of the fork, the pterygoid-bone is rough and hollowed out ; it is thin and lathy where it binds on the inner face of the quadrate. That region is only slightly ossified by the quadrato-jugal (*q.j.*) ; its condyle (*q.c.*) is large, elegantly bilobate, with the inner or hinder lobe much the larger of the two.

The annulus (*a.ty.*) is rather small and open above ; the stapes (fig. 5, *st.*) is oval, with an emargination antero-superiorly, and an external boss. The pistol-shaped columella is large and bilobate behind, but the unossified part is not distinct from the medio-stapedial (*m.st.*). The decurved extra-stapedial (*e.st.*) is spatulate, and it gives off a narrow supra-stapedial (*s.st.*), which degenerates into a ligament, above.

The stylo-hyal (*st.h.*), from its junction with the pedicle, runs into the cerato-hyal (fig. 4, *c.hy.*), without change of breadth; the hypo-hyal region (*h.hy.*) is narrower, and there is no projecting lobe. The great notch in front of the basal plate (*b.h.br.*) is wide; the plate is short and wide, and the fore lobes are uncinate and but little wider than the hind lobes. The thyro-hyals are long, a little more than half bony, and hooked inwards behind. There is an irregular patch of "endostosis;" between these bars, in front, as the sign of a tendency to form a basi-branchial bone. The mandible (fig. 3) is long and slender, with a pyriform condyle (*ar.c.*), a long dentary and mento-Meckelian (*d., m.mk.*), and a low coronoid ridge to the articulare (*ar.*).

The investing bones are extremely delicate, some of them being less than may be seen in the Tadpoles of other kinds, before the fore legs are free.

This is true, especially of the fronto-parietals (*f.p.*), which are delicate styles overlapping the girdle-bone in front, and only forming a *wall-plate* to two-thirds of the narrow marginal band of roof-cartilage. In the temples each bone projects, and binds upon the anterior canal; behind, it becomes pedate, not reaching, however, to the small ex-occipital; thus the hinder, complete "tegmen" is nearly all naked.

The nasals (*n.*) have a fuller development, but they are more than their own width apart, and are thin lunate shells. The premaxillaries (*px.*) are extended in front of the wide snout, and the main bar of each is long; their nasal and palatine processes are moderately developed. Over the junction of these bones with the maxillaries there is a small irregularly radiate septo-maxillary (*s.mx.*); the maxillary (*mx.*) is of great length, almost reaching to the hinge (*q.c.*), but it is thin and not high; the small tooth-like quadrato-jugal (*q.j.*) is partly joined to the quadrate (*q.*).

The squamosal (*sq.*) runs one-third further forward than the tegmen (fig. 1, left of *au.*); it overlaps that cave a little, and is split at its free postorbital end; the descending part is normal. The parasphenoid is thin and like that of a Tadpole, but it reaches from the front of the "cephalostyle" (*b.o.*) nearly up to the notch in the margin of the narrow girdle-bone (*eth.*); its processes are all normal, but the lateral wings are rather limited, and the hind lobe is irregular.

The vomers (*v.*) are like bony spicules; there is a *post*-narial, but not a *pre*-narial hooked spike; the "body" is in two arcs, the hinder ends in a thick lobe covered with teeth, and the front lobe is pointed and underlies the angle of the floor-cartilage (*s.n.*).

The main points to notice in this skull, as compared with the "norma," are its general arrest at a condition of growth equal to that of a young Common Frog of the first summer. The particular points are:—

1. Rudiments of a basi- and supra-occipital.
2. Its *one* large fontanelle.
3. The long orbital fontanelles.

4. The small girdle-bone.
5. The definite prenasal with its lateral projections.
6. The junction of the pedicle with the stylo-hyal.
7. The absence of a distinct inter-stapedial.
8. The absence of hypo-hyal lobes.
9. The extreme delicacy and narrowness of the fronto-parietal bones; this is repeated in the other investing bones, but not to the same extent.

44 (continued).—(B) *Acris Pickeringii.*—Larva, 1 inch 2 lines long; tail, ¾ inch; hind legs, ¼ inch; same locality.

The adult of this species is smaller than that of *Cumariotius*, but its Tadpoles are larger by one half; this specimen is more metamorphosed than the larva of that minute Australian "Oxydactyle Frog" (see Plate 15, figs. 6, 7).

In this larva the chondrocranium (Plate 30, figs. 6, 7) is oval in outline, for the hind part of the suspensorial bands and the fore part of the auditory capsules project, and the quadrate condyles (*q.c.*) are drawn in towards the mid-line. It has gone beyond the *Petromyzine* stage, for there are two bones, both of them axial and azygous: one intrinsic, the cephalostyle or rudimentary basioccipital (*n.c.*); and the other superficial, the parasphenoid (*pa.s.*). The first vertebra (*c.v*¹.) is drawn with the crown of its arch cut away; in it we see the ossifying neural arches, and the relatively huge notochord (*nc.*) becoming invested with a layer of bony deposit, principally on its upper surface (fig. 6). This is taking place in the mesoblastic sheath, and is two-horned in front. Entering the skull, the notochord suddenly becomes very small, reaches to the same transverse line as the Gasserian ganglia (V.), and ends in a rounded point, which is twisted to the right hand. This styliform apex of the notochord is invested—most above—with a deposit of bone, like that which in the vertebral region forms rudimentary "centra." The bony "cephalostyle" has already been described as existing in the adult. The occipital arch is fast finishing, for cartilage is breaking out, above, from the inner edges of each auditory capsule, and this has converted most of the membrane there into the same tissue. Also in front of the notochord there is a long *intertrabecular tract* of new, small-celled cartilage (fig. 6, inside *tr.*), and the orbital walls are beginning to grow over as a super-cranial (*tegminal*) rudiment; this is especially seen in front.

Between the narial passages (*i.n.*) the trabeculæ have coalesced, converging for this purpose, but there is no appearance here of a distinct mesethmoidal rudiment. The trabeculæ break free of each other again, to end as free, crescentic, decurved horns, from whose inner edge a crest is growing to fill up the re-entering angle between them. The nasal roofs are only beginning to chondrify, and with the eye-balls, were removed in making the dissection. The auditory capsules began to chondrify directly after the trabeculæ and suspensoria, and are now highly developed; their canals (*a.s.c.*, *h.s.c.*, *p.s.c.*) bulge largely, and the ampulla of the horizontal canal projects greatly.

The tegmen tympani is showing itself, and outside the *parachordal floor* the oblique fenestra ovalis is nearly filled with the oval stapes (*st.*), now well chondrified. The unfilled fore end of the fenestra ovalis is waiting until after most of the metamorphosis is done to receive the columella, which will articulate with the newer parachordal floor at its outer edge. Cartilage has spread from the tegmen tympani into the angular space between the capsule and the otic process of the suspensorium (*ot.p.*), just as cartilage spreads from the inner edge of the capsule into the membranous superoccipital tract—for *cartilage*, like *bone*, in many instances, does not heed morphological landmarks. The little cordiate cartilage (fig. 7, *sp.c.*) which is developing in the *opercular membrane*, over the 1st cleft, becomes free from the ear-capsule, coalesces with the otic process of the suspensorium, becomes detached again from that part, and then develops into the annulus tympanicus, and is permanently kept as the cartilaginous "operculum" of the cleft to which it belongs; the specialisation of both cleft and operculum failing to disturb the original morphological relation of the parts. In like manner the modified hyo-mandibular or columella, will be wedged in between the auditory capsule and the basal plate, exactly as in its simple and generalised counterpart in the Selachians.

The suspensorial arch has the normal form; the pedicle (fig. 7, inside *sp.c.*) is narrow, the otic process (fig. 6, *ot.p.*) is a thick crest, ascending from the main bar as it suddenly turns inwards to become the pedicle. The quadrate condyle is a small trochlea, and looks inwards as well as forwards, and the terminal or quadrate region (*q.*) is altogether oblique; it sends inwards an angular pre-palatine spur which is attached to the cornu trabeculæ in front of the narial passage (*i.n.*) by a ligament.

The orbitar process (fig. 6, *or.p.*) is large and rounded, with raised edges; the hyoidean condyle beneath it has the usual reniform outline, and is scooped. The wings of the ethmoid are very rudimentary at present, and I find no post-palatine rudiment outside them. The mandibles (fig. 7, *mk.*) are short, thick, ulniform rods, with the usual hooked angular process; they are articulated to the lower labials by a wide surface; these latter are each half a thick crescent with a raised lip. The *temporary* upper labials (*u.l.*) are large, thin, crescentic flaps fitted on to the end of the cornua trabeculæ. The hyoid bar (fig. 7, *c.hy.*) has the usual form; it has a large condyle and an unciform stylo-hyal free end; it has a narrow waist, obliquely ridged, externally, for muscular attachment, and then expands largely, most on its hinder margin, to contract again where it is joined to its fellow, ventrally, by the square, soft-celled, basi-hyal. The branchial arches were not figured in this instance, they were quite normal.

The parasphenoid (fig. 7, *pa.s.*) formed a floor, very exactly, to the newer cartilage of the intertrabecular space; it had, although it was very thin and young, developed the rudiment of its lateral and posterior angles; the last was directly under the (endoskeletal) bony sheath of the notochord.

Second sub-division.—Tree-frogs, with dilated sacral apophyses, and without parotoids.

Family "HYLIDÆ."

First genus. *Hyla.*

45. *Hyla Ewingii.*—Adult female; 1¾ inch long. Van Diemen's Land.

The lesser, glandless, flat-backed Tree-frogs form a very natural group; those of the genus *Hyla* a very neat group. The Australian Anura of various genera and families are very frail and delicate in their build, and their skulls especially are often extremely deficient even in cartilage, which, like the outer bone, is often used with the utmost economy.

This in the case of the *Hyla* is in perfect harmony with the life led by these insect-like Batrachians, and is very instructive as throwing light upon the influence of external conditions upon a most sensitively modifiable group. I shall return to this subject again when summarising the whole of this piece of work.

The skull of this species (Plate 31, figs. 1, 2) is an even half oval; the breadth is to the length as 9 to 8; the occipital condyles (*oc.c.*) project but little, are postero-inferior, of medium size, and wide apart. The condyles of the quadrate (*q.c.*) reach as far back as the fore edge of the stapes; this is a correlate of the arrested size of this type. The skull is very flat and wide, and very open; there is one large oval fontanelle, which reaches almost from the closing in of the cranial cavity in front, to some distance into the inter-auditory region, behind; at the sides, also, the tegmen is scarcely developed at all, and thus three-fourths of the roof is membranous. The whole occipital arch, and all but the edge of the tegmen tympani, and the rim of the fenestra ovalis, is one continuous (generalised) osseous tract, which reaches up to the optic fenestra (II.) in front. The flatness of this little skull is such as to throw the nerve-passages both before and behind the ears (fig. 2, II., V., IX., X.) on to the general plane of the gently convex lower surface. Thus this skull resembles that of a young typical Frog, artificially compressed; and that in spite of the intense ossification of both the fore and hind skull. In accordance with the general arrest as to size, the auditory capsules are relatively larger, and the parotics less extended, than in the Pelodryadidæ. So also the mid skull is wider; the cranial cavity is larger, in proportion to the part in front, and is twice as long as the nasal region. There is very little pinching in of the orbital region, which is widest behind, where the temporal region begins, and in front where a distinct cave of cartilage is left unossified (*s.ob.*). The girdle-bone (*eth.*) runs into the nasal region in front, transversely above, and as a spike below; it takes in also part of the proper territory of the anterior sphenoid, for more than half of the orbital region is ossified by it, and it also runs well into its own wings, stopping where the ethmo-palatine bar is segmented in *Bufo*. Here, as in *Pelodryas* (Plate 34), the nasal roofs are small (figs. 1 and 4), and the nasal floor large (fig. 2, right and left of *s.n.*). The end of the septum (*s.n.*) projects as a triangular rostrum (fig. 4, *p.n.*), and the paired trabeculæ have developed their new horn or pro-rhinal (*p.rh.*) in such a manner as to almost rival the outer angle of the first horn (fig. 4, *e.tr.*); that is pedate with the toe outwards: this is like an adze blade. The outer nostrils (*e.n.*) are only moderately wide apart; the inner (*i.n.*) are not so far apart as in many cases; they are very large and circular; the labials (*a.l^1.a.l^2.*) are normal. The palato-suspensorials are like those of a young Common Frog; they are slender and but little affected by the bony tracts; have a large cultrate pre-palatine; and end behind in two massive growths, the condyles of the pedicle and quadrate (*pd., q.c.*); these parts are much like those of *Pelodryas*. The Eustachian opening (*eu.*) in the angle is almost circular and of the average size—only half as large as the inner nostril. The palatines and pterygoids (*pa., py.*) are slender, but quite normal. The mandible (fig. 3) is normal.

The annulus (figs. 1 and 5, *a.ty.*) is of normal size, but its horns are wide apart.

The stapes (fig. 5, *st.*) is an oval plate obliquely truncated in front; on it, antero-superiorly, the arched columella fits; it is equally wide at both ends, has no segmentation, and no supra-stapedial cartilage—only a fibrous band (*s.st.*). There is no inter-stapedial segment, even marked by ossification, and the bony matter runs up nearly to the stapes. The extra-stapedial (*e.st.*) is two-fifths the length of the outer rod; it is a rounded spatula.

The stylo-ceratohyal is confluent above (fig. 2, *st.h.*), and gently widens to the middle (fig. 3, *c.hy.*); it then lessens again, and returns backwards into the basal plate without any hypo-hyal lobe. Only the sharp postero-lateral lobe exists on the short, small basal plate (*b.h.br.*); the thyro-hyals (*t.hy.*) are well developed, moderately diverging, and straight.

The investing bones are such as we should find in a young *Pelodryas* equal to this species in size. The fronto-parietals (fig. 1, *f.p.*) form a narrow, straight wall-plate, widened, like a foot, over the temporal fossa; they turn slightly outwards over the superorbital cave, and inwards, a little behind. The nasals (*n.*) are narrow crescentic shells, with a spike on the middle of their convex edge.

The marginal bones (*pœ., mx., q.j.*) are well formed and typical, but extremely thin; I find no septo-maxillary in this species. The squamosals (*sq.*) have a good upper bar, projecting as a free postorbital process in front.

The parasphenoid (fig. 2, *pa.s.*) is characteristic of this Family; it is well formed, wide in its main part, has all its processes pointed, but the basi-temporals are narrow. The vomers (*v.*) show nothing of the breadth of those of *Pelodryas* (Plate 34, fig. 2), nor the radiate character of those of *Phyllomedusa* (Plate 34, fig. 8); they are quite normal, and have a small obliquely-oval dentigerous plate.

This skull should be compared with that of a young Common Frog of the first autumn; then we can see that its divergence from the type is in these several peculiarities, namely:—

1. The extremely flat form of the skull, generally.
2. The intense ossification of the occipito-otic and ethmoidal regions.
3. There is but one, that a very large, fontanelle.
4. The prenasal is present, and the lesser cornua (pro-rhinals) are half the size of the outer angles of the snout.
5. The superorbital tract is distinct as a semi-oval cave.
6. The nasal roof is very narrow, as compared with the floor.
7. The annulus has its horns wide apart.
8. The pedicle is very solid, and the Eustachian opening is margined behind by a confluent stylo-hyal.
9. There is neither a separate inter-stapedial nor a supra-stapedial band.
10. There are no lobes on the hypo-hyals, and no antero-lateral lobe on the very small basal plate.
11. The roof-bones are extremely arrested and narrow.
12. There are no septo-maxillaries.

46. *Hyla phyllochroa.*—Adult female; 1 inch 5 lines long. Cape York, Australia.

This skull resembles the last very much, but is altogether frailer, and less ossified; I am not aware of having dissected a skull more light and delicate than this: its main rivals are some of those of its own country, viz.: *Rappia bicolor*, and *Camariolius tasmaniensis*.

The outline of this skull (Plate 31, figs. 6-9) is elegantly semi-oval; it is much shorter than the last, the breadth being to the length as 7 to 6, and the condyles of the quadrate have retreated much further back, being opposite the proximal part of the occipital condyles. These condyles project as little as in the last, are smaller, and differ in direction, being exactly posterior.

Both skulls are equally depressed, but this differs in shape in all the three regions. The auditory capsules are only three-fourths the size of those of *H. Ewingii*; the parotics project much further outwards, and are considerably narrower.

The tegmen cranii reaches more than half-way from the foramen magnum to the front of the cavity, and runs round the single fontanelle (*f.*) as a considerable band both right and left and in front. The greatest width of this membranous space is equal to that of *H. Ewingii* and to its own length, but the widening backwards of the marginal tegmen makes it lose what would otherwise be the circular form.

A large triangular tract of cartilage exists above, occupying most of the hinder tegmen, and a clear synchondrosis exists both above and below at the foramen magnum (*f.m.*); the floor of the vestibule and the tegmen tympani (fig. 7, *eb.*, and fig. 6, *t.ty.*) are also unossified. Above, there is a narrow tract of cartilage running from the horizontal canal to the junction of the anterior and posterior canals (fig. 6). In front, the prootic bone reaches to the foramen ovale (fig. 7, V.); above (fig. 6), the bone reaches the roof over the optic fenestra (II.) The girdle-bone (*eth.*) only occupies one-third of the mid skull, runs a little into the wings below, and above creeps along the septum nasi almost to the end of the snout (*al.n.*). The whole of the large ear-shaped superorbital tract (*s.ob.*) is unossified; it is thrice as large as in the last. The optic fenestra (II.) is twice as large as in the last. In that species the flat mid skull was widest behind; in this, the skull widens and bulges from behind forwards. With the exception of the top of the septum, all the nasal region is unossified; and in this kind the roof and floor are equal, and there is no rostrum. The pro-rhinals (*p.rh.*) are very wide, and have the shape of the angular lobes (right and left of *s.n.*).

Here the snout is almost as broad and transverse as in the Indian Polypedatidæ; the labials (*a.l¹.a.l².*) are normal. The palato-suspensorials are curiously different in these two kinds: here the ethmo-palatine (*e.pa.*) is broader; the pre-palatine a sharper adze blade, and the post-palatine cartilage retains a greater breadth; it ends in this, as in the other, in solid condyles—both of pedicle and quadrate, as in *Pelodryas*. But in this the palatine (*pa.*) is a mere needle of bone, whilst the pterygoid (*py.*) is broader; it is of necessity longer, on account of the greater retreat of the quadrate: this latter part is not ossified.

In this species the Eustachian openings (*eu.*) are large and nearly circular, as in the last; but the inner nostrils (*i.n.*) are still larger; they are oval, and look obliquely inwards and forwards.

The annulus (*a.ty.*) is large, but unclosed; the stapes (figs. 7 and 9, *st.*) large and sub-oval; it is but little produced outwards. The bony medio-stapedial (*m.st.*) has a very thick, solid, proximal part, from the core of which an ear-shaped inter-stapedial lobe (*m.st'.*) passes inside the stapes. The shaft is long, and in front the unossified extra-stapedial (*e.st.*) grows as a narrow tongue, bent on the core of the shaft, and having a lobate edge, but no distinct supra-stapedial.

The stylo-hyal (fig. 9, *st.h.**) is confluent above; it widens downwards into the cerato-hyal (fig. 8, *c.hy.*), which soon lessens again, and becomes a long, delicate, retral, non-lobate hypo-hyal (*h.hy.*). There is no front lateral lobe to the very short basal cartilage, which tends to develop a hinder segment (*b.br'.*); this exists as a very rare median lobe between the long, slender sub-arcuate thyro-hyals (*t.hy.*).

The mandible is normal, is very long and slender, and has a distinctly lobate coronoid process to the articulare (fig. 8, *ar.*).

The mento-Meckelians (*m.mk.*) are large; the feeble dentaries (*d.*) reach half-way along the ramus.

The investing bones are, on the whole, about as thin and delicate as in the last; but the roof-bones are wider by one-third. The fronto-parietals (fig. 6, *f.p.*) partly overlap the ear-capsule, and nearly reach the ethmoidal wings. They are sigmoid in outline, and bend inwards, first behind the fontanelles, and again at its fore end; then they lessen this space slightly, and end in a rounded lobe. The nasals (*n.*) are large, thin, crescentic shells. The premaxillaries, maxillaries, and quadrato-jugals (*px., mx., q.j.*),

---
* The top of the stylo-hyal is hidden in fig. 7.

and also the squamosals (*sq.*), are very similar to those of the last kind; there is a small club-shaped septo-maxillary (fig. 6, *s.mx.*) on each side under the second upper labial (*u.l*²·). The parasphenoid (fig. 7, *pa.s.*) is altogether more slender and pointed than that of *H. Ewingii*, especially in its basi-temporal spurs. The vomers (fig. 7, *v.*) are large; they have a curiously-arched form, a dentigerous lobe twice as large as in the last, and come nearer together. The postero-external edge of the bone embraces half of the great inner nostril, and the outer spike in front of this passage is but the hinder of a series of tooth-like projections of the bone.

The difference between this skull and that of the "norma" is much the same as in the last; it comes nearer, however, to the pattern-form, in having the ex-occipitals and prootics somewhat marked out, and in having the lateral masses of bone distinct; also in having septo-maxillaries. It diverges further in one thing, namely, in having a basi-branchial rudiment projecting beyond the basal plate. Both the annulus and columella, in these two kinds, come equally short of that which is typical. Here, also, we have the optic fenestra and the superobital, both twice as large as in *H. Ewingii*.

On the whole, there are about the same number of points in this and the last, in which these typical *Hylæ* disagree with the "norma."

47. *Hyla arborea*.—Adult male; 1½ inch long. South Europe.

This is a semi-oval skull, still shorter and broader than the last; the breadth is to the length as 6 to 5.

On the whole, this specimen comes very near to *H. Ewingii*, notwithstanding the distance between the homes of the two species.

The skull is not quite so depressed as in the two last, and is rougher and stronger in its general build; also it is more fully roofed in.

The occipital condyles (Plate 32, figs. 1 and 2, *oc.c.*) are larger and wider apart; they are postero-inferior, as in *H. Ewingii*.

As in that species, the occipito-otic ossifications are fused together, but in this kind there is a basioccipital tract of cartilage (fig. 2, *b.o.*).

The canals are larger, but the parotics are similar in both; there is very little unossified cartilage near the fenestra ovalis in this species; altogether the ossification is more intense above than below; the bone encircles the foramen ovale (V.).

The three regions of the skull are nearly equal; the single fontanelle (*fo.*) is exactly one-third the length from snout to occipital condyles; it is an irregular oval, for the edges of the tegmen cranii are sinuous.

The short, wide, depressed mid skull lessens a little from the temples, and enlarges again into the superorbital lobes, which are ossified, proximally, by the girdle-bone (*s.ob., eth.*), and are semi-oval, not projecting much more than in *H. Ewingii*.

The girdle-bone scarcely occupies a third of the mid skull; and the large optic fenestra (II.), which is intermediate in size between that of the two last, is well margined with cartilage.

The bone hardly affects the ethmoidal wing, but runs fairly up to the septum nasi (*s.n.*), below; above (fig. 1, *eth.*), it is very narrow, where it forms an arcuate margin to the fontanelle.

The nasal region is well developed, both as to roof and floor; it is wholly cartilaginous, it is broad, but arcuate (not transverse), in front, and ends in five large, well formed processes (figs. 1, 2, and 4). Each outer angle of the primary cornua trabeculæ (fig. 4, *c.tr.*), is a large, broad, decurved ear of cartilage, and the pro-rhinals (*p.rh.*), are long, large, and pedate; the premasal (*p.n.*) is spatulate, and one of the best, in shape and size, to be seen in the whole group. Between this bar and the out-turned prorhinals, the nasal-nerve openings (*n.n.*) are large. The palato-suspensorials are stouter than in the two last; the pterygoid and palatines (*pg., pa.*) are stronger even than in *H. Ewingii*, but the cartilage of the pedicle (*pd.*) is much less solid; that of the quadrate is equal to what is seen in that species; this latter part is not ossified. These condyles have retreated to a point opposite the hind margin of the stapes; this is intermediate between what we see in the two last kinds. So, also, are the internal nostrils (*i.n.*), they are sub-oval and oblique, but not to the same extent as in *H. phyllochroa*; the Eustachian openings (*eu.*) are very similar in all three; they are quite circular in this species. The external nostrils (*e.n.*) and the upper labials (*u.l.n.l.*) are very similar to those of *H. Ewingii*.

The annulus (*a.ty.*) is imperfect above, but it is of the average size.

The stapes and the other linked segments of the ear-chain (fig. 5, *st., i.st., m.st.*) are large; the proximal segment is oval and unossified, but quite distinct from the solid head of the medio-stapedial.

The extra-stapedial is a long, decurved, spatula, giving off a ligulate cartilaginous supra-stapedial (*e.st., s.st.*), which like the stylo-hyal (*s.th.*) is confluent above.

The rest of the hyoid band (fig. 3, *c.hy.*) is rather broad; it lessens before it turns back, has no lobe, and soon melts into the basal plate.

That plate (*b.h.br.*) is extremely small, both fore and aft, and has inordinately large thyro-hyals (*t.hy.*): its lateral lobes are but little pronounced, especially on the right side. The mandible (fig. 3, *m.mk., d., ar.*) is normal.

The investing bones are somewhat denser than in the small Australian *Hylæ*; the fronto-parietals (*f.p.*) are only one-third as long as the skull, and overlap the marginal "tegmen" very little; they are small, straight, and gently conchoidal, with only a slight temporal dilatation.

The nasals (fig. 1, *n*) are fine large stalked crescentic shells of bone, coming very near each other at their inner edge; these bones, and the pre-maxillaries, septo-maxillaries, maxillaries, quadrato-jugals, and squamosals, are all similar to what would be found in a young Common Frog of the same size as the adult of this species.

The parasphenoid and vomers (fig. 2, *pa.s., v.*) are normal and large; the former has its wings pointed, and not dilated, and the latter are thrown wide apart by the dilatation of the basi-nasal tract.

Divergence from the "norma" is seen in :—
1. The general breadth and depression of the skull.
2. The thorough continuity of the endo-cranial bones of the hind skull, above, and of those of the same side, below.
3. In the mid skull the great size of the optic fenestra, and the limited extent of the girdle-bone.
4. The superorbital cave, and large single fontanelle.
5. In the fore skull, the large, well made prenasal, and the unusual size of the prorhinals.
6. The stylo-hyals and supra-stapedials are confluent, above.
7. The fronto-parietals are arrested.
8. The basal plate is very small, and the thyro-hyals very large.

48. *Hyla albomarginata.*—Adult female; 2¼ inches long. Brazils.

This species, which is nearly twice as large as those whose skulls have just been described, cannot be far removed from *Phyllomedusa*, in spite of its want of "parotoids." Its skull is moderately strong in its outworks; the endocranium very strong, although very flat; the ethmo-nasal region takes up half its length, and this causes it to be almost a *long* skull, the breadth being only *one-fiftieth* greater than the length; the jaw-hinge reaches very nearly as far back as the occipital condyles.

Thus much of the divergence from the type is in this kind evident at first sight. Besides its likeness to the skull of the gigantic South American kind, it resembles very much that of an Australian sub-type, namely, *Litoria marmorata* (see Plate 19, figs. 11, 12; and Plate 32, figs. 6, 7); but the skull of that species is much less unlike the "norma."

Even here the *Ranine* skull is but thinly veiled, and not so much metamorphosed in any part as to hide its fundamental character; and yet this normal form is an acquired character, whilst the primary form was *Petromyzine,* or "Suctorial."

The occipital condyles (Plate 32, figs. 6, 7, *oc.c.*) are small, subreniform, and postero-inferior; they are separated by an interspace which is greater than their own width, and this is deep and crescentic; the epiotic eminences (*p.s.c.*) are nearly flush with them.

A line drawn athwart the middle of the hind brain, and cutting the junction of the anterior with the posterior canals (*a.s.c., p.s.c.*) would also run across the opening of the tympano-Eustachian cleft (*eu.*), where the columella pushes out its extra-stapedial process (*e.st.*); this point is a little behind the middle of the drum of the ear (*a.ty.*). The great parotic wings are half the breadth of the rather small proximal region of the auditory capsules; they turn a little forwards, externally, but they are dilated somewhat along the tegminal margin (fig. 6, under *sq.*).

The ossification of the hind skull is continuous everywhere, except a short oblong tract under the hind brain; its greatest width is in front. Another tract, twice as large, with its base in front, is seen under the pituitary body, and the triangular superorbitals (*s.ob.*) are soft. With these exceptional tracts, all the cranium is solid bone up to the middle of the proper nasal region. The fontanelle (*fo.*) is a long oval; it is half the length of the cranial cavity, and half the width of the interorbital region at its narrowest part.

The temporal shoulders of the hind skull are high and large; from thence the outline narrows suddenly, and less rapidly regains its inter-temporal width, so that the interorbital region is at first very narrow, with a deep, large, crescentic emargination right and left; a small round notch is seen between the superorbitals and the ethmopalatines. Then the wide girdle-bone is flat above and concave on each side below; moreover, the edge of the cranial boat is produced outwards, and is scooped below up to the optic fenestra. The girdle-bone takes up all the ethmoidal wings, but leaves the *facial region* of the ethmo-palatines untouched; in front it occupies full half of the partition wall, the roof, and the floor of the true nasal region, besides its own *ethmoidal* wall, roof, floor, and wings.

From below, the skull looks very much like the repetition, in front, of such a continuous vertebral tract as is seen in the neck of a Skate. The nerve-passages in this flat skull are almost *inferior* in position, and they form a double series, right and left, in a very orderly manner, and are almost equidistant; moreover, those for the trigeminal and facial (fig. 7, V.) are subdivided by a bony bar, like those for the glossopharyngeal and vagus (IX., X.); the optic passage is small.

The nasal roof (fig. 6, each side of *s.n.*) is wide behind and narrow in front; the floor (fig. 7, *s.n.*) is wide at both ends and rather contracted, by a crescentic retreat of the margin, in the middle. The internal nares (*i.n.*) are very large, turn inwards in front, and lie against the narrowing hind part of the floor. The outer nostrils (*e.n.*) are only half as large, and half as wide apart; they are well protected by the two upper labials (*u.l¹.u.l².*) right and left. The snout is not of great extent, but it is directly transverse, and has no rostrum; the pro-rhinals (*p.rh.*) are rather small, but the angles of the floor are large and fan-shaped, and lie well within the wide maxillaries (*mx.*). The ethmo-palatines (*e.pa., pa., py.*) are widely transverse, and end in an adze-shaped dilatation externally; the palatine bone (*pa.*) takes on the same form, and ends inwardly as a sharp point far from its fellow bone; the post-palatine cartilage is continuous with the pterygoid tract, which is strongly arcuate, but very narrow and slight; its bony correlate (*py.*) runs nearly to the palatine in front, and behind, forks at less than a right angle, in which space there is a very large egg-shaped Eustachian opening (*eu.*), the narrow end of which is in the sharp re-entering angle of the bone, and is therefore turned outwards. The outer fork binding the quadrate bar runs far back; the inner, growing over the stunted "pedicle" (*pd.*), clamps it and ties it down to the skull by sutural teeth, instead of allowing it to glide on the corresponding cartilage. The quadrate-condyles (*q.c.*) are twice as large as the occipital; the cartilage above them is ossified by the quadrato-jugal (*q.*, *q.j.*). The annulus (*a.ty.*) is large, broad, and open above; the stapes (fig. 9, *st.*) is thick and oblique; the medio-stapedial (*m.st.*) is a strong, arched rod, having a large ovoidal mass of cartilage on its upper fork, which, however, is not distinct as in the last kind. The extra-stapedial (*e.st.*) is tongue-shaped; it has a free rounded selvedge, but no supra-stapedial band. The stylo-hyal (fig. 7, *st.h.*) is confluent above; it is very narrow, and widens very little in the cerato-hyal region (fig. 8, *c.hy*), which is a very narrow tape all through, with only a slight hypo-hyal lobe. The basal plate (*b.h.br.*) is largely converted into the membrane of the great front notch, which is more than twice the extent of the solid tract.

The narrow ear-shaped front lobes run far along by the hypo-hyal, and the hind lobes are small spikes; in each of these, at their origin, there is an endosteal patch. The thyro-hyals (*t.hy.*) diverge moderately, are of the average strength, and are sigmoid in form; they spread well into the basal plate, proximally, and have a blunt unossified end.

The mandibles (fig. 8) are extremely long and slender; the dentary (*d.*) is only one-third as long as the ramus; the coronoid process of the articulare (*ar.*) is well formed, and the cylindroidal condyle (*ar.c.*) is very large; the mento-Meckelian (*m.mk.*) is small.

The fronto-parietals (fig. 6, *f.p.*) are falcate, less than a third the width of the narrowest part of the skull, pointed at both ends, and dilated a little behind, but far apart there; they just touch the ethmoidal wings in front. The nasals (*n.*) are long, narrow, angulate bones, with a small posterior and two large external emarginations.

The marginal bones (*px.*, *mx.*, *q.j.*, *sq.*) are all normal and well developed, but are thin; there is a good supratemporal plate and post-orbital spur to the squamosal; there is no septo-maxillary. The parasphenoid (*pa.s.*) is large, long, attenuated in front, gnawed externally, and triangular behind. The vomers (*v.*) are peculiar; the front part is a dilated lobe, giving off a short spike in front of the inner nostril; then the bone runs along, flat and thin, by the inner margin of that passage up to the ethmo-palatine. This thin part is flanked with a thick, arcuate, dentigerous crest, and the right and left crests running towards each other in front, meet within a distance of one-third their own length.

This large Neotropical Tree-frog has an intensely specialised skull, which, however, lies along a line diverging far from that of the typical skull:—

1. The whole skull is extremely depressed.
2. The ethmo-nasal region takes up half the length of the skull.
3. There is only one fontanelle which is totally uncovered by the roof-bones.
4. All but the small tracts below, and externally, and the fore end of the nasal region is one continuous bony box.
5. The snout is without a rostrum (normal), but is very transverse, and the roof-cartilages are rather scant.

6. The ethmoidal roof gives off large superorbital outgrowths.
7. The palatine bones are unusually developed, taking on the dilated form of the corresponding cartilages.
8. The pterygoid bone, like the cartilage on which it is grafted, is slender, but it articulates, by suture, with the bony skull, and fixes the pedicle.
9. There are neither inter- nor supra-stapedial.
10. The basal plate is largely membranous in front, and all the "lobes" are small.
11. There are no septo-maxillaries.
12. The fronto-parietals are narrow, arrested bands.

49. *Hyla rubra.*—Adult male; 1 inch 11 lines long. South America.

This is a lightly built, but strong, skull (Plate 33, figs. 6, 7); its greatest breadth is to its length as 11 to 10, and its general outline is seven-twelfths of a very neat oval; the incurving of the quadrato-jugal ($q.j.$) gives a sixth of the (supposed) other half of the figure. The condyles of the quadrate ($q.c.$) reach as far back as the upper edge of the foramen magnum ($f.m.$).

It is a flat, wide skull, the "parotics" beyond the horizontal canal ($h.s.c.$) doubling the width of the capsule, right and left. The axial extent of the hind skull is only two-thirds as great as that of the other two regions, which are equal.

Measured across the ethmo-palatine wings within the maxillaries ($e.pa.$, $mx.$), in front, and across the tegmen tympani of each side within the squamosals ($sq.$), behind, the chondrocranium has the same breadth. A considerable synchondrosis exists above and below, at the mid-line ($f.m.$); between the small, oval, posterior condyles ($oc.c.$) the basal outline is convex, and is equal to both the facets in breadth; the arch, above, comes short of the basal plate only moderately; the obliquity of the foramen magnum is not great.

The extent of the outstanding ear-capsules is twice as great against the skull as at the tegmina (inside $sq.$); each tegmen is ossified for one-third of its extent, and the bony tracts are not divided into a prootic and an ex-occipital ($pr.o.$, $e.o.$); the foramen ovale (V.) is not quite enclosed by this bony tract. Below (fig. 7), the bone only leaves a scooped cartilaginous space, margined by a bony balk, for the pedicle ($pd.$) and a small tract running from the setting on of the stylo-hyal ($st.h.$) to the fenestra ovalis ($st.$). The hinder third of the interorbital space is cartilaginous, and the large optic fenestra (II.) occupies its middle. That region is almost oblong—it dilates a little at both ends; it is only three-fifths the width of the sub-oval orbital spaces. The single fenestra is long-heart-shaped; the fore end of it has been filled in by periosteal growths from the girdle-bone; the lateral tegminal growths are wide. This is a very rare ethmoid; its superorbital region is unossified, and grows out and back into a narrow cartilage with a sinuous outline; this process carries a distinct oval superorbital ($s.ob^1.$) which is turned forwards. This free cartilage occurs again in *Phyllomedusa* and *Alytes*

(Plates 34 and 24), but I have looked for it in vain in most of the skulls of the Anura.

The girdle-bone reaches in front to the full extent of the ethmoidal region, but at the sides does not ossify the "alæ." From the axils to within a short distance of the hind margin there is a large oblong fossa: this is due to a very remarkable structure, viz.: the filling in by periosteal bone of a pair of large oblong fenestræ in the sidewalls of the skull, in front of, and similar to, the optic fenestræ (II.). In *Rappia bicolor* ("Polypedatidæ") (see Plate 19, figs. 6, 7) and in *Camariolius tasmaniensis* ("Cystignathidæ") these two spaces are continuous (Plate 19, figs. 1, 2); it is possible that the orbito-nasal nerve may have passed into the skull in the front part of this space, which is very anomalous. The septum nasi (*s.n.*) is thick and well marked, and ends in a distinct, but short, prenasal (fig. 6); the roof and floor (on each side of *s.n.*, above and below) are well developed, and the new cornua, or pro-rhinals (*p.rh.*), are miniatures of the primary cornua trabeculæ.

The sub-tubular nostrils (*e.n.*) have a definite raised rim; they are not very wide apart, and have the usual appendages (*n.l¹.n.l².*) which are well developed. The inner nostrils (*i.n.*) are much wider apart; the ethmoidal alæ are broad up to their outer edge, behind; there the cartilage, now ethmo-palatine (*e.pa.*), is much narrower and then expands outside into the pre-palatine plate (*pr.pa.*). The bone (*pa.*) only partly hides the cartilage; it is almost straight and dilates outside; the pterygoid (*py.*) does not reach it by a considerable distance; that bone and the correlated cartilage is very slender. The bone continues so, but the cartilage dilates considerably to form the swelling pedicle (*pd.*) and the oblique bilobate trochlea of the quadrate (*q.c.*); the cartilage above the joint (*q.*) is considerably ossified from the quadrato-jugal (*q.j.*).

In the right angle formed by the suspensorial forks the Eustachian opening (*eu.*) is seen to be large and crescentic. The "annulus" (*a.ty.*) is of normal size, but is open above. The stapes (figs. 10, 11, *st.*) is a thick oblique valve, hollow within and thickest at its lower edge; it has a crescentic emargination for the inter-stapedial (*i.st.*) and a definite, oblong "umbo" outside. The inter-stapedial (*i.st.*) is a large saddle-shaped segment, half the size of the stapes; the medio-stapedial (*m.st.*) has the average pistol-shape, and a nearly perfect joint divides its unossified end from the *symplectic* element. This latter (*e.st.*) is a rod of cartilage, duckbill-shaped, with a thin flange; this extra-stapedial gives off from the end of the flange a ligulate supra-stapedial (*s.st.*), which is confluent above.

The mandible (fig. 8) is perfectly normal.

The fixed stylo-hyal passes into the cerato-hyal region (fig. 9, *c.hy.*) without enlarging; the cartilage has a sharp inturned horn before it grows back as the hypo-hyal (*h.hy.*). The notch is very large and transversely oval; the body of the basal plate (*b.h.br.*) is extremely short, and ends behind in a small free rounded lobe, in front of which there is an endosteal basi-branchial (*b.br¹.*); outside this the thyro-hyals (*t.hy.*) are extremely long and bound a space which is half a long ellipse; their hinder third is unossified.

2 A 2

The lateral lobes are well developed, the foremost are stalked obovate leaves, the hinder pair are ligulate; altogether, this is one of the most remarkable and elegant hyo-branchial structures to be seen in the Order.

The investing bones are full of interest: the fronto-parietals (*f.p.*) are oblong, rounded, however, and scarcely dilated behind, and having a transversely dentated front margin; they barely overlap the endocranium (fig. 7). Contrary to rule, they cover in the fontanelle (under *f.p.*), as in the larger "Polypedatidæ."

Over the large fore skull the nasals (*n.*) are nearly as large as the fronto-parietals; they only touch the edges of the septum nasi (*s.n.*) and do not meet each other; they are crescentic shells, with a facial "handle."

The premaxillaries (*px.*) are narrow, and so is the palatine portion of the maxillaries (*mx.*), but the facial part, although extremely thin, is high: there are no septo-maxillaries.

The quadrato-jugals (*qj.*) are sharp and curved; they are well grafted on to the quadrate; the squamosals (*sq.*) are well developed, the supratemporal part is lozenge-shaped, and sends outwards and forwards a long narrow postorbital process. The parasphenoid (*pa.s.*) is not two-fifths the length of the skull, it is a very elegant dagger, with backwardly bent narrow processes for the guard, and a short triangular handle. The vomers (*v.*) are long and slender; the tooth-hillocks are sub-crescentic, the pre- and post-narial spikes are long, and the fore part a short pointed process. This skull is a very exquisite structure, differing from that of the "pattern" rather in a certain delicacy of the parts and the light airy character of the whole, than in anything essential; some things may especially be noticed:—

1. There is only one fontanelle.
2. The bones of the hind skull are confluent on the same side.
3. There is a cartilaginous process, and also a distinct cartilage, in the fore angle of the orbit, above.
4. There is a short prenasal rostrum.
5. There are no septo-maxillaries.
6. The girdle-bone has its sides partly formed of ossified membrane.
7. There is a short hinder projection from the basal plate, and some endostosis there; and the whole of that apparatus is very remarkable in the shortness of the plate, the depth of the front notch, and the great length of the arcuate thyro-hyals; but these things, and many more, which the figures will show, are almost undefinable modifications of the typical form, and this Tree-frog is just such a refinement, so to speak, of the typical Frog, as that is of the Common Toad.

50. *Hyla* ———? sp.—Tadpole, 1 inch long; hind legs, 5 lines. Rio Janeiro.

These, and the skulls of more advanced larvæ of *Nototrema marsupiatum* (Plate 30, figs. 8, 9, and 10–13), show that the modification of the larval Batrachian chondrocranium in the "Hylidæ" is very characteristic; the stoutness of the larval skull gives no promise of the delicacy of structure shown in that of the adult. The skull at this stage corresponds very closely with that already described (Plate 30, figs. 6, 7) in the larva of a narrow-backed Tree-frog, viz.: *Acris Pickeringii*, a type not very far removed from the Oxydactyle "Cystignathidæ."

Up to the quadrate condyles (Plate 30, figs. 8, 9, *q*.) this is a rather square skull, but those hinges are further back than in any other kind known to me, reaching very little in front of the inner nostrils (*i.n.*); the cornua trabeculæ, on the other hand, are so long as to make this really one of the longest of the Tadpoles' skulls.

In this skull the occipital arch is fully formed, and the notochord (*nc.*) has become very small; the condyles (*oc.c.*) are perfect, and the tegmen cranii (fig. 8.) runs forwards as far as to the exit of the 5th nerve (V.).

The large auditory capsules (*au.*) reach as far back, nearly, as the condyles; and their breadth is about as great as that of the intermediate basal plate. Externally, they reach nearly as far outwards as the suspensoria, for the tegmen tympani (*t.ty.*) is already developed outside the horizontal canal (*h.s.c.*). In front of the tegminal lobe there is a smaller lobule of cartilage; here the formation of this tissue is spreading over the opercular region of the 1st cleft; that lobule will be detached as the "spiracular cartilage" (*sp.c.*). The fenestra ovalis (fig. 9) is formed, but the stapes (*st.*) is not chondrified. The interorbital region is one-fourth longer than the inter-auditory; apparently the basal plate is of the same width, but the wings that form the auditory floor are not evident at first. From the point where the notochord was, to the projection on each side in front of the inner nostrils (*i.n.*), the trabeculæ (*tr.*) are very uniform in width. There is an evident tract of newer cartilage (intertrabecular) from the point where the notochord ended, to the most contracted part of the trabeculæ (fig. 9, *tr.*): and above (fig. 8) this median tract has risen in front of the great fontanelle (*fo.*) as a rudimentary "mesethmoid" (*p.e.*).

The fontanelle (*fo.*) is large and elegantly lanceolate; the walls are well developed, and the roof (*tr.*) exists as a marginal band from the hinder complete roof (above *nc.*) to the up-growing middle wall (*p.e.*).

The wings of the ethmoid (each side of *p.e.*) are very indeterminate at present, and there is no distinct elevation outside them forming a rudimentary post-palatine. The inner nostrils (*i.n.*) are wide apart—unusually so—and small; in front of them the trabeculæ are lobate to catch the pre-narial ligament, which arises from the blunt pre-palatine spur (inside *q.*). The cornua (*c.tr.*) are very long, narrow, and sinuous; they each send out a sharp external angle. The suspensoria have a thick pedicle (fig. 9, *pd.*), a thick curved otic process (fig. 8, *ot.p.*) and a short and broad orbitar process (*or.p.*).

The ethmo-palatine band (*p.py.*) is wider than long, the suspensorium coming very close to the trabecula; beyond this band the quadrate region (*q.*) is one-third wider than it is long, and the pre-palatine spur is very blunt. The mandibles (*mk.*) are extremely large and massive; the angular process is short and thick. The lower labials (*l.l.*) are thick scooped crescentic lunules; the upper (or temporary) labials (*u.l.*) are large, thick, and lozenge-shaped, with a deep notch outside. The hyoid (fig. 9, *c.hy.*) is very elegant. The lobate lower part is lozenge-shaped, and is joined to its fellow by the normal square mass of soft cells. The condyle and the condyloid cavity on the suspensorium (fig. 9, between *q.* and *sp.*) are very large, and the stylohyal process is shaped like a Tiger's claw, broadly uncinate with a definite point; this tract is separated from the hinder distal lobe by a deep notch.

Altogether, this is a very peculiar chondrocranium; the next stage, in another genus of the "Hylidæ" (*Nototrema*, Plate 30, figs. 10–13), will show still further the peculiarities of the skull in the Tadpoles of this group.

Second genus. *Litoria*.

51. *Litoria marmorata.*—Adult male; 1½ inch long. Australia.

The skull of this species (Plate 19, figs. 11, 12) differs considerably from that of the wide-backed Australian Tree-frogs already described, viz.: *Hyla Ewingii*, and *H. phyllochroa*. In some respects it comes nearer that of *H. albomarginata* (Plate 32, figs. 6, 7), a Neotropical species, and in others it is not so far removed from that of a typical *Rana*.

It is rather strongly built, and is not so flattened out as those I have been describing; in many things it shows characters that indicate affinity to other Families. The breadth is to the length as 11 to 10, and the quadrate condyles reach to the middle of the stapes (*q.c., st.*); hence it is a medium skull in these respects. So, also, in its flatness, in density of endocranial bone, and strength of investing bones, it is an average skull. The occipital condyles (*oc.c.*) are of the medium size; they are postero-inferior, reniform, and are separated by a convex line of basal bone greater than their own breadth. The opiotic eminences over the posterior canals (*p.s.c.*) reach nearly as far back as the occipital condyles; then the other canals (*a.s.c., h.s.c.*) are large, but the parotic wings, although widely extended, are narrow and run to a rounded end (*t.ty.*), which, at the point where the squamosal (*sq.*) binds on, is unossified.

But the whole hind skull, with this exception, and with the exception of the basisphenoidal region, is all solid bone up to the optic fenestra (II.). The hinder tegmen cranii is well ossified, but here was seen what is evidently rare in the Hylidæ, namely, two large secondary fontanelles; they are, however, filled in with periosteal bone, and are traceable by means of its thinness. The sides of the mid skull form a gently concave outline, inbent most at the middle; the great fontanelle (*fo.*), which is longer than either of the bony tracts above, is a long oval, has but narrow tegminal margins, and is a little roofed over by the fronto-parietals (*f.p.*). The optic fenestræ (II.) are of medium size, and have a tract of cartilage nearly of their own extent in front of them; from thence bone—the girdle-bone (*eth.*)—reaches above, to the middle, and below, along a third, of the proper nasal territory. This extensive bony tract takes in the ethmoidal wings up to the proper *facial* ethmo-palatine (*e.pa.*), and uses up the small superorbital projections (*s.ob.*).

The roof and floor of the nose have the usual width; the latter narrows, crescentically at the middle, and the roof is widest behind. The septum nasi (fig. 11, *s.n.*) is thick and well marked, it ends in a distinct prenasal, to the sides of which are fused the horns of the roof crescents (on each side of *s.n.*). As in *Hyla Ewingii* (see Plate 31, figs. 2 and 4, *p.rh.*), the pro-rhinals (fig. 12, behind *px.*) have the form and appearance of the angles of the primary cornua, and are half as large—a very unusual size. The appendages of the nostrils (*n.l¹.n.l².*) are well developed; these passages are at the average distances, outside the crescentic snout; the inner nostrils (*i.n.*) are large, circular, and their distance is one-fourth greater than that of the outer holes. The palato-suspensorials are in several respects varied from the norma; the palatine portion is evidently stronger than the pterygoid, and the arch is angulated where these regions meet. The ethmo-palatine (*e.pa.*) narrows as usual where it joins the ethmoidal wing, and both this lessening, and the cessation of the bony deposit, mark off the true facial part. The pre-palatine is the point and edge of a dilated blade, and under this a large palatine bone (*pa.*) binds, which is falcate and dilated externally, and has a cultrate ridge growing from its middle third—*like an old tooth-bearing crest*; as in *Bufo agua, Callula pulchra, Cystignathus ocellatus,* and some others.

The pterygoid bone (*pg.*) is slender, but strong; the re-entering angle of its fork is rounded, its inner fork (*pd.*) forms a suture with the skull, and ties down the cartilaginous pedicle, as in *Hyla albomarginata* (see Plate 32, fig. 7). The quadrate region is rather short, moderately retral, and considerably ossified; the condyle (*q.c.*) is a well-formed bi-cristate trochlea. There is a middle sized annulus (*a.ty.*); its band is wide, leaving a small central space; it is open above. The mandible (fig. 13) is perfectly typical, and the Eustachian opening (fig. 12) is oval and rather large. The stapes (fig. 15, *st.*) is large, oval, and knobbed; the medio-stapedial (*m.st.*) is pistol-shaped, with a heavy "handle" of cartilage, notched off from the bony part, and itself emarginate behind. The shaft is arcuate and very slender; it is followed by an extra-stapedial (*e.st.*) which is two-winged, and the wings are crenate. A bud grows from the middle of this oakleaf-shaped plate, which ends as a free knob, behind, but from which no ray protrudes, as a supra-stapedial.

The stylo-hyal (fig. 12, *st.h.*) is confluent above, and passes into a middle-sized tape (fig. 14, *c.hy.*) that ends in a straight hypo-hyal horn (*h.hy.*).

The retiring part passes quickly into the base, for the "notch" is shallow; so also the plate itself (*b.h.br.*) is of small extent, fore and aft, and has no front lobes, only long hinder processes. The large, expanding, highly ossified thyro-hyals (*t.hy.*), diverge considerably behind; in front they are anchylosed together, and there they form a "basi-branchial" bone (*b.br.*), whose wedge-like point nearly reaches to the selvedge of the emarginate basal plate.

The investing bones are slight, but rather strong; the fronto-parietals ($f.p.$) are more developed than in many *Hylæ*, and creep over some considerable tract of the temples, clamping the ridge of the "anterior canal" ($a.s.c.$). They cover in a little of the exposed membranous roof ($fo.$), and diverge only a little to their pointed fore end. The nasals ($n.$) are crescentic trowels, that stop in front with the bony deposit, and thus are far off from the nostrils; they are a moderate distance from each other. There are no septo-maxillaries, that I can find; the other outside bones ($px.$, $mx.$, $qj.$, $sq.$) have the typical development. The parasphenoid ($pa.s.$) is a very elegant dagger, at first very wide, and then running to a fine long point. The guard is composed of two long slender bars; the handle is broad, short, triangular, and anchylosed to the skull.

The vomers ($v.$) are very large, perfectly typical in form, and have an oval toothed boss, they are only separated by the septum, behind.

With an outer form more *Ranine* than that of the species of *Hylæ*, this skull yet shows a considerable number of specialisations that are different from what is seen in a typical Frog:—

1. The skull is more flattened out.
2. The occipito-otic region is one continuous mass of bone, running to the front of the optic fenestra.
3. The girdle-bone takes up half the anterior sphenoidal region, and half the nasal.
4. The superorbitals are present, but ossified.
5. The main fontanelle is open through the arrest of the roof-bones; and the lesser spaces are ossified.
6. The pedicle is tied down by the pterygoid.
7. The quadrate is ossified.
8. There is no septo-maxillary.
9. There is neither inter- nor supra-stapedial.
10. There is no fore lobe to the small basal plate, and the thyro-hyals are anchylosed together and ossify much of its middle part.
11. The palatines have a cultrate ridge.

Third genus. *Nototrema*.

52. (A) *Nototrema marsupiatum.*—Adult male; 1⅘ inch long. South America.

This skull (Plate 33, figs. 1-5) has all the massiveness of that of any stout Toad, and is in extreme contrast with what is found in some of the species of *Hylæ*—such as *H. phyllochroa,* and *H. Ewingii.* It is flat at the top and wide, but is also very high or deep, for a Batrachian.

The proportion borne by the bony, to the soft, tracts of the endocranium is not much greater than is common even in the genus *Rana,* but where bone is, there it is strong and rugged, and the investing bones might belong to the skull of a small Crocodile, both for their strength and sculpture.

The outline is semi-oval, and the length is to the breadth as 10 to 11; the hinge of the jaw has got no further back than the end of the upper bar of the squamosal (*sq.*, *q.c.*), in this it is in strong contrast with the skull of *Hyla albomarginata*. The obliquity of the parotic wings is much more evident than it is in that species (see Plate 32, figs. 6, 7); a line drawn along the hind margin of the auditory capsule, touching the end of the tegmen tympani and the epiotic eminence, would form, with a transverse line intersecting it, an angle only half as acute as would be obtained by the same measurement in the skull of that species of *Hyla*; I shall take that species for my model of comparison in describing this. In this skull the moderate degree of retreat of the jaw-hinge makes the suspensorium form a little less than a right angle with the basi-facial line; and there is a great obliquity of the squamosal (*sq.*) outlying it, the hammerlike head of which is tilted downwards to articulate with the jugal process of the maxillary.

Without any loss of elasticity—for soft tracts of cartilage alternate with hard territories of bone—this skull is one of the strongest, for its size, of any to be found in the "Order;" albeit, like that of a Crocodile, it retains not only much cartilage, but, also, nearly all its sutures.

We have, here, the normal proportion of the three (serial) regions of the skull; the auditory and nasal are about equal, and the orbital one-third larger; measured across the axis, the middle region has its cranial part very large, thus lessening the orbital vacuities.

The occipital condyles (*oc.c.*) are semi-ovoidal, and are postero-inferior; they are separated by a straight basal space one-third larger than their long axis. Over and below the large foramen magnum there is a clear cartilaginous supra- and basioccipital tract of cartilage (behind *f.p.* and at *b.o.*); the upper tract becomes the soft selvedge of the short hind skull, up to the single, very large, pear-shaped fontanelle (*fo.*), the broad end of which it encloses.

The lower tract passes into an extensive cross-shaped field of cartilage, not divided off from the unossified interorbital part (fig. 2). Thus we have the normal division into prootics and ex-occipitals (fig. 2, *e.o.* to 11.). These, however, are anchylosed together on the epiotic eminence (fig. 1 *p.s.e.*), but a narrow band of the roof-cartilage runs along almost to the end of this rounded balk. The parotics are ossified to their end, leaving but little cartilage even at the tegmen. The prootics enclose the foramina ovalia (V.), and almost touch the optic fenestra (11.). The ex-occipitals run up to the fenestra ovalis, but the outer part of the vestibule (fig. 2, *vb.*) is floored with cartilage, which is continuous with the soft part of the basal plate (*b.o.*). The girdle-bone (*eth.*) is equal to the unossified tract behind it, yet their junction is not in the middle of the orbits, but further forwards, for that bone runs a little into the proper nasal region (*s.n.*). Yet it does not harden the wings of the ethmoid, nor the narrow superorbital eave (*s.ob.*). The endocranium in the orbital region is, as to its cavity, very similar to that of *H. albomarginata* (see Plate 32, figs. 6, 7; and Plate 33, figs. 1, 2), but the intrinsic roof of the latter, and the secondary roof of *Nototrema*, dilate the top at opposite ends, so that whilst they have the same actual outline, this is recessed ; the skull in this kind is twice as wide at the temples as in front, and in the other twice as wide in front as at the temples.

Moreover, the skull is almost as completely roofed as in the narrow-backed Oriental Tree-frogs (Polypedatidæ), for only a little of the membranous fontanelle is naked, as the roof-bones converge rapidly, and for half their extent form a "sagittal suture." Somewhat hooked in front, the fronto-parietals (*f.p.*) insert their point under the nasals (*n.*), but leave a pentagonal tract of the *girdle* uncovered, in front of the fontanelle, and also expose, right and left, the small semi-ossified superorbital eaves (*s.ob.*). Above, in front, this bony part is hidden by the conterminous nasals (*n.*), and below, in some degree, by the inner end of each palatine (*pa.*), and by the toothed boss of each vomer (*v.*). The dilated end of each ethmoidal wing, and nearly all the nasal region, are left untouched by the girdle-bone. The nasal roof (*al.n.*) is formed by a pair of large, but narrow, crescents, that are confluent by their backs with the top of the septum (*s.n.*); the whole tract is narrow-waisted. So also is the floor (fig. 2, *s.n.*), but it is twice as wide, and at its angles spreads into leafy lobes, that grow well into the fore end of the maxillaries (*mx.*). The outer nostrils (*e.n.*) are not wide apart, although the snout is transverse ; yet the outline of the bones (*px.*) is crescentic. The inner nostrils are one-half larger, and their distance is one-half greater. The transverse snout-margin has three well-formed projections, namely, a teretc prenasal, and the two pedate, out-turned pro-rhinals (*p.rh.*). The appendices of the nostrils (*n.p.n.p.*) are of the typical size and form.

The palato-suspensorials are strong, the front bar (*pa., pr.pa.*) is directly transverse and cultrate, and the lateral part (*pa.* to *pg.*) gently arcuate. The palatine bone (*pa.*) is thin, falcate, and narrowest in the middle ; it is but little united to the cartilage. The pterygoid (*pg.*), as usual, runs up along the post-palatine bar to the curved end of the palatine bone ; it has largely affected the cartilage beneath, and has run outwards into the post-palatine lobe, thus making what is generally only seen during metamorphosis (see in *Pseudis*, Plate 12, fig. 4), a permanent structure ; for at one time all the arch which now stretches from the ethmoidal wing to the jugal lobe was *antorbital*, and it really belongs to the palatine. The pterygoid (fig. 2, *pg.*) seems, as seen from below, to have two sub-equal forks, but the side view corrects this (fig. 3), and the quadrate region is seen to be of considerable extent ; it is partly ossified. The fork of the bone (*pg.*) is sharp, and thus the Eustachian opening (*eu.*), which is of the medium size, is oval. The inner fork (*pd.*) is formed of the cartilaginous pedicle and pterygoid bone ; the latter *almost* obliterates the joint, but not so much as in *Hyla albomarginata*. The quadrato-condyle (*q.c.*) is a very finely finished trochlea, for the solid cylindroidal condyle of the mandible (fig. 3 *ar.c.*). The auditory outworks have but a middling development ; the annulus (*a.ty.*) is not small, but its horns are wide apart ; there is a good sized tympanic cavity. The stapes (fig. 5, *st.*) is large,

DEVELOPMENT OF THE SKULL IN THE BATRACHIA.

bossed, and sub-oval, with an oblique antero-superior emargination. The columella is pistol-shaped, large proximally, and small distally. The medio-stapedial (*m.st.*) is notched above, and carries a huge unossified inter-stapedial lobe (*i.st.*), not separate; the extra-stapedial (*e.st.*) is tongue-shaped, with a side wing which ends abruptly without an ascending process.

The stylo-hyal (*st.h.*) is confluent above, and passes into a band (fig. 4, *c.hy.*) which widens gently and passes back (*h.hy.*) into the basal plate without a lobe. The great front notch is a wide semi-circular space; the basal plate (*b.h.br.*) is one-third larger fore and aft than in most of the *Hylæ*; the front lobes are small, and the hind lobes sharp; the thyro-hyals (*t.hy.*) are long, gently divergent, and dilated behind. The mandible (fig. 3) is normal, the mento-Meckelian (*m.mk.*) moderate, the dentary (*d.*) large, the coronoid process of the articulare (*ar.*) small, and the condyle (*ar.c.*) solid, and sinuously rounded.

The investing bones must be compared with those of *Calyptocephalus* and *Pelobates* (Plates 21 and 25); they are thick, ornately sculptured, and sub-ganoid. Wherever these bones are directly sub-cutaneous there they are elegantly honeycombed; the pits are well cleared, and the network of ridges sharp.

The fronto-parietals (fig. 1, *f.p.*), like the other outer bones, are wide and strong, as in the species of *Bufo*. Their form is many-cornered, with the antero-external and postero-external margins concave, and the hind margin crenate, and notched; together, they reach, behind, over the rising parts of the auditory capsule, and each sends a round process backwards which clamps the front part of the epiotic eminence (*p.s.e.*). Towards the middle they cover the front third of the occipital arch. Binding on the anterior canal (*a.s.c.*) they send a sharp postorbital process outwards, and then run, with a concave margin, their sharp-edged superorbital plate up to the proper superorbital cave (*s.ob.*). Their pointed diverging fore part retreats and exposes the girdle and the fontanelle. There is a supraorbital fossa, beyond which the bone is strongly honeycombed, and also over the temporal edge behind; there is there, on each side, one large oval fossa, more strongly marked than the rest. The nasals (*n.*) form a pair of bony wings, whose long tips bend down to lie along the top of the maxillary (*mx.*); they overlap the frontals, the girdle-bone, and the ethmo-palatines, and leave the front part of the snout naked. They are only moderately wide, much angulated, sculptured, and they meet to form a nasal suture, and dip to reach the "girdle."

The premaxillaries, maxillaries, and quadrato-jugals (*px., mx., q.j.*) are normal in form, *Bufonine* in strength, and have the special fretwork, outside, of this species. I find no septo-maxillaries.

The squamosals (*sq.*) have only a moderately broad supra-temporal part; it binds by an irregular edge on the parotics. The postorbital process (fig. 3) is a broad, rough, oblique part, whose *depth* is continued with little diminution to the end of the "head" of the bone, which thus shows an *Otilophine* character. The "handle" of the hammer lies deep, and is therefore smooth; it is narrower than the head, and is set on behind the middle; the fore part of its head is bent, adze-like. The facial part of the maxillary is very high, so also, although lower, is the jugal part up to where it is united, by suture, with the squamosal.

The parasphenoid (*pa.s.*) is a very large, but truly *Hyline*, bone; the four rays all run to sharpish angles; the wide blade is swollen, and then rapidly attenuated; the guard is unusually wide, and the handle is a large, low triangle. The vomers (*v.*) are perfectly typical; they touch the sides of the septum, but not each other, and each carries an oblique, oval, rather large, dentigerous boss.

The difference between this skull and that of *Hyla albomarginata* can be seen better by reference to the figures than by any description. This skull is about equidistant from that of the typical *Rana*, and that of such a Neotropical *Bufo* as *B. chilensis*, or the Ethiopian *B. pantherinus*; it nevertheless keeps the *Hyline* characters intact, whilst diverging less in some things from the norma than the species of *Hyla*. The divergence of such a species as this from an oxydactyle Frog may have been gentle and gradual; there was no morphological stumbling-block to be got over of any importance whatever. The likeness of this skull to the "norma" is as instructive as its unlikeness.

1. The whole skull is short, solid, strong, and deep; and the osseous tracts of the endocranium, although well, and almost normally, limited by cartilage, are remarkably *Bufonine*.

2. The sculpturing of the very strong outer bones is another variation.

3. The bones of the hind skull are anchylosed above on the same side.

4. The quadrate condyles have only partially retreated.

5. The fontanelle is single, and partially open.

6. There is a distinct prenasal rostrum.

7. The septo-maxillaries are wanting.

8. There is a wide post-palatine lobe.

9. The horns of the annulus are wide apart.

10. There is no proximal segment of, nor ascending process from, the *small* columella.

11. The stylo-hyal is confluent above, there is no hypo-hyal lobe, and the basal plate is short, and has small lateral lobes.

52. (B) *Nototrema marsupiatum*.—Larva, $2\frac{1}{3}$ inches long; body, 1 inch; tail, $1\frac{1}{3}$ inch; hind legs, 13 lines; *right* fore leg free, $\frac{1}{2}$ inch long; *left* fore leg still under the operculum. South America.

This skull is at a very important stage; the larval structures are all present, and the permanent structures are fast coming in. Its shape (Plate 30, figs. 10, 11), now, is a short oval, for while the quadrate cartilage is where it was, right and left, the cornua trabeculae have dwindled down remarkably; nevertheless, all the external cartilages are of full size at present. Besides the odd outer bone below, two pairs more are now present above, and the post- and pre-auditory endoskeletal centres have appeared (*pa.s., f.p., pa., e.o.*, and the patch behind V.); three pairs of new cartilages, also, have appeared—the spiracular and the *permanent* upper labials (*sp.c., u.l¹., u.l².*).

The notochord (*nc.*) is fast shrinking; the condyles (*oc.c.*) are well formed; and a crescentic bony tract appears on each side, reaching to the foramen for the 9th and 10th nerves (IX., X.); these are the ex-occipitals (*e.o.*); they are climbing up behind, and flanking the auditory capsules in the opisthotic region; the prootics are small. The whole basis cranii is now well chondrified, and behind, has completely coalesced with the auditory capsules; they are obliquely placed, the horizontal canal (*h.s.c.*) and its fringing tegmen (*t.ty.*) pushing outwards. Outside the oblique, oval, vestibular convexity, below, we see a perfect oval stapes (fig. 11, *cb., st.*); and between the otic process of the suspensorium (*ot.p.*) and the capsule there is a thick biconvex mass of cartilage wedged in; this is the distinct spiracular cartilage (*sp.c.*)—the "annulus tympanicus" that is to be. The interorbital region, or mid skull, lessens considerably forwards, and bulges gently; its floor is perfect and so are its walls, but the roof is membranous from the narrow occipital arch (lettered *f.m.*) to the perpendicular ethmoid (the hind part of *s.n.*): this is the great fontanelle (*fo.*). The rudiment of an ethmo-nasal wall seen in the last stage (fig. 8, *p.e.*) is now complete, and yet has quite a clear outline above and below (figs. 10, 11, *s.n.*); this is the high fore part of the intertrabecula, which often, in this "Family" (the Hylidæ), grows forwards as a free prenasal rostrum. The broad bulbous hinder end is directly in front of the parasphenoid below, and of the fontanelle above; its fore end is narrow and is still growing, for it will run forwards between, and even in front of, the cornua trabeculæ (*c.tr.*). Here we see the distinction between the nasal roof-cartilage (*n.r.*), or ali-septals, and that growth of the skull itself at its closing in, which belongs to the true ethmoidal region. Here the hind part of the roof is the ali-ethmoid (*al.e.*)—an ox-horn-shaped spreading and curving growth of cartilaginous wings, enfolding the sides and hind part of the nasal sacs. The nasal roofs (*n.r.*) are broad lunules of newer cartilage, their deep concavity, or notch, margining the nostrils (*e.n.*).

The cornua trabeculæ (*c.tr.*) are quite like those of the last stage (figs. 8 and 9), but they have dwindled down to half their proper size. In front of these, but more than twice as large, are the temporary upper labials (*u.l.*); these have a deep cleft behind, more than half dividing each, and a long round outer angle; the outer part corresponds to the angulo-labial of the Lamprey, and the inner moiety to half its anterior dorsal cartilage, or main upper labial.

Behind this inner part, running obliquely backwards and outwards, there is a half-tube of thin bone, rather dilated at its base, in front; this is the premaxillary (*p.x.*), which thus develops its nasal process, first.

Each bone reaches the front of the outer nostril (*e.n.*), and has under it two little beads of *new* cartilage; the inner lobule lies under the bone and is discoid; the outer passes outside the front of the nostril and is a hollow shell; these are *permanent upper labials* (*u.l¹, u.l².*).

The lower labials (fig. 12, *l.l.*) are, as in the earlier stage, very solid; but now they form a very good half ring, segmented across the middle; these scooped, thick ridged cartilages are articulated with the mandibles (*mk.*), massive, swelling segments, with a concave condyloid face, and a squarish, hooked, angular process. The massive suspensorials (*sp.*) are nearly half as broad as long; their pedicle (fig. 11) is narrow at its root but soon widens; their otic process (fig. 10, *ot.p.*) is a high, thick, inturned rib of cartilage.

The large leafy orbitar process (*or.p.*) is half the size of the suspensorium, it is of good width, and very long, from its broad, sessile root, to its rounded apex. The condyle for the hyoid (*h.y.f.*) is the normal rimmed hollow, and the condyle of the quadrate is rather small, and selliform. The pterygo-palatine bridge (*p.pg.*) is rather small, and the cartilage in front of it very broad, sending inwards a sharp, but small pre-palatine hook (*pr.pa.*); I find no rudiment of a post-palatine over the bridge. The hyoid (fig. 13) is massive, its distal lobes are lessening; its condyle large, and its stylo-hyal end sharp and uncinate. The sharp, shell-like fronto-parietals (fig. 10, *f.p.*) are as much developed now, as in the average of the *Hylidæ*, in the adult; the parasphenoid (fig. 11, *pa.s.*) has the shape and proportions, now, that it has in the adults of many dwarf Toads (*Pseudophryne*, &c.), being broad, splintery, and with sharp, angular lateral processes.

The basi- and hypo-branchials are normal, and the cerato-branchials (*proper*) are evident, as distinct from the extra-branchial pouches, which are, also, quite normal; these parts were dissected and examined, but not figured.

This specimen has added considerably to the sum of our knowledge of the growth and metamorphosis of the Batrachian skull; I shall refer to it, especially, in my "Summary."

Third sub-division.—Tree-frogs, with dilated sacral apophyses and parotoids.

Family. "PELODRYADIDÆ."

First genus. *Pelodryas.*

53. *Pelodryas ceruleus.*—Adult male; 3 inches long. New South Wales.

This skull is of the average breadth, which is to the length as 9 is to 8; the quadrate condyles almost reach as far back as the occipital, and the general outline is a very accurate semi-ellipse.

At first sight this skull is seen to belong to a very different type to that of the Common Frog; it is intermediate in size between those of the Common and Edible kinds, and in detail differs much from them; the large Oriental "Polypedatidæ" help to fill in the space between the typical Frogs with pointed toes, and the typical flat-toed kinds.

The reasons for the great unlikeness of this skull to that of the type are not, at first, easy to find; the same elements are there, having the same relations, and the degree of ossification of the chondrocranium, and the form and size and density of the investing bones is very similar in both. The occipital condyles (Plate 34, figs. 1–4, *oc.c.*) are reniform, and posterior; they are of the average size, and are separated by an emarginate space more than equal to their own diameter. The post-auditory part of the skull is here unusually developed—there is thus a "neck" to the skull, if there is none to the body. The "parotic wings" stretch out to an unusual distance, are of a good breadth to their end, and are then dilated along the tympanic roof.

Considerable super- and basioccipital tracts of cartilage remain wedge-like between the lateral bones; those of the same side are perfectly confluent.

Each occipito-otic bony tract reaches forwards from the condyle to the middle of the space between the openings for the 2nd and 5th nerves (II. V.); laterally, the bone is arrested a little beyond the horizontal canal (*h.s.c.*).

There is, beyond the bone, an otic tract two-thirds the extent of the ossified part; this extended tegmen narrows gently, and then dilates over the ear-drum, sending outwards and downwards a pedate process, to which the hinder crus of the annulus is attached behind the squamosal (figs. 1, 3, 4, *t.ty.*, *a.ty.*, *sq.*). This postero-external process is the familiar "pterotic ridge" of the Fish, and is well seen, again, in *Rana pipiens*, and *Cystignathus ocellatus*.

I find no secondary fontanelles, but in that region there is a considerable tract of cartilage, triangular in shape, but with its apex truncated over the foramen magnum. From that edge to the ossified tegmen in front, the roof, ossified laterally, reaches nearly half way; the fore half is occupied by the long oval main fontanelle (*f.o.*) which is small, relatively, but almost entirely uncovered by the fronto-parietals (*f.p.*).

This space is rendered narrow by the marginal tract of endocranial roof, which also extends a good way back in the ethmoidal region. The mid skull has the hourglass outline, for there is a considerable expansion of both the post- and antorbital parts of the roof, and the skull is rather narrow where the girdle-bone (*eth.*) ends.

Where the superorbital (*s.ob.*) cartilage overhangs the orbit, there the breadth, above, is doubled; but this is only where this ear-shaped flap projects; it lessens again and then is continued, roof-like, into the ethmo-palatine, as in *Rappia bicolor* (Plate 19, figs. 6, 7).

Above (fig. 1), the girdle-bone keeps to the narrow width of the mid skull; but below (fig. 2), it ossifies all but the superorbital flap, and takes up all the cartilage belonging to it, both in the middle, and into the wings, right and left.

In the fore skull the nasal region is evenly rounded in front, and does not form a square snout, with the nostrils (*e.n.*) wide apart, and at the edge, as in the Polypedatidae; but these passages are superior, and not very wide apart; they are also large, and have a strong rim of uncovered cartilage.

The roofs (*al.n.*) are narrow crescents and only form a tract half the width of the floor (fig. 2); the septum (*s.n.*) is thick, and ends as a blunt mass, not projecting; the pro-rhinals (fig. 2, *p.rh.*) are rather small and inturned; the angles of the original cornua pass well into the maxillaries (fig. 2, *mx.*).

From the upper surface of these angular growths a thick and widish crescent of cartilage grows up as a limited nasal wall (figs. 1, 3), behind the nostril (*e.n.*), and confluent with the roof (*al.n.*).

The two pairs of labials (*u.l¹.u.l².*) that defend the nostril in front, are rather small.

The palato-suspensorial arch is normal but very strong, and the usual bony plates, namely, the falcate palatine (*pa.*) and the triradiate pterygoid (*pg.*), are normal, and leave an average amount of cartilage untouched. But the condyles of the pedicle and the quadrate (*pd., q.c.*) are very large and solid, and the latter region is partly ossified.

The cartilage is left uncovered towards the end of the pedicle, which is not only very solid, but is also almost *externed* in position. Moreover, the pterygoid bone forms a right angle by its fork, and in the space thus half enclosed there is an extremely large, circular Eustachian opening (*eu.*), whose boundary is finished by the hind process of the pterygoid, by the stylo-hyal, and by the binding web of fibrous tissue tying these together.

The mandible (fig. 3) is normal; the dentary (*d.*) is half the length of the ramus; the mento-Meckelian (*m.mk.*) is large, and the coronoid process (*ar.*) low. The annulus (*a.ty.*) is relatively almost as large as in *Rana pipiens* (Plate 8, fig. 1), but it is widely open above (fig. 3); yet the drum, altogether, is large and well specialised, and its additional structures are nearly typical.

The stapes (figs. 4, 5, 6, *st.*) is oblique, truncate, and lobate; it is gently convex externally, and gently concave within.

The columella has a reniform proximal osseous centre, which hardens most of the dilated upper end; this is the inter-stapedial (*i.st.*); it is segmented from the long medio-stapedial rod (*m.st.*) by separate ossification, and not by separation of the primary cartilage. The distal cartilage is as long as the bony part (*m.st.*); the extra-stapedial (*e.st.*) is a compressed tongue, which gives off a ligulate supra-stapedial (*s.st.*) that is confluent with the ear-mass above. So also is the stylo-hyal (*st.h.*) confluent above; it becomes first broader, and then narrower again as it passes round, without lobulation, to become hypo-hyal (see Plate 15, fig. 8, *c.hy., h.hy.*).

The body (*b.h.br.*) of the basal tract is only two-thirds the extent of the notch in front; it has no anterior lobe, and the posterior lateral lobe is long, pointed, and diverging. The bony thyro-hyals (*t.hy.*) are normal, but in front of them the basal plate is occupied in its centre by a large basi-branchial endostosis (*b.br.*). Also on each hypo-hyal there is a small oval extra-hyal piece (*e.hy.*).

The investing bones are normal as to strength, but they differ very much in certain regions from the norma: the fronto-parietals (*f.p.*) cover two-thirds of the inter-auditory region, the right overlapping the left.

They then become narrow bands, and in front turn outwards, over the superorbital, which they partly cover; their end is truncated. By this divergence the girdle-bone is left naked, and the fontanelle is scarcely covered at all by them.

The nasals (*n.*) are rather small shells; they have a falcate outline, and are notched, behind; the hollow blade overlies the very limited nasal roof. The marginal bones are well developed; the nasal process of the premaxillary (*p.x.*) is thick, wide, and high; it is capped by the small 1st labial.

The maxillary (*m.x.*) is very high in front, and rather high behind; where it shelves downwards, over the premaxillary there is a small, notched septo-maxillary (*s.m.x.*); the quadrato-jugal (*q.j.*) is normal in size; it is largely continuous with the quadrate (*q.*).

The squamosal (*sq.*) has a long falcate upper part, and a long, broad, retreating lower part.

The parasphenoid (figs. 2, 3, *pa.s.*) is large, smoothly convex below, and has all its four rays pointed.

The vomers (fig. 2, *v.*) are at their highest development; they occupy all but the fore margin of the very wide subnasal cartilage, and as the large circular inner nostrils (*i.n.*) are wide apart, there is a large inter-narial space for the two vomers, which fit, by "harmony," along the mid-line.

Each bone sends a narrow sharp snag out in front of the passage, and in front of this a broader falcate process.

But each bone has a large oblong body, with a rounded front, and a nearly straight hind, margin; this part is thickened into a smooth, rounded mass, whose hind margin bristles with retral teeth.

This toothed lobe projects a little further backwards on the outer, than on the inner, edge.

At first sight it would seem difficult to make out many points of difference between this skull and that of the two European species of the typical Frog (*Rana*), for the size is intermediate between the two, and the elements are essentially the same in this and in them.

Nor are the relative proportions of the three regions of the skull different in any remarkable degree; nor the proportionate quantity of bone to be found in the endocranium.

Carefully compared with the type, this skull is seen to differ in the following respects, namely:—

1. There is a very distinct post-auditory arch.
2. The parotic wings are of great extent, and give off a long "pterotic" process postero-externally.
3. Each lateral mass of bone has lost all signs of division, above.
4. The mid skull is very flat, has wide terminal margins, and has one moderate fontanelle.

5. The superorbital cave is very large and projecting, and is continued on to the lateral ethmoid.

6. The nasal roof is only half the width of the floor, and the post-narial wall is very thick and crescentic.

7. The inner nostril passages are very large, wide apart, and circular.

8. The Eustachian openings are still larger, and are circular.

9. The condyles of the quadrate and of the pedicle are very large, and the quadrate is partly ossified.

10. The annulus is very large, but its horns are wide apart.

11. The stylo-hyal and supra-stapedial are both confluent, above; the inter-stapedial is only segmented by osseous distinction; and the stapes is lobulate.

12. The mento-Meckelian is very large, and the coronoid process very low.

13. There are no hypo-hyal nor lateral basal lobes; there is a distinct basi-branchial bone; and an extra-hyal on each side.

14. The roof-bones—fronto-parietals and nasals—are very narrow, and cover but a small part of the mid and fore skull.

15. The vomers are extremely large, covering most of the nasal floor, and have a very solid dentigerous lobe.

In this species, and in the next, we see, to the utmost degree, the specialisation that is correlated with discoid toes and fingers, and a flat, wide sacral region.

Second genus. *Phyllomedusa*.

54. *Phyllomedusa bicolor*.—Adult female; $3\frac{1}{2}$ inches long. Santarem, River Amazon, lat. 2° 20′ S., South America.

This (Plate 34, figs. 7–11) is a longer skull than the last, the length being to the breadth as 13 to 14; this greater length is due to the unusual extent of the nasal region, for, if that were normal, it would be a short, broad skull.

Here the *Petromyzine* embryo has metamorphosed its skull into that which is extremely *Raiine*,—*Raiine*, that is to say, in its endocranium, considered as free from bony tracts in itself, and from bony investments, outside. It fails, however, in one character, viz.: in not possessing a prenasal rostrum, a part well developed in some of the "Hyloids." Its quadrate condyles are behind those of the occiput.

This flat, broad skull, with a smallish, single fontanelle and much tegminal growth laterally, as well before as behind, from which grows an *ethmo-nasal* region one-third longer than the *spheno-occipital*, is a curious renewal in Nature of the Skate's skull. The cavity of the skull is scarcely longer than the closed-in tract in front of it; and this shallow, boat-like skull is much too large for the enclosed brain.

The occipital condyles (figs. 7, 8, *œ.c.*) are posterior, project but little, are rather small, and are separated by a straight space equal to one condyle. The occipital ring projects but little, and the arch, above, is rather cut away; the whole region, including the auditory capsules, is continuous bone. Above, the cartilage persists for a small extent behind the fontanelle; and, outside, the tegmen (*t.ty.*) is unossified. This soft tract is less than half the huge parotic wing, which extends beyond the horizontal canal (*h.s.c.*). The three canals (*a.s.c.*, *h.s.c.*, *p.s.c.*) are small, but well marked; beyond them the parotic projects as far again from the middle of the skull, and increases in width so as to be equal in size to the hind skull. In this kind the tegmen is rounded behind, and sends outwards, and downwards, a projection in front, like the process that appears, behind, in *Pelodryas*. The whole outline of this two-winged hind skull is strongly sinuous, behind and before. The fontanelle (*fo.*) is heart-shaped, and is open over a third of the cavity of the skull; it has a considerable tegminal margin. The narrowing in front of the temporal region is followed by a continual increase in width up to the superorbital cartilages; and the skull (figs. 7-9) is both wider and shallower than in *Pelodryas*.

The prootic bony tract runs round the foramen ovale (*pr.o.*, V.) up to the optic fenestra (II.); this latter opening is very large, as in some of the lesser Australian *Hylae*. A definite margin of cartilage (*o.s.*) bounds that fenestra in front, and then the girdle-bone (*eth.*) begins; it does reach the edge of the skull-wall, but does not occupy the all proper ethmoidal territory either in front, or antero-laterally. Here, as in *Pelodryas*, the ethmoidal region has two pairs of wings, the front pair confluent with the nasal region, whilst the hind pair are the projecting superorbital caves.

Here we have, as in *Alytes*, a separate, but smaller, superorbital cartilage (*s.ob¹.*); it is finger-shaped, and turns downwards (fig. 3). The outspread cranial roof runs forwards into the nasal roof, and outwards on to the ethmo-palatine bars (*e.pa.*). The distance from the foramina ovalia (V.) to the front of the girdle-bone is equalled by the cartilage in front of that bone. The trabecular floor (*s.n.l.*) is immense; the nasal roof is only a little smaller; but, from the narrowing of the snout, which is transverse, in front, the outer nostrils are only half as wide apart as the inner. These latter (*i.n.*) are as large as the Eustachian openings of *Pelodryas*; they are oval, and their direction is inwards and forwards. The openings for the nasal nerves (inside *s.n.l.*) are large, far backwards, and wide apart, the floor is narrowed between them and the inner nostrils; it is definitely notched at its sudden narrowing.

The pro-rhinals (*p.rh.*) are well-developed, near together, and out-turned; but the angles of the nasal floor converge in a remarkable manner in front; they are bilobate (*c.tr.*, *s.n.l.*), and the hinder lobes diverge. The septum nasi (*s.n.*) is moderately thick, and is rounded off in front, so that there is no definite rostrum. The nasal roof and wall (fig. 9, *al.n.*) are well developed, especially the latter, which is an unusually large, wide, ear-shaped band. The labials (*u.l¹.u.l².*) are normal, and are larger than in *Pelodryas*.

The palato-suspensorial arch diverges outwards to the postorbital region: it is of great breadth throughout, and so are the bones applied to it; these, however, do not greatly affect the cartilaginous pith, which is nowhere obliterated.

The palatines (fig. 8, *pa.*) are wide apart, are sub-falcate, and have a dilated outer end, which is notched in front. The pterygoid (*pg.*) fails, by a third of the length of the jugum, to reach the palatine; it clamps the under side and inner edge of the cartilage, arches suddenly round to bind on the pedicle, and turns inwards to bind the quadrate, making an elegant sigmoid curve. The pedicle, as in *Pelodryas*, is very large, and its thick condyle is obliquely external to the ear-mass, with which it articulates.

The almost transverse direction of the arch at this part, and the inbinding of the quadrate, also contracts the space for the Eustachian passage. The stylo-hyal (*st.h.*) has coalesced, contrary to rule, with the outer angle of the condyle of the pedicle (fig. 8, *e.pd.*), and arising as a hook bent forwards, this boundary bar also helps to lessen the passage, which is crescentic, with its convex margin looking forwards and outwards.

Here, owing to the depth of the retreating quadrate, the pier of the mandible—after the absorption of its dorsal end, half-way across the wide postorbital region—is still equal in length to its pterygoid outgrowth (fig. 9, *pg., q.*). Above the reniform condyle (*q.c.*), which is less than in the last kind, there is a considerable tract of bone.

The annulus (*a.ty.*) is more than equal in size to that of *Pelodryas*, and its fore horn is better developed, for it has formed a lobe above which almost touches the lesser horn; the front of the tegmen, clamped by the squamosal (*t.ty., sq.*), being thrust outwards, has made the position of the annulus very oblique (fig. 9).

The epi-hyal element (figs. 9 and 10), although specialised for a new function, is well-nigh as long, relatively, as in the Teleostei, and much longer than in the *Rays*. Looking at it from a simply morphological point of view, we see in it both the pharyngo-hyal of the Chimæra, as well as the epi-hyal proper, and also the segmented "symplectic" of the Sturgeon and the Paddle-fish. The top of this three-jointed epi-hyal fits into a most remarkable periotic piece—the stapes (*st.*). This part of the paraneural pouch is sub-quadrate, bilobate behind, and cupped, antero-externally, with a thick crescentic half-rim to the hollowed part.

The semi-osseous inter-stapedial (*i.st.*) is bulbous above to fit into this hollow, and then narrows to articulate with the slender terete sub-arcuate medio-stapedial (*m.st.*), which has its distal end soft. That unossified knob is nearly cut off from the extra-stapedial or "symplectic" (*e.st.*) by a segmenting notch. This distal part is compressed into a high supra-stapedial angle which gives rise to a mere ligament (*s.st.*); then the bar flattens in the other direction to form the normal spatulate manubrium.

Here the apex of the lower element of the hyoid arch (stylo-hyal) has coalesced, not, as usual, with the floor of the tympanum, but with the stunted apex of the *upper element* (pedicle) of the arch next in front of it; a very rare state of things.*

* In my earlier researches into the development of the Batrachian skull ("Frog's Skull," Phil. Trans. 1871, Plate 7, fig. 4, *sy.*), I supposed that part of the hyoid to have become confluent with the mandibular pier; that this never takes place in the Common Frog, Professor HUXLEY showed me, and I soon found the truth of the matter. Here, however, is an instance of this sort of thing, and in the larger and lower

This bar is very slender all the way to the basal plate, but it has a small hypo-hyal lobe (Plate 15, fig. 9, *e.hy.*, *h.hy.*); the front open space between the hypo-hyals is less than the basal plate, which has no front lateral lobes, has sharp hinder lobes, and normal thyro-hyals (*t.hy.*).

The mandible (fig. 9) is long, sinuous, and very similar to that of *Pelodrytes*, having a large mento-Meckelian and a low coronoid process (*m.mk., cr.*).

The investing bones are very similar, on the whole, to those of the last kind; the fronto-parietals (*f.p.*) scarcely meet behind; they have the same divergence forwards, and cover a little more of the fontanelle.

The nasals (*n.*) are twice as large; they have a greater surface to cover, and they do this better; yet much cartilage is left naked; their form is normal. The premaxillaries (*px.*) are almost transversely placed across the narrowed truncated snout; both their nasal and palatine processes are normal. The maxillaries are of the average strength and size, the ascending facial plate is enlarged and rough under the point of the nasal. Under the nostril there is a rough suborbicular septo-maxillary (fig. 9, *s.mx.*); it is notched in front.

The quadrato-jugal (*q.j.*) is slender, but it is well united with the quadrate. The squamosal (figs. 7, 9, 11, *sq.*) is a large T-shaped bone, with its upper bar diverging forwards, and scooped outside for the annulus; its descending or main bar is unusually long, and of a good width. The parasphenoid (*pa.s.*) is only two-fifths the length of the skull, but its main part is broad; it is but little convex (fig. 9); its processes are all pointed, and the basi-temporals are slender. The vomers (*v.*) are much less than in the last, and yet not small; their form is very peculiar. They are a wide distance apart, and diverge rapidly forwards; their main part is a long oblong, and ends behind in a transverse toothed part, which has a triangular elevation. There is no post-narial spike, but the pre-narial spike is forked, and the proximal part of this external process is almost notched off from the body of the bone; the outer fork is the longer of the two. The shorter hind spike would appear to be the homologue of the ordinary post-narial process, and the whole bone is evidently *dislocated forwards* through the hypertrophy of the trabecular nasal floor (*s.n.l.*).

Long as these vomers are, they are half their own length both from the front of the snout and from the fore edge of the girdle-bone.

The dentary (fig. 9, *d.*) is slender, and ends in the middle of the ramus.

Here we have an assemblage of characters which, in the aggregate, put this type far away from the " norma," and yet most of these differences are in reality gentle modifications of a *Ranine* skull.

1. The cranial cavity is both short and shallow.
2. The auditory capsules are small, but have a huge tegminal parotic growth.

Urodeles (Menopoma, Siren, Cryptobranchus, &c.) the epi-hyal becomes fused with the quadrate. In *Proteus* it is large, very similar to that of a Shark, and remains free.

3. They are ossified thoroughly, and the occiput, with its small condyles, is one with them.

4. There is only one fontanelle.

5. The ethmo-nasal region is as large as the cranio-auditory.

6. There are *distinct* superorbitals, as well as the projecting "eave."

7. The palato-suspensorials are very wide and are curiously elbowed behind, have very solid condyles (of pedicle and quadrate); the latter is partly ossified.

8. The annulus is very large, but its lobate front horn does not meet the lesser horn.

9. The stapes is sub-quadrate and cupped externally.

10. The columella is composed of three segments, all sub-equal in length, with no supra-stapedial.

11. The stylo-hyal is confluent with the pedicle of the mandibular pier.

12. The basi-hyo-branchial plate is nearly void of lobes.

13. The fronto-parietals are arrested, and leave the fontanelle bare.

14. The vomers are multiradiate, are almost divided into two bones on each side, and are displaced forwards.

These are most of the peculiarities; there are others, namely, the large optic fenestræ; the dilated palatines; the converging angles of the nasal floor, &c.; but I have enumerated enough to show that we have here the *Hyline* type of skull in its utmost perfection, and at the greatest degree of divergence from the skull of an average typical *Ground-frog*.

1. B. *a.*—*Toothless* "Anura." *Without digital disks. Oxydactyle Toads.*

First Family. "BUFONIDÆ."

Typical Toads, with parotoids, and processes of the sacral vertebræ dilated. Ear perfect; toes webbed; skull generally strongly roofed; an open fontanelle, exceptionally, in *Bufo calamita*.

First genus. *Bufo.*

55. *Bufo pantherinus.*—Adult female; 4¼ inches long. Africa.

The skull of this species (Plate 35. figs. 1–4) may be taken as typical of what is to be seen in this sub-division of the Anura; it is short, high, and coarsely strong in its bony outworks; the length is only five-sixths of the breadth, and the zygomatic arches are very convex. The occipital condyles (*oc.c.*) are rather small, postero-inferior, and are separated only by a narrow concave tract of cartilage, so that they might almost be described as one bilobate condyle.

The quadrate condyles reach as far back as the hind edge of the fenestræ ovales, and the epiotic eminences are half way between the two pairs of hinges. An arcuate line touching all these points of the hind skull would form part of a very large circle, and this arc would be sub-equal to that formed by the cheek and jaw bones, right and left.

The muzzle is very wide, but its arc would form part of a circle only half the size of the one just supposed.

A semi-circle finishing the lateral outline would make the form into an ovoid.

A rather narrow tract of cartilage divides the right and left bony masses of the hind skull, but the prootic is confluent with the ex-occipital (*on., e.o.*). This common bony mass reaches beyond the horizontal canal, externally, and in front runs up to the optic fenestra (II.), enclosing the foramen ovale (V.). Most of the cartilaginous tegmen tympani (*t.ty.*) is covered by the large ear-shaped temporal plate of the squamosal (*sq.*); its rounded hind part, however, is naked.

The epiotic eminences (*ep.*) are wide, they are not covered by the roof bones, but much of the anterior, and the ampulla of the horizontal, canal, are covered, for the post-orbital process is broad, and unites by suture with the squamosal. The endocranium in the orbital region is much overlapped by the roof slabs; it is of nearly equal breadth before and behind, and is gently narrowed behind the middle.

The girdle-bone (*eth.*) takes up three-fifths of this territory, and does not finish its own wings; but it sends forwards a short me.lian and a pair of lateral bony growths under the nasal region (fig. 2, *eth., s.n.l.*); its margins are sinuous. The unossified wings of the ethmoid are segmented from the ascending process of the palatine cartilage (fig. 3, *e.pa.*); they are very solid, and in front are continuous with the broad subnasal laminæ (fig. 2, *s.n.l.*). These laminæ end in front in dilated horns; the inner cornua, or pro-rhinals (*p.rh.*), are conical, and turn inwards. The snout is transverse, gently emarginate, and moderately broad; from its angles a narrowish band of cartilage runs backwards on each side, these are confluent, behind, with the wings of the ethmoid: they are the arrested nasal roofs (*n.r.*); between them and the widened top of the septum (*s.n.*) there is a large crescentic fontanelle (or fenestra).

The first and second labials (*u.l¹.u.l².*) are well developed; the second nostril-valve is very large and solid. The palato-suspensorials are so modified as to form the type of a sub-group—the *Bufonine*, as distinguished from the *Ranine*. Primarily, as I have shown in a former paper ("Batrachia," Part II., Plate 54, p. 607), the development of these parts takes place as in *Rana*, but their metamorphosis is much modified. The bony plates here use up much of the cartilage, which remains as an ethmo-palatine (fig. 3, *e.pa.*) jointed off from the ethmoid above and free below, exactly, now, like that of a Skate or a Salamander. The pterygoid cartilage also (fig. 3, *py.*) is now, as in those types, quite free from the ethmo-palatine, and is, what we find in Selachians, in young Teleosteans, and in Urodeles—merely a *symplectic* process of the suspensorium of the lower jaw.

The palatine bone, also (*pa.*), is composed of two parts: the normal ectosteal plate, and a sub-distinct knife-like crest, which is very sharp and steep; the old remnant, undoubtedly, of a dentigerous plate, like the palatine bone of a Urodele.

The fore part of the pterygoid bone is a large trowel (fig. 1, *pg.*), and it has used up the fore half of the cartilage which is its endoskeletal correlate.

The hinder, forked part, whose forks form a right angle, clamps the pedicle and the quadrate pier; the former (*pd.*) is, by it, tied down to the skull, and would be found very small and not quite segmented from the basis-cranii. The latter (figs. 1–3, *q., q.c.*) is a strong retreating bar, with a very large trochlea below, and over this some bone derived from the quadrato-jugal (*q.j.*).

The annulus (*a.ty.*) is large, thick, strong in the rim, broad, and complete.

The mandible (fig. 3) answers to the strength of the upper face, and is quite normal.

The stapes (fig. 4, *st.*) is a thick, oblique mass of cartilage, projecting externally in a boss; it is hinged to the columella by a round tooth in the middle of its oblique scooped fore edge. The columella, without a proximal joint, is unossified at that part, which is nearly as long as that in front of the stapes; it is very oblique, is scooped, and has three teeth-like projections for its hinge in the stapes, so that these parts fit together like the valves of a "Lamellibranch." The shaft, altogether (*m.st.*), is very gnarled and irregular, with a gentle arch; its fore end is unossified and is cut off, by segmentation, from the extra-stapedial.

That part (*e.st.*) is spatulate, with a thin flange; this edge is notched off in front, and ends above in a small supra-stapedial (*s.st.*), that soon becomes a mere ligament.

The hyoid band is confluent above (fig. 2, *st.h.*); it is broadish, and widens out below before it turns backwards. Both the "notch" and the basal plate (Plate 38, fig. 5, *b.h.br.*) are large—both long and wide; it has large ear-shaped front, and small styloid hind, lobes; the thyro-hyals (*th.h.*) are strong, and well bent, upwards, as they embrace the larynx.

The investing bones are strong and thick, and scabrous externally; the fronto-parietals (*f.p.*) are both thick and wide; they have a square postorbital process, which articulates with the squamosal (*sq.*), are very wide behind, and become narrow from before, backwards. Their orbital part overlaps the endocranium, and doubles its width behind. The fronto-sagittal suture is perfect, the fronto-nasal suture almost transverse. The nasals (*n.*) have the normal form and the *Bufonine* solidity; they meet along the middle, and only leave the end of the snout, and the ethmoidal wings, uncovered.

The marginal bones (*pr., mx., q.j.*) are normal in form, but very strong and steep (fig. 3); the maxillary (*mx.*) has a wide palatine plate, which is widest where it articulates with the pterygoid.

There is a large septo-maxillary (fig. 3, below *e.n.*) beneath the nostril and inside the maxillary. The squamosal (*sq.*) has a large, rough, ear-shaped supratemporal portion, and a long, retreating, descending bar. The parasphenoid (figs. 2, 3, *pa.s.*) is large and well-formed; its fore part is attenuated for some distance, the wings are splintery and angular, the hind part broad, and the median part thickened at the cross.

The vomers (v.) are rather small and edentulous, but are spiked both before and behind the inner nostrils; their inner lobe is ear-shaped.

This average Toad's skull differs from an average Frog's skull in several particulars:
1. It is shorter, deeper, stouter, and altogether rougher; the endocranium is more solid, but scarcely more ossified, than in the Frog.
2. The nasal region has the roof deficient, and the pro-rhinals pointed and turned inwards.
3. The cartilaginous palatines are segmented, both from the ethmoidal wings, and the pterygoid cartilages.
4. The palatine bones are cultrate, and the pterygoids are very broad, and bind down and fix the joint of the pedicle.
5. The quadrate is partly ossified.
6. The columella has an imperfect supra- and no inter-stapedial.
7. The roof bones meet the squamosals over the temples.
8. None of the bones bear teeth.

56. *Bufo melanostictus.*—Male, half-grown; $2\frac{1}{4}$ inches long (old specimens measure 5 inches). India.

The skull of this young individual comes very near to that of an adult *B. vulgaris*, and the skulls of old specimens almost rival the very large kind next to be described, viz.: *B. agua*; they are, however, very variable as to the degree in which the bony ridges are developed, some having their skulls much more crested than others.

This is a very short, broad skull (Plate 35, figs. 7, 8), its length is only three-fourths of its breadth; the condyles of the quadrate (*q.c.*) reach as far back as the exit of the vagus nerve (X.). The occipital condyles (*oc.c.*) are large, infero-posterior, and are separated by a narrow notch, which is the end of the basi-occipital synchondrosis. The prootics and ex-occipitals (*pr.o., e.o.*) are confluent, and the bone half surrounds the optic fenestra (II.). The orbito-sphenoidal region (*o.s.*) is cartilaginous, and occupies two-fifths of the orbital region. The auditory capsules are ossified as far as to the outside of the horizontal canals; but the tegmen (*t.ty.*) is cartilaginous, as also is the floor of the vestibule (*vb.*). The girdle-bone (*eth.*) occupies its own (ethmoidal) region very exactly; the nasal territory (*n.r., s.n.*) is quite free from true bone. The form of the various parts is quite like what is seen in *B. pantherinus*. The palatine bones (*pa.*) are sub-arcuate, and on the left side there are two sub-equal bones.

The pterygoids (*py.*) bind down on the pedicle (*pd.*); the quadrate region is soft, the annulus (*a.ty.*) perfect, and the stylo-hyal (*st.h.*) confluent above. The stapes (figs. 9, 10, 11, *st.*) is oval and plano-convex; there is a distinct reniform interstapedial cartilage (*i.st.*); the medio-stapedial (*m.st.*) is a thick bony rod above, and is very narrow and decurved below. The cartilage proceeding from it has been almost segmented off, and is also very narrow; it soon dilates into a sub-orbicular convex extra-stapedial (*e.st.*), which gives off a supra-stapedial band (*s.st.*) that is confluent above.

The labials (*n.l¹.n.l².*), septo-maxillaries (*s.mx.*), and the marginal bones (*px., mx., q.j.*) are all like those of the last kind, but smoother and less solid in this young specimen. The ear-shaped top of the squamosal (*sq.*) does not reach the roof-bones, and is less broad in its postorbital region. The roof-bones (*f.p.*) are already anchylosed to the endocranium behind; the fronto-nasal suture is more triangular. The cerato-hyal (Plate 38, fig. 6, *c.hy.*) is broad in the middle without any sudden enlargement; the basal plate (*b.h.br.*) is narrower than in the last kind; the anterior lobes are broader, and the right hind lobe is ear-shaped also.

The differences here noticed are partly due to age, but some of the modifications would be found in the skull of an old individual; it is more *Ranine*, having a separate inter-stapedial piece.

57. *Bufo agua.*—Old female, 6½ inches; young do., 5 inches long. South America.

This large kind has a skull which is the *Bufonine* counterpart of that of the Bull-frog (*Rana pipiens*); they are the largest of the Order, or have scarcely any rivals, and in both the skull has much that is archaic or generalised. In the great Frog the skull is smooth and neat, and very narrow in the interorbital region (Plate 8); here the roughness and strength, and the breadth of the mid skull, are all exaggerated: the one is a caricature, so to speak, of a Frog's skull, and the other of a Toad's.

In this (Plate 36), as in the last, the length is only three-fourths of the breadth; here the condyles of the quadrate go as far back as the root of the occipital condyles, but they fall short, very much, of what is seen in the larger Frogs.

The ossification of the endocranium is much greater than what we have seen in the Ethiopian *B. pantherinus;* and here, with no pretence to the sub-Ganoid condition such as is seen in some Frogs, the investing bones are very extensive, solid, and coarse.

There is a great approach here to the triangular form in the general outline; this will be seen still more in the species yet to be described; the hinder margin shows all the projections to be almost flush with an imaginary transverse line drawn across the skull behind; the occipital condyles project beyond, the quadrate condyles reach, and the epiotic eminences just come short of, such a line. The occipital condyles (*oc.c.*) are sub-oval, *supero*-posterior, they are of moderate size, and are separated by a crescentic notch larger than themselves. The whole hind skull is one mass of bone up to the covered tegmen tympani; in a younger individual (fig. 4) this part is largely cartilaginous; the ossification reaches in front to the optic fenestra (II.). Only one-fourth of the interorbital region is unossified, and only the front half of the nasal; the girdle-bone takes up all its own region and half of the nasal and orbito-sphenoidal.

DEVELOPMENT OF THE SKULL IN THE BATRACHIA.

In the orbital region the endocranium forms scarcely more than a third of the expanse; the rest is due to the roof bones (*f.p.*); *they* increase in size from before, backwards; the cranial box, from behind, forwards. Behind, the endocranium is exposed very little; the edge of the occipital ring is bare, and so are the epiotic eminences, and the bony ear-capsules for a small space further outwards, and forwards. The paroccipital wings, ending in the "tegmina," are very much outspread, but they are roofed over and hidden by the huge top of the squamosals (*sq.*).

The endocranium is not wanting in height (fig. 3), its ethmoidal axillæ are large and shallow (fig. 2, *eth.*). The snout is very broad, but rounded; it is unusually steep (fig. 3). The winged portion of the girdle-bone is largely hidden, below, by the enormous palatine bones (fig. 2, *eth., pa.*). The subnasal laminæ (fig. 2) are wide, and are half occupied by the extended ossification of the ethmoid. Above, the nasals hide all but the end of the snout (fig. 1, *n.*); when these are removed it is seen that the roof-cartilages (*n.r.*) merely form a widish ring round each outer nostril (*e.n.*), and perhaps run a little along the side of the septum (fig. 1). These rimmed nostrils are large and wide apart, and the second labial (*u.l².*) forms a very large valve in front of the passage. The first labial (*u.l¹.*) is small and lenticular; it supports the first, lying inside the rod-shaped vertical nasal process of the premaxillary (fig. 3, *n.px.*).

The roof cartilage swells (fig. 1) out on each side of the bevelled end of the septum; below (fig. 2, *p.rh.*), the pro-rhinals are seen as inbent-spikes. The cartilage of the palatine and pterygoid regions is broken up into an ethmo-palatine and a pterygoid, the bones corresponding to these regions having devoured most of this sub-ocular arch.

The palatine bones (fig. 2, *pa.*), together, form a strong cross-bar, as thick and almost as long as the mandible (fig. 3). They are sigmoid, flattened within, where they meet and form a strong suture; they are dilated and convex, and, externally, become falcate. A bony ridge, sub-distinct, and half their length, grows from their middle—at an equal distance from each end; it is sharply serrated, and the *serræ* (fig. 7, *pa., pa'.*) might easily be mistaken for teeth; they are undoubtedly the atavistic marks of teeth.

The hooked end of the palatine slightly overlaps the corresponding pterygoid (*pg.*); this latter is an extraordinary bone, composed of three nearly equal rays, each of which strongly clamps the contiguous parts of the skull.

A small epipterygoidean rod of cartilage still lingers in the outer part of the front ray (fig. 3, *pg., sp.*); this runs to a point in the *fixed*, bony pedicle (*pd.*); below, it widens, and then is stopped by the bony growth of the quadrate (*q.*), borrowed from the quadrato-jugal (*q.j.*).

A rounded angle, less than a right angle, lies between the jugal part of the pterygoid and the bony pedicle (*pd.*). The former bar becomes a style in front, being aborted by the huge maxillary (*mx.*); it swells outwards against the space for the temporal muscle, and then becomes vertical, applying itself to the inner face of the quadrate, as it descends, growing also both backwards and outwards. The inner ray is a very massive and wide bar, it is sigmoid, bending backwards towards the oval Eustachian tube (*eu.*), and then growing directly inwards to articulate with the basitemporal bar of the parasphenoid (*pa.s.*), whose end is, contrary to rule, most extended at the hinder angle. These binding bars hide all but the hinder half of the vestibular floors below; they are sub-parallel with the palatine bars.

The condyle of the quadrate (*q.c.*) is a trochlea, grooved obliquely; the bar above it is largely ossified (fig. 3, *q.*); its height, from the otic process, which is covered by the squamosal (*sq.*), is very great for a Batrachian; this Toad has almost the deepest face of any of its Order.

The mandible (fig. 3) is normal; none of its processes are large, but it is of considerable thickness, and only moderately ossified; the rod (*mk.*) is continued forwards undiminished from the reniform condyle which lies in the grooved articulare (*ar.*). The mento-Meckelian rod (*m.mk.*) is rather large, and the dentary (*d.*) is three-fifths the length of the entire ramus.

The "annulus" (fig. 4, *a.ty.*) is only of a medium size, and narrow; but it is complete. The stapes (figs. 4, 6, *st.*) is lozenge shaped, and has a boss for muscular attachment; it is followed by an equally large segment, the distal half of which is ossified; this is the inter-stapedial (*i.st.*). The next segment is the medio-stapedial bone (*m.st.*); it fits obliquely under the first, is slightly arcuate and slender, and ends in a thick spatulate cartilage, the extra-stapedial (*e.st.*), which sends upwards a supra-stapedial band (*s.st.*), that is confluent above.

Both the stylo- and hypo-hyal regions are absorbed, (figs. 2 and 5) and the cerato-hyal (fig. 5, *c.hy.*) is a sigmoid band, sharp at both ends.

The front notch of the basal plate (*b.h.br.*) is deep and the plate itself rather narrow, with an angular projection on each side, behind; the thyro-hyals (*t.hy.*) are large, bent, and divergent.

The edges of the basal plate are thickened, and run in front into sharp horns; the "anterior lateral lobes" are large, distinct cartilages, narrow in front, and dilated behind; the hind lobes are absent.

The roof is almost as complete as in *Calyptocephalus* (Plate 21), but it is formed of thick rough slabs, the large fronto-parietals, which are thrice the breadth of the top of the endocranium.

Their edges, like the edges of all the surface-bones, are very thick; they reach within a short distance of the foramen magnum, and send out on each side a square postorbital process to articulate with the corresponding squamosal (*sq.*).

The whole dorsal region of the skull is a shallow trough, through the raising of the edges; this is divided by the sutural line along the middle, which is complete from end to end. More than a third of this suture is nasal, the rest fronto-sagittal; the fronto-parietals project furthest at the middle, and the nasals form almost half the orbital rim.

Their pre-orbital process (figs. 2, 3, *n.*) is immense, and an ingrowth there would make them correspond to those of the Tortoise. Besides forming a wide foot to rest upon the steep maxillary (figs. 1 and 3, *n.,mx.*), the bone swells out to finish the front of the orbit; these bones are blunt over the wide snout, which lies beyond them; their antero-external edge is straight.

The palatine processes of the premaxillary (fig. 2, *px.*) grow well inwards near the mid-line; on the right side the outer angle forms a separate bone (*px'.*).

The nasal processes (fig. 3, *px.*) are thick "uprights," propping up the snout, and padded with the first labial (*a.l¹.*); together, these bones occupy a large part of the fore face.

The maxillaries (*mx.*) are high rough slabs; their lower margin is gently concave, and along it and along the dentary edge of the premaxillaries (fig. 3) there is a row of sharp denticles growing from the epidermic sheath; these *ataristic memorials* of teeth might easily be mistaken for the organs they so closely imitate. The upper margin of the maxillary is sinuous; it rises four times, in front, under the nasal, where the pterygoid joins it, and as it binds upon the quadrato-jugal. The palatine margin (fig. 2, *m.x.*) is well developed, and of equal breadth up to the fore edge of the cavity for the temporal muscle; the bone then lessens into a sharp curved style, which binds the quadrato-jugal up to the hinge—nearly. The latter bone (*q.j.*) is a flat style, and is continuous with the bony half of the quadrate (*q.*). In front, over the joining of the premaxillary and maxillary, there is a semi-circular septo-maxillary (fig. 3, *s.mx.*) obliquely set on to the latter, and raising it nearly to the height of the bone in front (*n.px.*). Behind this part there is a fenestra between the nasal pouch (fig. 3) and the concave edge of the maxillary.

The squamosal (*sq.*) is quite Batrachian, but it is exorbitantly large; it is a roughly cruciform bone, with a sigmoid stem, and a large over-lying temporal plate, which is lozenge-shaped (fig. 1). This oblique plate is thick in front, and unites by suture with the postorbital projection of the roof bone (*f.p.*); the scooped part lies like a scale on the parotic region. The postorbital process of this bone looks downwards, it is blunt and short; it only reaches half-way to the jugal crest; the pre-opercular portion is long, curved backwards, and carinate inside where it binds upon the quadrate.

The parasphenoid (figs. 2, 3, *pa.s.*) is only half as long as the basis cranii; and is very peculiar. The fore part, for a fourth of its length, is a narrow style; then the bone widens at once, and is notched on both sides. The hind part is triangular, and the side bars are oblong and widest behind, and articulate by a toothed suture to the pterygoids (*py.*).

The vomers (fig. 2, *v.*) stand, behind, against the huge palatines; they are sub-crescentic shells of moderate size, and are separated by a space more than equal to their own width. There is only a *pre*-narial snag; on the *left* side this is a distinct bone like the distinct palatine ossicle detached from the premaxillary on the *right* side (*px'.*).

In comparing this type of skull with the "norma," a good proportion of the discrepancies will be seen to be divergences, also, from the typical Bufonine skull, such as that of *B. vulgaris*, or the larger skull of *B. pantherinus*.

It differs from that of the typical Frog in the following particulars:—
1. It has no true teeth.
2. There are horny imitations of the Ranine teeth.
3. The whole skull is as massive and rough as the " norma " is light and smooth.
4. The endocranium is much more ossified, and is shorter and broader.
5. The nasal roofs are imperfect.
6. The pro-rhinals are small and inbent.
7. The palato-suspensorial arch is very large, and the pterygoids bind upon the pedicles, and fix them, and articulate with the parasphenoidal wings.
8. The massive palatines are united by suture at the mid-line and also bear a serrated carinate crest; the core of cartilage is divided.
9. The quadrate is largely ossified.
10. The supra-stapedial is confluent above, and the inter-stapedial is largely ossified.
11. The stylo- and hypo-hyals are absorbed.
12. The fore lobes of the basal plate are detached.
13. The premaxillary on the *right*, and the vomer on the *left*, side, have a supernumerary bone.
14. The whole skull is of great height, as compared with the depressed " norma."
15. The orbits are largely roofed over, and the auditory regions almost hidden, by the investing bones.

58. (A) *Bufo chilensis.*—Adult male; 3 inches long. Arequipa, Peru.

On the whole, the skull of this species is so much like that of *B. pantherinus* that I have not figured it; there are some points, however, in which it differs from it. The palatines (Plate 35, fig. 6, *pa.*) are not so much crested as in that kind, but each bony tract is composed of two pieces, obliquely overlapping one another. The outer piece on the left side is twice as large as the other; the converse of this takes place on the right side.

The stapes (fig. 5, *st.*) is more oval, and has an equally large boss; the columella is shorter, and more regular in form, there is no inter-stapedial segment, proximally, and the medio-stapedial is widely emarginate; its two oblique lobes are unossified.

The extra-stapedial (fig. 5, *e.st.*) is a broad spatula, and sends up a strong supra-stapedial band (*s.st.*), which is confluent, above.

58 (continued).—(B) Tadpole of *Bufo chilensis.*—Total length, $\frac{5}{6}$ inch; tail, $\frac{1}{2}$ inch; hind legs, $\frac{1}{20}$ inch. Arequipa, Peru.

This is the youngest of the "fry" of this large Toad examined by me; the recently metamorphosed individuals were no larger than the Common House Fly.

This skull (Plate 38, figs. 7, 8) is only two thirds the length, but the same breadth, as the one next to be described, viz.: the Tadpole of *B. lentiginosus*, whose legs were only a line long, and in which (see Plate 38, figs. 9, 10) the parasphenoid was beginning. There is a remarkable want of uniformity in the time as to which certain parts appear.

From the root of the pedicle (*pd.*) to the quadrate condyle (*q.*) we have the generalised counterpart of the *quadrate* of the Sauropsida; the metapterygoid and quadrate regions (together) of the Osseous Fishes.

The free mandible (*mk.*) is only a short, thick, ray, with a notched condyloid tract, like that of the "ulna," ending in an angular process, like the "olecranon."

Another, much larger, cartilage is articulated to the side plate by a rounded condyle; the hollow for it is under the orbitar process (*or.p., hy.f.*): this is the stylo-cerato-hyal (*c.hy.*), it is a flat phalangiform piece, enlarged both proximally and distally, and united to its fellow by *simple cartilage*.

The succeeding arches (branchial) have all been removed, and will not be described. Over the curling, pointed, horns of the trabeculæ there is a pair of semi-lunar cartilages—notched behind; these are the upper labials (*u.l.*) not divided into two pairs.

Below, between the mandibles, there is another pair (*l.l.*), arranged crescentically; they unite with each other in the middle, and are scooped, above.

Above, and below, these labials are covered with the serrated horny plates that form the primary dental apparatus or "odontophore" of the Tadpole, which vanishes away, or is moulted off, during metamorphosis.

This skull is unlike enough to that of the adult Toad, and would be even if no investing bones appeared, nor any bony tracts in the endocranium; the passage of the larval into the permanent skull has, however, in this group, already been described. I shall now compare this and the next together.

59. (A) Tadpole of *Bufo lentiginosus*.—$\frac{3}{8}$ inch long; hind legs, $\frac{1}{12}$ inch long. Penekese Island, Mass., U.S.

The skull of this Tadpole (Plate 38, figs. 9, 10) (which was the same length as the last) is nearly a third longer, but very little wider; its breadth is only three-fourths of its length, and not nearly equal to it, as in the last kind.

The difference between this skull and the last, and of both of them from that of *B. vulgaris*, is very remarkable, showing that the variations in the adult do not arise out of a uniform larval "model;" the *species*, even, begin to vary as soon as they are hatched.

In this specimen the hind legs are apparent, the Tadpole is only $\frac{3}{8}$ of an inch long, and both this and the last are smaller larvæ than those of the Common Toad.

In *Pseudis* (A), as we have seen, the legs have relatively less development, but the skull is largely ossified, and as large as a *crown-piece*, the whole length of the Tadpole being $10\frac{1}{2}$ inches, and its tail 4 inches across (Plate 1, fig. 1). Here, the skull is the size of a mustard-seed, and is no larger directly after it is metamorphosed. I mention these things to show how sensitive the Batrachians are to their surroundings; how much their transformations are modified by the chances of their life; and also, how modifiable these creatures are—*Protean* in their changeableness.

The skull at this stage[*] (Plate 38, figs. 9, 10) is truly *Petromyzine*, yet it has a modification of the cranial structures not seen in the Lamprey; viz.: the "orbitar processes." These parts are free above; but in the Tadpole of *Bufo vulgaris* (Phil. Trans., 1876, Plate 55, fig. 3, between *q.* and *eth.*) they are confluent with the ethmoid, as Professor HUXLEY pointed out to me; suggesting, at the same time, that they might be the homologues of the anterior crus of the Lamprey's suspensorium: we neither of us hold this view, now.

The chondrocranium of the larval *Bufo chilensis* (figs. 7, 8) is almost circular; the whole length, including the labials, is but little greater than the breadth across the subocular arches (*sp.*); here the skull is oblong.

The chondrification is nearly perfect, but the lines of union of the various elements are all visible, being made up of younger, more elongated, and crowded, cells. From the occipital condyles (*oc.c.*) to the internal nostrils (*i.n.*), the basis-cranii is of nearly uniform size; it is then lessened by a notch, on each side, half the size of these passages, and gains this breadth again, gradually. The hinder part is, as it were, cut away for the ovoidal ear-capsules (*au.*); the interorbital part is straight-sided in the last and pinched in this, and the fore-part is in two diverging bands, with a deep notch between them.

Behind the orbital space, the pedicle of the suspensorium (*sp.*) passes into the basal plate; in front, the ethmo-palatine bands run into it; behind the ear-capsules the basal moieties curve round, to enclose the 9th and 10th nerves (IX., X.). From the middle of the inter-auditory space, to the great notch in front, there is a spindle-shaped space of *new* cartilage, leaving the thicker, and older, marginal bands of the same size as the free bands in front; these latter are curled over, in the frontal wall, pointed, externally, in the last, and blunt in this.

The basal tracts are as follows :—

The investing mass.—This reaches to the middle of the hind skull, and is separated by the cranial notochord (*nc.*), which is shrinking, and is invested with a thin (mesoblastic) layer of long cells, scarcely cartilaginous. The rest of the paired cartilages are due to the rapid growth of the trabeculæ (*tr.*), and the free fore parts are the "cornua" (*c.tr.*). The spindle-shaped, thin, new tract along the middle is the first part of the "intertrabecula" (see fig. 8, *i.tr.*); it is very long and wide.

A second region of this element is seen from above, in the last (fig. 7), as a short, thick wall of cartilage, the rudiment of the perpendicular ethmoid (*p.e.*).

The internasal or fore part of this element is membranous at present in both

---
[*] I have already described the earliest state of the Batrachian chondrocranium (see p. 16) in that of *Bufo vulgaris*.

species; it becomes the nasal septum afterwards, the space between the cornua becoming filled up, and often a projecting prenasal spike grows forwards from it. Over the foramen magnum, the occipital ring is completed by the investing mass in the last, but not in this. On each side, save where the ear-balls are intruded, the trabeculæ have developed a steep wall, with the rudiment of a roof ("tegmen cranii"); the two rudiments run up to the short ethmoidal wall.

Thus there is left one large fontanelle (*fo.*), oblong for the most part, semicircular behind, and more pointed in front in the last, and unfinished behind in this. The nasal capsules are not cartilaginous yet; the eye-balls have been removed; the ear-capsules (fig. 9, *au.*) are fused partially with the chondrocranium; they are losing the simple ovoidal form, and taking a shape in which the curves and swellings of the canals (fig. 9, *a.s.c.*, *h.s.c.*, *p.s.c.*) are seen.

Below (fig. 10, *f.o.*), there has been dehiscence of the fruit-like capsule, and the spindle-shaped space is the new fenestra ovalis; its occluding (indifferent) tissue will be the stapes.

On each side of the orbits, in front, an oblique squarish lobe of cartilage grows out of the cranial wall: these two processes are the "wings" of the ethmoid (*al.e.*), and answer to the "antorbital" of Birds and the "pars plana" of Mammals; close to these we see the rudiment of the post-palatine (*pt.pa.*). Under these a band passes outwards—the pterygo-palatine, and in front of the ear-capsules a similar, but longer, band passes outwards to join (or become) the large facial bar or suspensorium (*sp.*); this is two-thirds the size of the whole basis cranii.

This outer plate runs forwards, curving round the inner nostril (*i.n.*), and ends opposite the middle of the cornu trabeculæ (*c.tr.*); each band ends in an oblong, emarginate, obliquely inturned condyle—the quadrate condyle (*q.*).

The hinder conjugating bar curves and twists round the front of the ear-sac; its root is the pedicle; its rounded elbow the "otic process."

Half the upper edge of the side plate is occupied by a large, sessile, decurrent leaf of cartilage, which turns inwards and just touches the ethmoidal wing (*or.p.*, *al.e.*). I cannot find that it has, as yet, become fused with it; under it, the conjugational band is the pterygo-palatine.

Here the basal plate shows signs of sub-division; the apices of the trabeculæ (*tr.*), from which the notochord (*nc.*) is retreating, are obliquely marked off from the newer investing mass behind, which has not yet finished the occipital condyles nor even the occipital ring. Hence the fontanelle (*fo.*) reaches from the foramen magnum to the ethmoidal region, where the intertrabecular tract is not evidently raised at the mid-line.

The auditory capsules (*au.*) are longer than in the last, and the interorbital region of the skull is narrower in the middle; the cornua trabeculæ (*c.tr.*) are not so pointed externally, nor so definitely decurved.

The part taken by the intertrabecula in the mid skull is not half as great as in the last; and this part shows, on its under surface, threads of new membrane-bone—the parasphenoid (fig. 10, *pa.s.*) is appearing.

The notches for the inner nostrils (*i.n.*) narrow the skull very much; and in this, as in the last, the quadrate cartilage (*q.*) shows a slight enlargement, which is attached by a membranous ligament to the point of the cornu (*c.tr.*) in front of the notch; the projection on the quadrate part of the suspensorium is the rudiment of the pre-palatine spike.

The rudimentary ethmoidal wing (*al.e.*) is a more angular projection at present, and is much smaller than in the last kind; the elevation on this band is the rudiment of the post-palatine bar (*pt.pa.*). The mandibles (*mk.*) are like those of the last, but not so massive; the same may be said of the lower labials (*l.l.*).

But the upper labials (*u.l¹.u.l².*) differ from those of the last Tadpole; they are formed as two on each side, at the first, and do not become two by segmentation. The two inner pairs, if melted together, as in the larva of *Rana pipiens*, would answer to the anterior dorsal cartilage of the Lamprey; the styliform outer cartilages answer exactly to the external pieces in that Fish.

Here the orbitar process (*or.p.*) is still further from the ethmoid than in the last, and yet it possibly may unite with it afterwards.

The hyoid bar (*c.hy.*) is narrower in its shaft, and dilates into two still more projecting angles distally; the hinder of these is the stylo-hyal rudiment, the large cells at the lower end form part of the basi-hyal tract. Part of the first "extra-branchial" (*ex.br¹.*) is shown in relation. In neither of these instances is there any rudiment as yet of an "epi-hyal" element.

A comparison of these crania with those of the larval Frogs, already described, will show to what degree the skull may differ in its initial (*Petromyzine*) stages. I now come to the description of the newly metamorphosed skull of this Toad, to be used as a standard of comparison in describing the skulls of the small, arrested *glandless* Toads ("Phryniscidæ," "Engystomidæ"). These small kinds will be found to have skulls arrested at various points that correspond with what is seen at the different periods of life in the higher and more developed kinds. They also show very instructive instances of a sort of relapse into *old ichthyic* conditions; they resume characters that have been suppressed in the more normal and better developed *Anura*.

59 (continued).—(B) Skull of *Bufo lentiginosus*.—Recently metamorphosed; ⅜ inch long.

In the skull of the young Toad of the first summer (Plate 39, figs. 7, 8, 9) the length and breadth are equal, and the quadrate condyles (*q.c.*) reach, as yet, no further back than the Eustachian openings (*eu.*). Up to that part the form of the skull is a very neat semi-ellipse, and behind the round edges of the tegmen tympani (right and left) (*t.ty.*), the epiotic eminences (*p.s.c.*), and the occipital condyles (*oc.c.*), all may be said

DEVELOPMENT OF THE SKULL IN THE BATRACHIA.   211

to project from an arc of a large circle; thus the outline of the whole is like that of a short egg.

The pattern of a Batrachian chondrocranium is here perfect; and this has, already, been modified by two pairs of bony centres that are enclosing the 9th and 10th nerves, behind (IX., X.), and the 5th and 7th (V.) in front of the ear-sacs (*au.*). The occipital condyles (*oc.c.*) are large, sub-reniform, near together, and postero-inferior.

The foramen magnum (*f.m.*) is very large; each ear-sac projects far outwards as a parotic process, whose outer edge is the tegmen tympani (*t.ty.*). The floor is perfect (fig. 8), the roof as complete as in the "norma," there being one large, and two small, oval fontanelles (*fo., fo'.*). Infero-laterally, there are the nerve passages, and that for the optic nerve (II.) is a large fenestra—not a mere foramen.

The ethmoidal wings pass as yet right into the antorbital bars (*e.pa.*), and then, again, into the pterygoid fore-growths of the quadrate cartilage.

The roof of the nose (fig. 7) is not clearly divided off from the septum (*s.n.*), and the floor (fig. 8) is about equal to it.

The prenasal end of the septum is a small bud, and the pro-rhinals (*p.rh.*) are very small points of cartilage that look inwards; the angles of the floor are sub-falcate.

The labials (*a.l'.a.l².*) have their permanent form and relations. The *f*-shaped palatine is applied to the ethmo-palatine bar, which is dilated in front, externally, as the adze-shaped pre-palatine. The post-palatine cartilage is dilated to a less degree and ends, regionally, but not by division yet, in front of the great opening for the temporal muscle. There the pterygoid region begins, which narrows, somewhat, and then dilates to make its forks; its bony plate (*pg.*) is already broad and Bufonine. Under the main foot of this bone the cartilaginous pedicle (*pd.*) is dilated, but does not lose its continuity (as far as I can find, here, certainly not in *B. vulgaris*) with the basis cranii.

The quadrate condyle (*q.c.*) is large and reniform; the Eustachian passages (*eu.*) are small; the stylo-hyal (*st.h.*) is uniting with the skull, and passes outwards under the projecting, ear-shaped, tegmen tympani. The stapes was not chondrified in the larva just described, but that process took place very soon afterwards; now, there is a columella (fig. 9, *co.*).

This new cartilage, which appears in the beginning of summer, is a thick oblique wedge, helping the solid stapes (*st.*) to fill the fenestra ovalis; it is arched above and concave below, and ends in front as a fine thread of long cells.[*]

The small "spiracular cartilage," which had undergone various fortunes in the larval state—sometimes fixed at one end and sometimes at the other, and then free—is now

---

[*] This is the state in which this organ was shown to be in young specimens of *B. vulgaris* by Professor HUXLEY in the early summer of 1874. My own dissection of this stage is shown in my second paper on the "Batrachian Skull" (Phil. Trans., 1876, Plate 55, fig. 8, *co.*); it is somewhat more advanced than this which I am now describing in another species.

2 E 2

a small U-shaped annulus (*a.ty.*), over which the subcutaneous stroma is becoming arranged radially, and towards which the pointed end of the new columella is growing.

In *Pelobates fuscus* (Plate 25) such a rudiment ossifies throughout; it stops growth in a form soon to be described, viz.: *Rhinoderma* (Plate 39, fig. 5, *co.*), it is arrested at this identical stage and does not ossify; whilst in *Pseudophryne*, some kinds of *Phryniscus*, and in *Bombinator igneus* it never appears; in these kinds we have the counterpart of the newly metamorphosed Common Toad and Frog, and of many of the adult "Urodeles."

The mandible in this stage (Plate 38, fig. 4) is formed, the mento-Meckelian (*m.mk.*) is forming, and the whole bar is very similar to what it will be in the adult.

The investing bones are already well developed, but they are very thin shells at present, and do not cover the endocranium nearly as much as they will do; I shall refer, again, to these structures in describing the skull of cognate, but pigmy, or arrested Toads, of the Families "Rhinodermatidæ," "Phryniscidæ," "Engystomidæ," and "Brachycephalidæ."

60. *Bufo vulgaris.*—Adult. England.

In describing the Bufonine skull I shall have to refer again and again to my published description both of that of the adult and larva (see Phil. Trans., 1876, Plates 54 and 55, pp. 605–625).[*]

The skull of the adult Common Toad is short and wide, the length is only five-sixths of the breadth.

In the degree of its ossification, and in the possession of three fontanelles, it agrees with the Common Frog; but it is shorter and altogether a coarser and stronger skull; and there are no teeth, either on the vomers or the jaw-bones. By referring to my published figures the reader will see that, besides these general differences, there are several very important morphological modifications in the skull of this species; these are as follows:—

1. The nasal roof is a narrow, jagged cartilage, confluent with the fore end of the septum, in front, and with the ethmoidal wings behind, but separated by a large "olfactory fenestra" from the main part of the septum.

2. The palatine cartilage is a T-shaped, distinct piece, segmented from the ethmoidal wing, above, and from the pterygoid cartilage, behind.

3. The pterygoid bone ties down the pedicle, which has not lost its dorsal portion, and thus has never become free.

4. The pro-rhinals are very small and turned inwards.

5. The "annulus" has a short posterior horn, which does not unite with the anterior horn.

---

[*] In the figures the outer labial is lettered *u.l*[1]. instead of *u.l*[2].; and the deficient nasal roof is lettered *u.l*[2]. instead of *u.r.*; the description of these parts is erroneous, as I have discovered since, for I failed then to find the first upper labial, and mistook the nasal roof for the second.

6. Both supra-stapedial and stylo-hyal are confluent.
7. The medio-stapedial is a very short bone and the inter-stapedial takes up most of the rod.
8. The extra-stapedial is peltate.

61. *Bufo calamita.*—Adult female; 2¾ inches long. England.

This skull (Plate 40, figs. 1–5) is, on the whole, about as strong as that of *B. vulgaris*, but it differs greatly from it in several things, notably in having what is quite exceptional in the genus, viz.: an open fontanelle (*fo.*). The length is to the greatest breadth as 6 is to 7; the condyles for the lower jaw reach very little further back than the condyles of the pedicles; the outline of the face is a very exact semi-oval. There is in every part a remarkable Bufonine coarseness of structure, in spite of the arrest of the roof bones: the ossification of the endocranium is normal, as in *B. vulgaris*.

The occipital condyles (Plate 40, figs. 1, 2, *oc.c.*) are large, postero-inferior, and separated by a shallow notch less than half their own width. I can find no secondary fontanelle: the main space (fig. 1, *fo.*) is oval, and is half the length, and more than half the width of the roof, most of which is uncovered. The upper and lower synchondroses (*f.m.*) are wide, and widen rapidly so as to become large cruciform tracts between the prootics and ex-occipitals (*pr.o., e.o.*). The latter merely form a flange to the epiotic eminence, above (*au.*); below, they well enclose the double foramen for the 9th and 10th nerves (IX., X.).

The prootics reach to the facet for the pedicle, right and left, and to the optic fenestra in front (fig. 2, II.), quite enclosing the hole for the trigeminal (V.).

Above (fig. 1, *pr.o.*), they reach far over the capsule, nearly up to the squamosal (*sq.*), but the narrow tegmen tympani and the whole of the eminence over the posterior canal are unossified; they are confluent with the temporal wing of the roof-bones (*f.p.*). The mid-skull narrows steadily up to the ethmoidal ala and axilla; it is half cartilage and half bone, for the girdle-bone (*eth.*) has its normal development, affecting the septum nasi a little, above, not quite reaching it below, and only partly ossifying the alæ. The whole of the nasal territory is cartilaginous; the floor is wide, its angles in front are not large, yet the pro-rhinals (fig. 2, *p.rh.*) are unusually large, but straight, as in most Toads; there is no prenasal rostrum. The roof (fig. 1, *n.r.*) is very narrow, and I cannot find the band which, at a distance from the septum, runs back to the ethmo-palatine in most species of *Bufo*; here the true nasal "paranoural" cartilage is mainly confined to the fore part of the snout. The oblique external nostrils (fig. 1, *e.n.*) are well defended by the second labial (*u.l².*); the first (*u.l¹.*) is a mere pad inside the nasal process of the premaxillary; the internal nostrils (*i.n.*) are round, and not much wider apart than the outer.

The palato-suspensorial structures are large and strong, but the whole arch is continuous, above, with the ethmoid, and, behind, the dilated post-palatine is not segmented from the pterygoid cartilage as it is in other species of this genus.

The whole of the cartilage of the palatine region is a large L-shaped band, it narrows at its starting from the ethmoidal wing (*e.pa.*), dilates into the pre-palatine, and forms a rounded projection, behind, where it passes into the pterygoid band (*pg.*). The palatine bone (*pa., pa'.*) is composed of two pieces, imperfectly soldered together; the main piece is the ectosteal plate, the normal *Anurous* bone, it is large and falcate. The superficial bone is one-fourth the size of the other; it is an irregular rod, with a *distinct suture* in front, and lies somewhat obliquely on the middle of the fore edge of the main bone. This is the true counterpart of the dentigerous palatine "parostosis" of the Urodeles, and belongs to the same category as the vomers. The pterygoid bone (*pg.*) is very large, falcate, and two-membered, behind; contrary to rule it overlaps the pre-palatine, *above* (fig. 1), where, abnormally, it is more developed than below, and throws the remains of the unused cartilage on the ventral aspect, instead of the dorsal. Also, unlike the other species, it has a large, *free, Ranine* pedicle, the pterygoid only ossifying the surface of the thick stump of the original pedicle, and not binding it down; hence the joint is free; it is an oblong condyle moving in a shallow glenoid cavity.

Below the confluent otic process, hidden by the squamosal (fig. 1, *sq.*), the outer member of the suspensorium forms the partially retreated quadrate condyle (*q.c.*); this is very oblique, long, reniform, and has the quadrato-jugal a little confluent with the cartilage of which it is the base. Between these short thick forks we have the large semi-oval Eustachian opening (*eu.*), the hinder boundary of which is formed by the large confluent stylo-hyal (*st.h.*).

The "annulus" (*a.ty.*) is three-fourths the average size, and like that of *B. vulgaris*, is widely open above, and not a perfect ring as in most of the species of this genus.

The stapes (fig. 5, *st.*) is large, thick, umbonate, and notched in front; the medio-stapedial (*m.st.*) has an upper rounded, and a lower lip-shaped, cartilaginous process, but no separate inter-stapedial. The shaft is twisted and carinate in front, and is bent outwards to reach the outer edge of the suspensorium; I do not see here, what in old specimens of *B. vulgaris* can be plainly seen, namely, a *partial segmentation* of the bone at this outward bend. The extra-stapedial (*e.st.*) is sub-peltate, and the supra-stapedial band (*s.st.*) is perfect and confluent, above.

The mandible (fig. 3) is normal, but the articular bone (*ar.*) is feebly developed; the condyle (*ar.c.*) is long and reniform, and from it the rod (*mk.*) is scarcely lessened. The dentary (*d.*) is three-fifths the length of the ramus, and the mento-Meckelian (*m.mk.*) has only ossified half of the inferior labial.

The cerato-hyal (fig. 4, *c.hy.*) is broad, and doubles its width in the lower half; there is a small hypo-hyal lobe, and the notch of the basal plate (*b.h.br.*) is deep. The plate itself is of good width and of great length; the fore lobes are large and ear-shaped, the hind lobes ligulate and uncinate, inwards.

The thyro-hyals are long and straddling, with much cartilage at the end, and on the *left* side there is a round nucleus of cartilage (*t.hy*.) outside the bone, proximally. Just behind the front lobes a band of endostosis runs across the plate—a faint attempt at the formation of a bony " 1st basi-branchial."

The investing bones are coarse and strong but limited in their superficial extension.

The fronto-parietals (fig. 1, *f.p.*) are roughly radiated, grooved externally, and pitted in the hind part, where they very imperfectly cover the hind skull.

Their oblique temporal groove is inside the combined arches of the anterior and posterior canals (fig. 1, *au*.), and their ragged expansions are anchylosed to the prootics within. They are mere thick bars on each side of the naked fontanelle, and overlap the girdle-bone (*eth.*) in front. The nasals (*n.*) are large, long-handled below, and do not quite meet over the septum-nasi.

The premaxillaries (*pr.*) are gently arcuate, of great extent, and have well developed nasal and palatine regions.

The maxillaries (*mx.*) are also well developed, with a considerable palatine edge that broadens behind before it gives off its jugal process; there is between these bones a small angular septo-maxillary (*s.mx.*). The quadrato-jugal (*q.j.*) is a short curved spike, partly confluent with the quadrate (*q.*).

The squamosal (*sq.*) has a rough, short upper, and a gently expanding lower, limb. The parasphenoid (*pa.s.*) has all its processes well developed, it is quite *Bufonine*, narrowing rapidly in front, and with very extended lateral processes that run to a point; the handle is large and notched behind; and the opposite process, under the bone at its basi-temporal part, is an extended oblique toothed ridge.

The vomers (fig. 2, *v.*) are small, with a prenarial hook, a hooked fore part, and a notched hinder process.

In this skull the Bufonine characters are arrested and modified: this is especially seen in the deficiency of the roof bones, in the continuousness of the palato-suspensorial cartilages, and in the perfect freedom of the pedicles. It, therefore, is not so different from a Frog's skull as that of its congeners.

From the "norma" it differs in—

1. The general strength and coarseness of its structure.
2. There are no secondary fontanelles.
3. The main fontanelle is uncovered.
4. There are no dentigerous bones.
5. There is a superficial ("parosteal") palatine.
6. The pterygoid is most developed below, and not above.
7. The mandible is unusually cartilaginous.
8. There is no inter-stapedial; and the supra-stapedial and stylo-hyal are confluent above.

62. *Bufo ornatus.*—Adult female; $2\frac{1}{9}$ inches long. South America.

This species should be put into the next genus, viz.: *Otilophus*; the difference between the skull of this and of a young *O. margaritifer* is very much less than what may be seen between several kinds of *Bufo*.

The general form of the skull is that of a triangle with the apex off; the length is nine-tenths as great as the breadth.

Some of the characters of *B. agua* are exaggerated in this; although its size is only a third of that large species.

That which especially strikes the eye in this skull, and in the exaggerated form seen in the next genus, is the projection of the square snout forwards over the marginal bones of the face.

This form of the nasal region occurs amongst the Ganoids in the "Palæoniscidæ" (TRAQUAIR, Trans. Palæont. Soc., 1877, plate 1, figs. 1, 2, 11), and is exactly repeated in the half-ripe embryo of the Pig (Phil. Trans. 1874, Plate 34, figs. 1, 2, 6). In this, as in many other things, the Anura prefigure the Mammalia.

The occipital condyles (Plate 37, figs. 1-3) are large postero-inferior, and reniform; they are separated by a shallow notch less than their own width. The epiotic eminences (fig. 1, *e.o.*; fig. 3, *ep.*) are almost flush with the occipital condyles; the quadrate condyles (*q.c.*) only reach back to the inter-stapedial, the supra-temporal plate (*sq.*) as far as to the middle of the stapes (*st.*). This top of the squamosal, growing so far backwards, gives the squareness and breadth to the outline of the skull, behind. Like that of many of the Caducibranchiate Urodeles, the endocranium is almost entirely ossified into one continuous bony box: there may be a little cartilage at the tegmen tympani; there is a tract in front, for the fore half of the nasal region is soft. The parotic parts are large but do not reach out beyond the middle of their squamosal roof. The rim of the occipital arch, the superoccipital region, and three-fourths of the ear-capsules are naked above. Also a small lozenge of the ethmoid (fig. 1, *eth.*) is seen in front of the frontals, and the cartilaginous snout is bare at its fore margin. Through the broad roof (fig. 1, *f.p.*) the three fontanelles shine: the first is large and heart-shaped, the two small spaces are sub-oval, and lie at a moderate distance behind the emargination of the main space.

Below, as in *B. agua*, the outline of the endocranium is very irregularly sinuous; it is very broad behind, then lessens up to the axillæ and also of the ethmoid (*eth.*), where it expands suddenly; in front, the bone runs up to the points of the palatine processes of the premaxillaries (fig. 2, *pv.*).

The outline of the fur overhanging roof is sub-parallel with the edge of the endo-cranium, being dominated by it. Each "eave" is more than half the width of the cranial chamber. Seen from the side (fig. 3), this chamber is extremely shallow—as in other Neotropical Anura—much more so, relatively, than in *B. agua*. The ethmoidal wings are ossified to their proper end, where, in the genus *Bufo*, the segmentation

DEVELOPMENT OF THE SKULL IN THE BATRACHIA.   217

of the antorbital bar takes place. The passages for the emerging nasal nerves are large, and lie on each side the square end of the far-extended ethmoid bone. They lie close behind the facial edge of the snout, and on their outside the pro-rhinals spring ; these are hidden by the premaxillaries, and are small.

The sinuously-transverse snout overhangs this part obliquely ; the large external nostrils (*e.n.*) are in its sides, and are wide apart. The roof is feebly developed, but there is a well-formed pouch in the outer wall (figs. 1, 3, *al.n.*, *n.w.**) ; in front of the nostril there is the large, oval, hollow second labial (*a.l².*) ; below it, inside the vertical nasal process of the premaxillary, the small, lenticular first labial (Plate 36, fig. 8, *u.l¹.*). The palato-suspensorial cartilage is broken up into an adze-shaped ethmo-palatine, to the under surface of which the large falcate bone (*pa.*) is grafted ; whilst the suspensorium has a remnant of the pterygoid cartilage.

The pedicle (*pd.*) is completely covered with bone, and its original apex has been lost in the ossifying cartilage close to the foramen ovale (V.). The cartilage breaks out from beneath the inner foot-like ray of the pterygoid (*py.*), which has used up nearly all the front process that was originally continuous with the palatine cartilage (fig. 3, *sp.*).

The suspensorium, itself, or quadrate region (*q.*), is very large, and remains unossified (figs. 2, 3). The condyle (*q.c.*) is oblique, and reniform ; and above it a shell-like flange of cartilage grows from its outer edge, clamping the hind margin of the long, descending bar of the squamosal ; this crest of cartilage is the old "orbitar process," (fig. 3, *or.p.*) once under the antorbital, and now beneath the auditory, region. The stylo-hyal (*st.h.*), carried back by the retreating suspensorium, but loosened from it, now has grown inwards, and caught hold of the tympanic floor of the vestibule (*eb.*) on which it is grafted. The front fork of the pterygoid bone (*py.*) is a sharp style ; the hind bar is flat and vertical, binding the inner face of the quadrate ; between it and the large inner process (*pd.*) the Eustachian opening (*eu.*) is seen to be large and circular.

The articular region of the mandible (fig. 3) is hollowed out for the reniform trochlea of the quadrate (*mk.*) ; the cartilage is but little ossified by the articulare (*ar.*) ; the dentary (*d.*) is three-fifths the length of the ramus, and the mento-Meckelian (*m.mk.*) is of the normal size. The "annulus" (*a.ty.*) is large and perfect.

The stapes (Plate 36, fig. 10, *st.*) is large and sub-oval ; it has an ear-shaped boss and some ossified cells ; the inter-stapedial (*i.st.*) is oval and semi-osseous ; it fits in like a wedge between the stapes and the next segment—the medio-stapedial shaft (*m.st.*). This part is pistol-shaped and large, it is especially thick, proximally ; joined on to it is the broadly-spatulate extra-stapedial (*e.st.*) with its fastened, strap-like process, the supra-stapedial (*s.st.*).

The hyo-branchial structures (Plate 37, fig. 4) show that this *lesser Neotropical Toad* is approaching its small, "glandless" relatives of the same region, viz. : the "Phryniscidæ" (Plates 40 and 41), which of all the Anura have the narrowest basal plate. I shall return to this comparison when I come to that genus. Here the

* The *lower a.l².* in fig. 2 should be *n.w.*

cerato-hyals (fig. 4, *c.hy.*) gradually enlarge and contract again before they turn back into the non-lobate hypo-hyal bands. The anterior lateral lobes of the basal plate (*b.h.br*). are but little freed from the main sheet of cartilage; they end in a blunt point far in front of the deep, narrow notch. Then the plate is very narrow, and gives off two small hind lobes. The thyro-hyals (*t.hy.*) are rather short, bent strongly upwards to embrace the larynx, diverge at a right angle, and end in a blunt piece of cartilage.

The investing bones are intermediate between those of *B. agua* and *Otilophus* (Plates 36–38); belonging to a species so much smaller than the former, they are far less solid; their surface is smooth, except at their free sub-cristate edges, which are serrate, but not beaded and ornate as in *Otilophus*; on the whole, this skull is like that of either of the two kinds between which I have placed it, and links them together.

The broad fronto-parietals are (*f.p.*) more than double the width of the inner skull, they are separated by a fronto-sagittal suture, and by an interrupted transverse suture from the nasals (*n.*); they are lobate behind, exposing the superoccipital region, and articulate largely with the dilated squamosals behind the orbits. A lozenge of the girdle-bone (*eth.*) is seen between the nasals and frontals; the former (*n.*) are large, triangular shells with a broad facial plate (fig. 3); they are jagged and pointed in front, and almost reach the end of the snout. As in the embryo of the Pig, the premaxillaries (figs. 2, 3, *px.*) are quite under the snout, and the long nasal processes are tilted forwards; the palatine processes are small and sharp. The maxillaries (*mx.*) are deep, but their palatine edge is not extended far inwards; they reach nearly to the hinge of the jaw, and are overlapped by the small quadrato-jugal styles (*q.j.*) which keep distinct from the quadrate piers. The squamosals (*sq.*) are almost entirely like those of *B. agua*, but they have sharper and more serrated edges, especially along the temporal region; the descending part is small, long, and sigmoid. In front (Plate 37, fig. 3, and Plate 36, fig. 8), old generalised *ichthyic* characters break out; there is a large, shell-like septo-maxillary (*s.mx.*), and behind this a thin, sub-crescentic "pre-orbital" (*p.ob.*); it is perched upon the ascending inner lamina of the maxillary (*mx.*). The parasphenoid (fig. 2, *pa.s.*) is more normal than in *B. agua*, but its fore part is broader; the pterygoid does, however, overlap the basi-temporal wing, obliquely; there is a transverse crest between the two wings. The vomers (*v.*) are thin, toothless shells, semi-circular in form, but with *post*- and *pre*-narial spikes, and a lobulated fore part; they throw the inner nostrils (*i.n.*) very far apart; the outer (*e.n.*) are wide apart, but these are at nearly twice their distance.

This is a very instructive skull, looked at in its Bufonine aspects; as compared with the Anurous " norma " it is very remarkable; it differs as follows :—

1. In the triangular form of the skull; its breadth above, the height of the face, and the shallowness of the cranial cavity; also in the tendency to a crested and sculptured condition of the outer bones.

2. In the intense ossification of the endocranium, blotting out all landmarks, except the necessary foramina.

DEVELOPMENT OF THE SKULL IN THE BATRACHIA. 219

3. In the curious *porcine* form of the snout, broad and overhanging the facial bones.
4. In having no teeth, either marginal or submarginal.
5. In the fixity of the pedicle.
6. In the breaking up of the palato-suspensorial cartilage.
7. In the fixity of the stylo-hyal and the supra-stapedial, and in distinctness of the stapedial " boss."
8. In the narrowness of the basal plate, and the absence of hypo-hyal lobes.
9. In having a second *pre-orbital* besides the septo-maxillary.

Second genus. *Otilophus*.

63. (A) *Otilophus margaritifer*.—Half-grown female; 1½ inch long. Venezuela.

This skull is about one-fourth less than the last; it resembles it very much, but is feebly ossified (at present), has a wider roof, and more ornate and crested bones, externally. If this skull had been arrested at this stage there would have been no reason for the sub-generic distinction *Otilophus*, as distinct from *Bufo*. I agree with that distinction, but would put *B. ornatus* into the *sub-genus*.

We have here the same triangular broad roof, and high skull (Plate 37, figs. 5–7), as in the last, but the occipital region projects more, behind, and the snout is much narrower, and more rounded; moreover the investing bones are very ornate with small beads or pearls of clear bone. The figures (Plate 37, figs. 1–4 and 5–10) will give a clearer idea of the great likeness and small unlikeness of these two exquisite little skulls. The occipital condyles (*oc.c.*) are large, reniform, and postero-inferior; they are separated by a rounded notch half their own width, and they show, more than those of the last kind, that the motions of the head on the atlas are very free, and worked by strong muscles.

The ex-occipitals (*e.o.*) are less than half the size of their region; and both above and below are separated by wide tracts of cartilage. The floor of the vestibule (fig. 6, *vb.*) is naked cartilage; the top of the ear-sac is all covered, except the epiotic region (*ep.*), so that the prootics are only seen where they surround the foramina ovalia (V.). The prootic region is wide, unossified, and forms a large tegmen (*t.ty.*) under the squamosals (*sq.*). The large optic fenestra (II.) are surrounded by cartilage which occupies two-thirds of the orbital region (*o.s.*); this part is of almost equal width, but widens at both ends; its depth is very small (fig. 7), but it bulges in the middle. The limited girdle-bone (*eth.*) is complete below (fig. 6), and takes up its own proper cartilage: above (fig. 5), it only appears just where the roof-bones partly expose it as a circle of bone reaching the septum nasi in front, and the fontanelle behind, but it has much cartilage on each side. The main fontanelle is rather small, and elegantly heart-shaped; the secondary fontanelles are large and oval; the " tegmen cranii " is largely developed and the endocranium is rather massive.

The narrower and more rounded snout is extremely *porcine*; the outer nostrils are very large and not far apart; the large round inner nares (fig. 6, *i.n.*) are twice as far apart, but as the skull narrows in faster than in the last, these passages are nearer together than in it. The nasal roof (*p.n.*) is not large, and appears to be imperfect; the floor (*s.n.l.*) is very wide; the pro-rhinals (*p.rh.*) are small and turned inwards; and the angles of the floor are but little dilated, as the maxillaries (*m.x.*) stop their outward growth. The palatine bones are rather short; they are falcate, and hide the small ethmo-palatine cartilage, which cannot be traced into the half-used pterygoid cartilage, growing inside the pterygoid bone (fig. 7, *pg.*), which is part of the suspensorium (*sp.*). The bone (figs. 5-7, *pg.*) is very small and pointed in front, has a large foot lying on and fixing the pedicle (*pd.*); and bending suddenly on the inner process, the hind part half encloses the smallish, oval Eustachian opening (*eu.*), and then runs down the inside of the quadrate (*q.c.*). That part is somewhat ossified by the quadrato-jugal (*q.j.*); is very high (fig. 7) and forms more than a right angle with the basis cranii, but the condyle (*q.c.*), which is a large sulcate trochlea, only goes as far back as the setting on of the stylo-hyal. A flange-like out-growth of the suspensorium is all that remains of the leafy "orbitar process" (*or.p.*), it binds on the squamosal (*sq.*). As in the last, the "annulus" (*a.ty.*) is large and perfect; the stylo-hyal (*st.h.*) is confluent, above.

The stapes (figs. 9, 10, *st.*) is sub-oval, and has a boss; the mediostapedial (*m.st.*) has a large, cartilaginous lobe, but no free segment; the extra-stapedial (*e.st.*) is spatulate, and has a membranous supra-stapedial (*s.st.*).

The hyo-branchial structures (fig. 8) are similar to those of the last kind, but the cerato-hyals (*c.hy.*) are narrower; the lateral lobes are larger and freer; in the base of the hinder pair there is a fenestra; the thyro-hyals (*t.hy.*) are much smaller. The mandible (fig. 7) is like the last, but the bony tracts are feebler in this young individual; the hinge is deep and large. The labials (*a.l¹.u.l².*) are similar to those of *B. ornatus*, but the second is smaller.

The investing bones are like those of a Chameleon; they are crested and ornate with "tears" of bone; these are especially developed over the inter-auditory region; the roof bones (fig. 5, *f.p.*) are very extended and polygonal, and leave a space in front where they narrow out and join with the nasals, obliquely. Behind, they and the squamosals (*sq.*) only leave the epiotic region naked, and their temporal is larger than their orbital edge; that free edge is elevated and grooved radially; the temporal suture is sinuous. The nasals (*n.*) are conchoidal, with a sharp retral facial stem; they are especially perlate on their thickest part in the prefrontal region. The feeble premaxillaries (*px.*) stretch in an arcuate manner under the projecting rounded snout, and the nasal processes (fig. 7, *n.px.*) have their axis almost coincident with the axis of the long deep maxillaries (*m.x.*). These latter bones are straight up to the hinge, whilst they almost completely overlap the small quadrato-jugals (*q.j.*). There is a rough notched septo-maxillary (*s.m.x.*) under the outer nostril.

The squamosals (*sq.*) have a huge supra-temporal plate, four-sided, irregularly, and with the largest, or postero-internal, angle sharply notched. The side view (fig. 7) shows a large triradiate upper part, and a long (deep) sinuous preopercular process bound by the "orbitar process," behind. The parasphenoid (fig. 6, $pa.s.$) is unlike that of the last, it is shortened in front and is there made up of spines; its basitemporal wings are very long, and the pterygoids bind over them so as to cut away their fore edge, outside; they then run to a point, almost as far as the Eustachian openings ($eu.$). The hind part is a broad triangle, and in front of it the middle of the bone is raised into a triangular apophysis, which looks forwards. The vomers ($v.$) are formed of three curved rays, the two hinder rays half surround the inner nostril ($i.n.$) and the front ray curves in the other direction.

63 (continued).—(B) *Otilophus margaritifer.*—Adult female; $2\frac{3}{8}$ inches long (HYRTL.'s prepn., Mus. Coll. Surg., Eng.). Brazils.

This skull, kindly lent to me by Professor FLOWER, shows to what a degree of external modification the Bufonine type may undergo without losing any essentially Anurous character.

In the side view (Plate 38, fig. 1) we see that the height of the squamosal, measured at right angles to the basis cranii, is equal to the *gape*: in *Rana pipiens* and *Cystignathus ocellatus*, it is only *one-third*; and that is the normal proportion.

This is due to two things—to the height of the lateral crest of the squamosal, and to the depth of the grooved sigmoid pre-opercular region (figs. 1, 2, $sq.$); it is the upper part which has developed so largely since the creature was half grown (see fig. 3, $sq.$).

The endocranium is now completely ossified behind (fig. 2); but even now half the orbital region is cartilaginous (fig. 1, $o.s.$); this part is extremely shallow. The sides of the skull are raised all along up to the snout, and the nasal, frontal, and squamosal crests ($n., f.p., sq.$) are all frosted with small bony spikes and knobs; the latter crest, as a large "ear," reaching further backwards than the occipital condyle.

Behind (fig. 2), the parietal part of the roof-bones ($f.p.$) has lost the suture, and the whole plate lies flat across the wide audito-occipital region, partly covering the epiotic eminences ($ep.$). Where the squamosals join the roof-bones there is a considerable valley, on each side, in the hind skull. The mid skull and the nasal region are concave.

The orbital part of the squamosal (fig. 1, $sq.$) projects but little from the shaft, only reaches half way to the jugum, and points towards the hind part of the facial plate of the nasal ($n.$). The nasal, frontal, and squamosal form three-fourths of an orbital ring, which is finished by membrane, below, far above the edge of the maxillary ($mx.$); all the edges are produced or limbate, and concave. A little more ingrowth of the huge nasal would have made it equal to the prefronto-nasal of the Chelonian. The premaxillary is a strong bone like a phalangeal segment, and is nearly all nasal process; its top carries the valvular labials ($u.l^1.u.l^2.$*) Behind this bone, in an interspace formed

* In fig. 1 the second labial is lettered $u.l^1.$ by mistake, and the line from $u.l^2.$ points to a part of the nasal wall.

by it, the nasals and the maxillaries (*mx.*), there is a roughly orbicular, notched septo-maxillary (*s.mx.*). The maxillaries (fig. 1, *m.x.*) are large, deep, sculptured bones, with both margins somewhat sigmoid; the quadrato-jugals are unusually long, thick, and broad; they are strongly grafted on to the quadrate (*q.*). The suspensorium curves forwards towards the large sulcate trochlea (*q.c.*). The inner process of the pterygoid (fig. 2, *pg.*) binds the pedicle, the hinder process binds the inside of the quadrate, and the front part, which is very wide, behind, shows a large tract of cartilage (*sp.*), and is strongly bound to the inner face of the maxillary. The mandible (fig. 1) has a large and deep articular process, the Meckelian rod (*mk.*) is scarcely affected by the half sheath of bone—the articulare; the dentary (*d.*) is only a third the length of the ramus; the mento-Meckelian (*m.mk.*) is of the average size; the whole lower jaw is rather weak and very flexible, quite unlike the upper jaw. As is shown also in the half-grown specimen, the "annulus" (*a.ty.*) is perfect, and sends a small flange down upon the intruding, down-turned columella (*c.st.*). The medio-stapedial (fig. 2, *m.st.*) is long; the stapes and the intermediate inter-stapedial lobe are partly ossified.

As compared with *P. ornatus* this type is, as it were, the same exaggerated; it is less ossified by far, even in the adult, in its endocranium, but this is made up by the enormous development of its investing bones. It comes nearer the skull of the Common Frog in the retention of large tracts of cartilage, but is less like that "pattern" in having no inter-stapedial. I find no *second preorbital* besides the septo-maxillary, and in this it is more normal; the fenestræ in the basal plate are quite peculiar.

One thing to be noticed is this, namely, that this *Chamæleonoid* skull has not been specialised at the expense of any essential Batrachian character; the skull of the Chameleon is a much greater modification of the Lacertilian type of skull, as I have shown in a monograph on the cranium of that Family elsewhere (Trans. Zool. Soc., 1881).

In the small Toads that form the Families most related to the typical kinds, we shall find, with far less change of outer form, skulls that are much fuller of exceptional characters than that of this species, whose abnormality lies mainly in outward form, the deeper characters being on the whole true to the "norma."

In the next Family of Toads, "Rhinodermatidæ," the parotoids are absent, and the ear is less perfect than in the "Bufonidæ."

Second Family. "RHINODERMATIDÆ."

First genus. *Rhinoderma.*

64. *Rhinoderma Darwinii.*—Adult male; 1 inch long. Chili.

The skull of this species (Plate 39, figs. 1–6) is peculiarly *Ranine*, both in form and in strength; it is sub-triangular; but the skull of newly metamorphosed Toads ("Bufonidæ") show this, also, more or less (figs. 7–9).

As in the immature skull lately described (pp. 210-212), the length and breadth are equal, but the quadrate condyles (*q.c.*) reach further back, viz.: up to the fore end of the stapes (*st.*); a position soon attained in the young Toad.

The occipital condyles (*oc.c.*) are large, moderately wide apart, and postero-inferior, and the foramen magnum is large, as in the last (fig. 7). The auditory capsules are also, relatively, large, and have a small, oblique tegmen (*t.ty.*), the whole parotic region being limited to an ear-shaped process of cartilage projecting beyond the horizontal canal (*h.s.c.*), and nearly covered by the squamosal (*sq.*). The canals project considerably; in the epiotic and prootic regions (*p.s.c., a.s.c.*) equally.

Below (fig. 2), scarcely any cartilage is left in the ear-capsules, except at the sides (fig. 3); but there is a distinct tract for the facet on which the pedicle (fig. 2, *pd.*) glides. Above (fig. 1), the oblique, bevelled, outer margin is soft, and, in its deficiency, shows the stapes (*st.*) and mouth of the vestibule (*vn.*) from that aspect. A wide and rapidly widening space of cartilage remains both in the basi- and supra-occipital regions (figs. 1, 2), but the bony tracts are thoroughly continuous, right and left; this is a generalised character.

The prootic region of the bone (*pr.o.*) reaches half way between the 5th and optic nerves (V., II.), and, above (fig. 1), the bony matter only leaves a cartilaginous selvage round the large, elliptical, single fontanelle (*fo.*), which reaches, behind, to the middle of the anterior canal (*a.s.c.*), and in front ends nearly opposite the ethmoidal wings.

The mid skull is widish, almost oblong, widening before and behind, and slightly bulging in the middle; it is of moderate depth (fig. 3) and very long; for the skull is long, and the bulk of the nasal region short.

The girdle-bone (*rth.*) is less than the cartilaginous part behind it (*o.s.*), in which is seen a very large optic fenestra (II.); as in many of the arrested types, the wall of the mid skull opens, as it were, to the setting of the eye-balls, as the hind skull does for the ear-balls.

The ethmoidal wings are only partly ossified, right and left; below (fig. 2), the ethmoidal region is ossified to its end; above (fig. 1), the bony growth creeps along a third of the true nasal region.

The tegmen cranii is very limited before, behind, and at the sides, so that the fontanelle is very large and long.

The nasal roof (*n.r.*) is relatively very wide, as in most young Anura; the floor is about three-fourths as wide (figs. 2 and 4, *n.f.*). The roof overhangs the fore face, as in many of the edentulous types, and here there is a very generalised and very *Raiine* prenasal rostrum (*p.n.*), which has, evidently, shrunk from the cavity of a (*once*) long dermal beak, *now* an oval leaf of skin projecting from the nose.*

The pro-rhinals (*p.rh.*) are small; the valves (*n.l*[1].*u.l*[2].) large; the whole region, stripped of the investing bones (fig. 4), is seen to be very much like that of the Skate;

---
* I believe that this type has become dwarfed, and its rostrum shrunken as a correlative of the new bony tracts that have appeared *in and on* the chondrocranium.

the palato-suspensorial arch is rather feeble; the pre-palatine (fig. 4) is blunt; the pedicle (*pd.*) is short, and the quadrate region (fig. 3, *q.c.*) of the average length; it is partly ossified. The pedicle glides on a disk, which has a thick rim; the part of the pterygoid (*py.*) which invests it is at right angles with the main part. The palatine bone (*pa.*) is a feebly falciform lamina lying loose under the cartilage; the condyle for the jaw (*q.c.*) is large and reniform.

The "annulus" (figs. 3 and 5, *a.ty.*) is one-third less than the average size, and is an oblique, open crescent; the stapes (*st.*) is a very regular ellipse, of medium size and slightly calcified. In front of the stapes, and passing a little inside its antero-superior edge, there is another ovoidal cartilage (*co.*), from the outside of which there grows a thick, short spike, also unossified; this ends in a fibrous thread which passes over the hind limb of the annulus and under the facial nerve (VII.). This "columella" is an accurate counterpart of that of the young Toad (fig. 9); the inter-stapedial region is here, with a rudiment of the medio-stapedial. The stylo-hyal (*st.h.*) is small above, and only partially confluent; the cerato-hyal band (fig. 6, *c.hy.*) is narrow and uniform; the hypo-hyals have a lobe, growing forwards, and are straight and long, bounding the very deep, semi-elliptic notch. The basal plate is long and wide, the anterior and posterior lobes are part of the same expansion, and the front part has a larger outer and a smaller inner projection; the thyro-hyals (*t.hy.*) are long, straight, and but little divergent. The Eustachian openings (*eu.*) are of moderate size and circular. The investing bones are very much like those of the young Toad (figs. 7, 8); the fronto-parietals (*f.p.*) leave the superoccipital region bare, behind, and do not cover all the fontanelle, and but little of the girdle-bone in front. The nasals (*n.*) are wide apart, they do not overlap much of the nasal region, and cover but little of the roof; they are small sub-crescentic shells. The premaxillaries (*px.*) are well under the snout, and are tilted forwards, above; the maxillaries, quadrato-jugals, and squamosals (*mx., q.j., sq.*) are normal, but feeble; I find no septo-maxillaries; the parasphenoid and vomers are like those of a young Toad. The mandible (fig. 3) is normal, but rather feeble; the dentary (*d.*) is more than half the length of the ramus.

As compared with the "norma," this skull has the following modifications:—

1. Its general shape is rather triangular than semi-elliptical; and its prenasal is very large and projects in front, like that of an "Elasmobranch."
2. It has only one fontanelle.
3. Its hind skull has only one ossification on each side.
4. The whole nasal region is very generalised and *Ranine*.
5. The annulus is small and very open.
6. The columella is only a cartilaginous rudiment.
7. The lateral lobes of the basal plate are confluent, and give off three spurs on each side.
8. There are no dentigerous bones.
9. The investing bones generally are very feeble and arrested.

10. Those of the fore face are thrown quite beneath the snout.
11. There are no septo-maxillaries.

With regard to the classification of the lesser Toads, without parotoids, I incline to something intermediate between Dr. GÜNTHER's system ("Bat. Sal.") and Professor MIVART's (Proc. Zool. Soc., 1869, p. 280).

The latter has put several of Dr. GÜNTHER's Families together to make up *his* "Engystomidæ" (see p. 289); and in the "Batrachia Salientia" the genus *Diplopelma* is put with *Rhinoderma*, *Atelopus*, and *Uperodon*, to form the "Rhinodermatidæ."

Again, Professor MIVART (p. 287) puts *Pseudophryne* and *Micrhyla* with *Phryniscus*, to form *his* "Phryniscidæ."

But the skull of *Pseudophryne* comes much nearer to that of *Diplopelma* than to that of *Phryniscus*, and the cranial characters of *Engystoma* and *Diplopelma* are almost the same; these two genera are very closely related.

The form of skull just described—*Rhinoderma*—is very unlike any of them, and I have examined none, as yet, with which it can be put; it may stand at the head of a Family, with *Diplopelma*, or the latter might be introduced (by a modification of the language used by Dr. GÜNTHER in his group-characters) among the "Engystomidæ."

I shall, for the present, keep *Pseudophryne* where that author puts it ("Bat. Sal.," p. 45), viz.: among the "Brachycephalidæ;" unfortunately, I have not worked out the skull either of *Brachycephalus* or *Hemisus*.

I am doubtful of the propriety of bundling up *Callula* (*Hylædactylus*) with the "Engystomidæ."

But the whole subject bristles with difficulties; in passing from species to species, in the same genus, some new and unexpected variation is always turning up.

Second genus. *Diplopelma*.

65. *Diplopelma ornatum*, vel *rubrum*.—Adult male; 11 lines long. India.

This, like the last, was a male; it was less than an inch in length; the skull (Plate 42, figs. 8 and 9), also, has its length and greatest width equal; but this skull agrees much more closely with that of *Pseudophryne* (Plate 42, figs. 1, 2), than that of the different species of the same genus that could be given in many instances.

Like that of *Pseudophryne* this is a very arrested skull, and shows some curious analyses of the Batrachian cranial elements. On the whole, the ossification is about equal in both; but, notwithstanding their close kinship, these two small Toads have several instructive cranial differences.

The hind skull is less massive, and is altogether proportionately less (figs. 8, 9; and 1, 2); and as the quadrate condyles reach nearly to the stapes, the facial outline is longer, and becomes arcuate in its hinder third; two-thirds of it are very straight, and the rather broad snout projects more in the middle, for the prenasal rostrum (*p.n.*) is more definite than in that type. The occipital ring projects less; the condyles are separated by their own width only, and they are large and not posterior, but postero-inferior. The synchondroses above and below (*s.o.*, *b.o.*) are equal in both, and have an endosteal deposit. The roof is less wide behind, shows only the large fontanelle, and the tegminal growth is altogether less, so that the fontanelle takes up more than three-fourths of the roof. The sides of the occipital ring and the auditory capsules, with their convex canals, are all continuous in their bony investment, right and left (*e.o.*, to V.); and there is the same small headland unossified at the very limited tegminal edge (*t.ty.*). There is cartilage also where the pedicle fits on, and where the stylo-hyal unites beneath, with the skull (*pd.*, *st.h.*); also in front, below, the large optic fenestra (II.) is surrounded by cartilage, and the foramen ovale (V.) by bone.

The gradually narrowing orbital region is more overlapped by the roof-bones (fig. 9, *f.p.*), and three-fifths of the wall is unossified, and nearly all the floor. The lateral rudiments of the girdle-bone (*eth.*) are less, and run into their own alæ partially; yet an endosteal deposit, with scarcely any perichondrial bone (ectosteal palatine), runs up to the cheek. These ethmoidal wall-bones reach the roof externally (fig. 8), but are separated by their own width below (fig. 9). Here we have a much more perfect median bone than in *Pseudophryne* (Plate 42, fig. 2), for the intertrabecular bar is ossified for the fore half of its extent. In front, the base of the septum nasi (*s.n.*) is ossified almost up to the rostrum (*p.n.*), and between the lateral ethmoidal centres (*eth.*) the bony deposit is continued for some distance in two tracts (*p.e.*). Here we have the continuous "mesethmoid" of the Ostrich Family foreshadowed, in which the bony deposit also takes place right and left of the cartilaginous wall.

The unossified roof and floor (below *n.*, and above *s.n.l.*) are moderately wide; the prorhinals (fig. 9, *p.rh.*) are small and sharp; the prenasal (*p.n.*) is rather large, the front is sinuously transverse and of medium width, and the sub-tubular nostrils are defended by well-developed valves (*u.l².u.l².*). The ethmo-palatine (*e.pa.*) is slender, and the external part adze-shaped; the bone (fig. 9, *a*) is mainly endosteal, and the post-palatine tract is either separated from the pterygoid cartilage, or united by a very fine thread. The pterygoid (*pg.*) is typical, and partly ossifies a *free* pedicle (*pd.*). The quadrate is unossified; the condyles (*q.c.*) reniform and oblique; and the Eustachian openings (*eu.*) are only half the average size. The same may be said of the annulus (*a.ty.*), whose horns are not united.

The stapes (figs. 9 and 10, *st.*) is not so large as in *Pseudophryne*, and only half as convex; its margin is cartilaginous, the rest is thin bone; its oval form is modified by a slight emargination before and behind. The columella (fig. 10) is almost of the average size; the whole rod is continuous; there is a semi-osseous enlargement of the medio-stapedial (*m.st.*, *i.st.*) behind, but no joint, and the bone is arched and geniculate; the extra-stapedial (*e.st.*) is a small oval shield, with a free suprastapedial spike.

The mandibles (Plate 43, fig. 6) are normal, the mento-Meckelian (*m.mk.*) of good size, and the coronoid crest (*cr.*) large.

The stylo-hyal (Plate 42, fig. 9, *st.h.*) is thick and confluent, but not so thick as in *Engystoma*; the cerato-hyal (Plate 43, fig. 6, *c.hy.*) is broad, and in bending back has no lobe, but it has a distinct and large extra-hyal (*ex.hy.*) as in some other kinds. The rest of these growths are very similar in the two kinds of *Diplopelma* (figs. 4 and 6), but the notch is less, the lateral lobes larger, the thyro-hyals (*t.hy.*) diverge more, and in the space between these roots there is a larger and more solid basi-branchial bone (*b.br¹.*); the keel in front of it is more or less calcified.

The investing bones are more developed above and below than in the next, and the roof bones are complete along the inner margin. The fronto-parietals (*f.p.*) are slightly convex, and fairly overlap the hind skull; they also overhang the walls in the orbital region; altogether they are more normal than in the other species. They scarcely meet in front, and this interspace becomes wider between the nasals (*n.*) which are, however, large conchoidal plates of the normal form. The marginal bones (*px., mx., q.j.*) correspond with those of the smaller species of *Bufo*, but the squamosal (*sq.*) is very feebly developed. The parasphenoid (fig. 9, *pa.s.*) is much more normal than in the next instance; its main bars are sub-equal, and its hind part triangular. The vomers (*v.*) are Bufonine, and not very small; they are crescentic shells, notched both before and behind. There are no septo-maxillaries. This type of skull differs from the "norma" on the whole as much as that of *Pseudophryne*: in some things more, as in the larger mesethmoidal bone; in other things less, as in the greater retreat of the condyles of the quadrate, and the more normal form of the parasphenoid. These two species may be said to belong to the same group, and to lie on the same morphological and zoological level, and they might, with a little cutting and contriving, be put into the same genus.

66. *Diplopelma Berdmorei* (?).[a]—Adult female; 1 inch 1 line long. Moulmin, Tenasserim.

This skull is of the same length as the last, but *its* greatest breadth was the same, in this it is much greater than the length. The main figures (Plate 43, figs. 1, 2) are only magnified three-fourths as much as those of the skull of *Engystoma* (figs. 7, 8); thus the greatest breadth of these figures is the same, or nearly; the smaller, more magnified figures (figs. 7, 8) have both measurements equal as in the last, whilst those of this species (figs. 1, 2) show the length to be only nine-tenths of the breadth.

This small skull has been metamorphosed very unequally; in some things it only

---

[a] This specimen, the gift of Jas. Wood-Mason, Esq., had lost its colour in the spirits; Dr. Günther considers it to be most probably *D. berdmorei*: if not, to be a closely allied species; it is larger than any of my specimens of *D. ornatum* vel *rubrum*, and differs much more from it in the structure of the skull than that species does from *Engystoma carolinense*.

equals that of a typical Batrachian whose tail is rapidly shortening; in others it nearly equals the skull of the adult in those high kinds; the ossification especially is defective here, as in some of the dwarfed kinds of Frogs. The foramen magnum is large, and the roof covers it well; the occipital condyles (*oc.c.*) project scarcely at all, and are postero-inferior. The roof runs half way to the ethmoidal region, and has no secondary fontanelles; the great space (*fo.*) is as long as the hinder "tegmen," and is very elegant in shape, like the cordate leaf of *Nymphæa alba*.

The tegmen cranii is no wider in front than at the sides, where it is unusually well developed. The ovoidal auditory capsules are well turned outwards, and are large; beyond them the tegmen tympani (*t.ty.*) is a small squarish lobe, as in the last two kinds. There is a little calcification above (fig. 1) but more below (fig. 2, *b.o.*), and that tract is wide, and rapidly widens forwards, for the ex-occipitals (*e.o.*) are only large reniform patches that enclose their own nerves (IX., X.); they creep up between the arch and the capsule so as to show a little above (fig. 1). The prootic also (*pr.o.*) is a curved band just margining the foramen ovale (V.), and creeping along the capsules for some distance above and below (figs. 1, 2).[*]

The mid skull is broad, lessening gently forwards, and bulging; it is moderately high, and is *wholly unossified*, as is also the fore skull and nasal region; this is equal to the hind skull in axial extent; the mid skull is one-third longer. The large baggy nasal roofs lying back over the ethmo-palatine bar (fig. 1), and the wide trabecular floor (fig. 2, on each side of *s.n.*) are quite *juvenile* in character; the nostrils (*e.n.*) are sub-tubular and projecting; they are wide apart, and protected by the normal valves (*u.l².u.l².*). The rostrum (*p.n.*) is also in its *first stage*, a mere decurved lip of cartilage; the pro-rhinals (*p.rh.*) are small and rudimentary.

The palato-suspensorial arch is delicate, and continues to the pedicle and quadrate (*pd., q.*), the pre-palatine is sharp, the post-palatine (*pt.pa.*) marked off from the pterygoid by a sharp lobe (a juvenile character), and the ethmo-palatine has a fine thread of ossifying perichondrium under it (*e.pa., pa.*).

The delicate bifurcate pterygoid (*pg.*) runs along inside the post-palatine bar, gives off a pedate fork for the short pedicle (*pd.*), and binds itself to the suspensorium (*q.*) behind; in its fork we see the small, round Eustachian opening (*eu.*).

The quadrate condyle (*q.c.*) is a perfectly normal trochlea, oblique, with the front elevation small; above it the "orbitar process" (*or.p.*) is retained, as an ear-like lobe, which is nearer the condyle (*q.c.*) than the otic process (fig. 1, above *or.p.*); this is a rare character. The annulus (*a.ty.*) is large (quite of the typical size), and is very perfect above, as a ring. The mandible (fig. 3) is long and very perfect in all its parts.

The stapes (fig. 5, *st.*) is thick, oval, with an oblique, antero-superior, *grooved* emargination for articulation with the columella; it is also umbonate. The columella (fig. 5) only wants segmentation of the large cartilaginous proximal lobe (*m.st.*) to

---

[*] The continuous bony tract right and left, in the last, is due, as this skull shows, to coalescence of the two normal centres.

DEVELOPMENT OF THE SKULL IN THE BATRACHIA.   229

make it normal; it is large, well developed, and has a spatulate extra- and a ligulate, fixed, supra-stapedial (*r.st., s.st.*).

The stylo-hyal (*st.h.*) is confluent; it is not enlarged above, as in the other kind; the rest of the bar (fig. 4, *e.hy., b.hy.*) is narrow, without a lobe.

The lateral lobes of the basal plate (*b.h.br.*) are well developed, the notch in front and the main plate are of the average size, and the thyro-hyals (*t.hy.*) are long, and moderately divergent. Between them there is a thick, ossified mass, with a free rounded fore margin; this is the last basi-branchial (*b.br.*); its hinder margin is grooved. In front of this bony ridge the plate is keeled, below, and this keel expands behind the front notch into a triangular mass, the first basi-branchial; the whole keel is more or less calcified.

The three characteristic hyo-branchial plates of this and the two other kinds (Plate 43, figs. 4, 6, and 10,—*Engystoma*) evidently belong to three closely related species.

In the main skull (Plate 43, figs. 1, 2) the investing bones are as feeble as the centres in the chondrocranium. The fronto-parietals (fig. 1, *f.p.*), thin shells pointed in front, dilated postero-externally, and nowhere meeting at the mid-line, exactly correspond to their counterparts in metamorphosing Common Tadpoles. So also the broadly-crescentic nasals (*n.*) with their facial handle, and the frail marginal bones (*px., mx., qj.*), and, running along the suspensorium, the squamosal (*sq.*). The parasphenoid (fig. 2, *pa.s.*) has its basi-temporal wings larger than it cochleariform rostrum; the hind part is a large triangle.

The small ragged vomers (*v.*) have all the *four* normal processes, in size and development they come between those of this and the next kind; they protect the inner edge of the rather small, round, inner nostrils (*i.n.*), which like the outer (*e.n.*) are extremely wide apart.

In some things this skull comes nearer the "norma" than the two last, viz.: in its perfect annulus, well-developed columella, and distinct prootics and ex-occipitals. It is farther from it in the extreme feebleness of all the bony tracts in the main skull, and in the total absence of the girdle-bone; also in the retention of the "orbitar process;" the clear regional mark between the post-palatine and pterygoid cartilaginous tract; in the more perfect development of the thick, bony, uncinate *last basi-branchial;* and especially in having a rudiment of both a supra- and a basioccipital.

These minute forms are well worthy of study in their irregular, and as it were, halting metamorphosis; and this is to be noted, namely, that generic groups may be made according to the taste of each individual Zoologist; no two species agree in all things, and in some existing genera each species might be put by itself, and have its own *generic*, as well as *specific*, name.

Third Family. "BRACHYCEPHALIDÆ."

Ear very imperfect; no parotoids; sacrum dilated.

Genus *Pseudophryne*.

67. *Pseudophryne Bibronii*.—Adult female, 1 inch long; and adult male, ¾ inch long. New South Wales.

The skull of this *Bombinator* Toad is another example of arrested metamorphosis combined with relapse, so to speak, into general vertebrate characters that are, normally, suppressed in the Anura.

Whether what we see in skulls like this is due to *relapse*, or the retention of more or less unchanged, old, ichthyic, pro-Batrachian characters, the interest of the matter is equal, for the transforming power which has wrought so mightily in the higher kinds to set them on high above the fishy tribes, generally, has in these southern dwarfs found some check, or has never, as yet, come into full play.

The length and the breadth of this little skull are equal (Plate 42, figs. 1, 2), and up to the hinges of the mandible the outline is a neat semi-ellipse; but these hinges are only opposite the scarcely open Eustachian cleft (*eu.*). Instead of getting some distance behind the small occipital condyles, as in some types, the condyles of the quadrate (*q.c.*) only reach along two-thirds of the length of the skull. The arrest, generally, corresponds with what has been done in typical kinds by the time the tail is well absorbed, or at most up to the first summer.

The three regions of the skull are about equal in length, and the auditory capsules are relatively very large, obliquely oval, with prominent canals (*a.s.c., h.s.c., p.s.c.*); and the parotic outgrowth is nothing but the small unossified selvedge which forms the tegmen (*t.ty.*). The occipital ring is very distinct and protruding, but the foramen magnum (fig. 1) is very oblique, and open above. The ossification on each side is generalised, for there is no distinction of prootic and ex-occipital (*au., rh., e.o.*), and yet it is very complete, except at the edge and the middle, above and below. There we see the cartilage is wide and widening; above (fig. 1), it runs forwards to the fore-third of the hind skull, and ends in a peak which converts the large, single fontanelle into a heart-shaped space.

Below (figs. 2 and VII., *b.o., nc.*), the *permanent* cephalic notochord is covered with bone (a "cephalostyle"), and this bony matter has run into the investing mass, right and left, so that here we have a true, but arrested, "basi-occipital" bone.

The broad, short mid skull lessens quickly up to the ethmoidal alæ; it is rounded, or swelling as in young Anura. The bone scarcely reaches the foramina ovalia (V.), which are large, and two-thirds of the orbital wall remains unossified. Nearly half that space is occupied by the large oval *fenestra optica* (II.), through the back part of which the optic nerve escapes. At the fore edge of this space we have a bony tract on each side; these are the lateral rudiments of the girdle-bone (*eth.*). These cochleate tracts reach the top of the cranial wall for a small extent, and also run into the ethmoidal alæ, but they are far apart, above and below. These are the true "lateral ethmoids," but between them the "perpendicular plate" is in rudiment; this is a triangular tract of imperfect bone at the middle of the floor, just over the front of the parasphenoid (fig. 2, *pa.s.*). The fore part of the chondrocranium is well developed, but quite unossified; the subnasal laminæ (*s.n.l.*) are broad, with large falcate angles, and well formed pro-rhinals (*p.rh.*); the roof (*n.r.*) is well developed, and has a distinct ring of cartilage round the outer nostril (*e.n., n.w.*); these passages are at a moderate distance, and are well protected by the labials (*u.l.u.l².*). The palato-suspensorials are largely developed in front, for the ethmo-palatine bar (fig. 2) expands into a large adze-shaped plate, the pre-palatine part of which (fig. 4, *e.pa., pr.pa.*) almost reaches the angle of the nasal floor. There is just a thread of bone answering to the palatine ectostosis (figs. 2 and 4, *pa.*).

The partially retreated hinder part of this arch is still continuous with the basis cranii by the unabsorbed pedicle (*pd.*); the joint-cavity is there, but is not complete— as in the species of *Bufo* and other types, where the strong pterygoid binds this part down. Here, however, that is not the cause of the unfinish of the joint, for this bone (*pg.*) is very feeble as in young Toads and Frogs.

The condyle of the quadrate (*q.c.*) is a large bilobate trochlea; the body of the suspensorium (fig. 6, *sp.*) is not ossified. There is a small ligulate sub-crescentic "annulus tympanicus" (figs. 5 and 6, *a.ty.*), but no columella.

The stapes (figs. 5, 6, *st.*) is large, oval, and apiculate, behind; the stylo-hyal (*st.h.*) is loosely attached to the capsule.

The rest of the band is wider (fig. 3, *e.hy.*) and is definitely dilated before it turns back as the hypo-hyal (*h.hy.*).

The notch in front of the basal plate (*b.h.br.*) is shallow, the plate itself short, the fore side lobes large and stalked, the hind side lobes very short, and the thyro-hyals (*t.hy.*) large and moderately divergent.

The mandibles (fig. 3) are quite normal, but the ossified labials (*m.mk.*) are very large, and so are the articular condyles (*ar.c.*) ; the articular bone (*ar.*) rises directly in front of the condyles, but very little in the coronoid region; the dentary (*d.*) is small and feeble, and the rod of cartilage (*mk.*) is not much affected by the bone, outside.

The investing bones are all in a *quasi-juvenile* condition; the fronto-parietals (*f.p.*) remain distinct, right and left; they more than cover the large fontanelle, and overlap the auditory capsules and super-occipitals moderately.

The nasals (*n.*) are thin shells of bone, imperfectly covering their own region; the premaxillaries, maxillaries, quadrato-jugals, and squamosals (*px., mx., q.j., sq.*) have all the same feeble arrested character.

So, also, the parasphenoid (*pa.s.*); it has all its processes but is only two-fifths the length of the skull, and is less developed than that of the Tadpoles of many kinds.

The vomers (v.) are very small shells, bifid behind, and protecting the circular inner nostrils (i.n.), which are wide apart.

Very much of what is peculiar in this skull may be summed up in the word "arrested;" but to this general character must be added such things as are not often seen even in young specimens of higher kinds:—

1. There is a small basioccipital, and a persistent cranial notochord.
2. The prootics and ex-occipitals are continuous on the same side.
3. There are no secondary fontanelles.
4. The optic fenestra is very large.
5. The girdle-bone is arrested, and there are three rudiments answering to the lateral and perpendicular ethmoids.
6. The palato-suspensorial arch goes no further back than the postorbital region.
7. The pedicle is not absorbed above.
8. The palatine bone is a mere thread.
9. The Eustachian opening is nearly closed.
10. There is only a lignlate, imperfect annulus.
11. There is no columella.
12. The mento-Meckelian rods and the condyles of the mandible are very large.
13. The hyo-branchial plate has its processes feebly developed, as in a young Common Frog or Toad.

I shall take the liberty to modify Dr. GÜNTHER's classification, somewhat; but I see no advantage in bundling together several of these groups—as Professor MIVART has done (Proc. Zool. Soc., 1869, p. 289).

Diplopelma might go with *Engystoma* and help to form the "Engystomidæ." Dr. GÜNTHER himself ("Bat. Sal.," p. 50) says that in *Diplopelma* the toes are only "one-third webbed," and in *Engystoma* free (p. 51). Moreover, the columella is well developed in both these genera, whereas in *Rhinoderma* it is a mere rudiment. The author says of another member of his "Rhinodermatidæ," viz.: *Atelopus*, "I have never seen the animal" (note to p. 48), and of another genus—*Uperodon*—that "the tympanum is hidden;" and his description of the skull (p. 49), although short, is enough to show that it is very similar to that of *Hylaplesia*, and extremely unlike that of either *Rhinoderma*, or of the species of *Diplopelma*, or that of *Engystoma*, soon to be described.

By their skulls these small species must be judged, not keeping out of sight other characters, especially the suppression of the *claricula* ("pro-coracoid"), and of the *manubrium sterni* ("omosternum").[*]

I shall follow Dr. GÜNTHER's example and limit the next group to the one genus *Phryniscus* (see "Bat. Sal.," p. 42).

[*] These pro-coracoids and omosternums are absent in *Engystoma*, *Callula*, and *Diplopelma* (MIVART, Proc. Zool. Soc., 1869, p. 289).

Fourth Family. "PHRYNISCIDÆ."

Toes webbed; sacrum broad; no parotoids; ear generally very imperfect.

Genus *Phryniscus*.

68. *Phryniscus cruciger.*—Adult male; 1⅜ inch long. Interior of Brazils.

The small skull of this species (Plate 41, figs. 1–5) has its breadth one-thirtieth greater than its length; its outline, up to the hinges for the lower jaws, is nearly triangular, but the extreme end of each hinge (*q.c.*) is opposite the centre of the Eustachian opening (*eu.*), so that much of the skull—all the hind skull—lies behind the facial margin, as in newly metamorphosed Anura of a high type.

The whole skull has the appearance of an abortively developed Bufonine structure, with an abnormal amount of ossification; it is like a badly developed skull of such a form as *Bufo ornatus*, or of the half-grown *Otilophus* (Plate 37); moreover, it is asymmetrical to a degree very seldom seen in the group; more so than that of *Siredon* among the Urodeles. The foramen magnum (*f.m.*) is large and obliquely superior; the occipital condyles (*oc.c.*) are large, sub-pedunculate, and directly posterior. Outside them, the epiotic and tegminal projections (*p.s.c.*, *t.ty.*) are a little and a little further forwards, and yet, on the whole, the large hind skull is a broad transverse tract, the "canals" standing out of it well, and the wide tegmina, bounded by the squamosals (*sq.*), enlarging gently, forwards. Two-thirds of the tract beyond the horizontal canal (*h.s.c.*), on each side, is cartilaginous; from the hind margin of the optic fenestra (II.) to a distance twice the extent of that space there is cartilage; and the snout (*p.n.*) in front of the nasals (*n.*) is also soft.

The rest of the cranium proper is bony, and in some parts these extensive ossifications are anchylosed to the investing bones.

Measured along the axis, the three regions are sub-equal; they are all very broad; and, taking in the face, the outline becomes just less than a right angle; for the proper cranial margins run inwards rapidly from behind, forwards. The temporal region is of great breadth; the orbital edge is concave; the main fontanelle (fig. 1), contrary to wont, is widest across—like a Tortoise's heart; near it, behind, the other two spaces are large and circular; the larger space is not quite covered.

The girdle-bone (*eth.*) leaves only a small orbito-sphenoidal tract (*o.s.*), behind, and in front runs to the fore edge of the nasals (*n.*) above (fig. 1), and to the premaxillaries below (fig. 2, *px.*); it takes in the "wings," above, and the palatine flap, also, below; moreover the palatines are anchylosed with it (figs. 1, 2, *ep.a.*, *pa.*), the bony "alæ" touching the pterygoids (*pg.*); the vomers (*v.*), also, are confluent with the large girdle-bone.

The nasal roof and septum (figs. 1–3) are formed into an over-hanging decurved beak, with a sinuous outline; this pre-nasal (*p.n.*) is a mere thickening of the middle part, in front, as in the embryos of many Vertebrata at an early stage; this arrested (or generalised) condition of a beaked snout is found in this genus *Xenophrys* (Plate 23, figs. 5-7), among the toothed Anura. From the snout to the ethmoidal region the ossified subnasal laminæ (fig. 2, *p.n.*, *eth.*) form a concave plate, in front of which the nerve passages (*n.n.*) are seen wide apart, and outside of them, the small pro-rhinals grow out, and are imbedded in the premaxillaries (*p.x.*). The nasal roof (fig. 1, *p.n.*) is moderately convex; the skull is very shallow at this part (fig. 3), which condition is increased by the hollow form of the fore-palate.

The subocular spaces are an almost perfect oval, and the bars surrounding them are strong. The nostrils (*e.n.*) are wide apart, their valves (*u.l².u.l².*) are normal. The fore skull is formed of the combined ethmo-nasal cartilages and bones, which are continuous with the palatines. The pre- and post-palatine regions of the latter are adze-shaped; there is some cartilage left above, in this part (fig. 1, *e.pa.*), and also along the whole palato-suspensorial bar, into the quadrate (*q.c.*) and pedicle (*pd.*), which has a free joint on the skull; the quadrate is but little affected by the quadrato-jugal. The forks of the pterygoid bone (*pg.*), which enclose the small oblique Eustachian pouch (*eu.*), run backward to the same transverse line, for the quadrate hinge is arrested in its retreat. Its condyle (*q.c.*) is long-reniform, and its front edge is opposite the exit of the optic nerve (II.), an extremely forward position in an adult Anuran; also the *left* is not in symmetry with the *right*, and is not nearly so far back.

Considering that the pedicles (*pd.*) once were continuous with the basis cranii under the outgoing trigeminals (V.), their present position speaks of a large amount of metamorphosis, after all; they are very wide apart now, and far from the skull-base.

In the obtuse angle formed by the suspensorium and its splint (fig. 3, *sy.*) there is a very small semi-lunar annulus (*a.ty.*), and in its inner rim the fore part of the columella (*e.st.*) fits. The stapes (figs. 1 and 5, *st.*) is an oblique half oval; it is convexo-concave, and umbonate.

The columella has no proximal joint, but the medio-stapedial (*m.st.*) is very thick, above, and the most solid part is oblique, emarginate, and unossified; it wedges itself within the oblique part of the stapes. The shaft is bent on the clubbed end, and ceasing to be bony, below, soon dilates into the broad trowel-shaped extra-stapedial (*e.st.*), which has no ascending ray.

The narrow stylo-hyal (*st.h.*) is uniting with the tympanic floor, and does not enlarge in the cerato-hyal region (fig. 4, *c.hy.*) until near to the hypo-hyal loop.

The basal plate has three rounded notches in front; the median space is deep and narrow between the hypo-hyal bands; there is, right and left, a small ear-shaped front "lateral lobe." There is no hind lobe, and the basal plate is very long (narrow beyond all precedent in this group, the "Anura"), and ends in two strong, upbent, widely diverging thyro-hyals (*t.hy.*). The mandible (fig. 4) is normal, the dentary (*d.*) is half as long as the ramus, the articular surface is large and obliquely reniform; the articulare (*ar.*) has not ossified the rod (*mk.*) appreciably, and the mento-Meckelians (*m.mk.*) are large.

The want of symmetry mentioned above is best seen in the roof-bones (fig. 1, *f.p.*, *n.*). Large as are the fronto-parietals, they barely cover the two lesser fontanelles, behind, and there is an open chink in front. In their hind third they are anchylosed; their angular postorbital projections are not opposite, the *left* is in front of the *right*, and their jagged fore part shows the same asymmetry; their orbital edge is sharp and separated from the rest by a sub-marginal fossa; there is a good space between these bones and the nasals, leaving the girdle-bone naked: the orbital plate (fig. 3, *f.p.*) is definite, but narrow, less than a third the depth of the shallow skull-basin. The *left* nasal (*n.*) does not go so far back or so far inwards as the *right*; it is only three-fourths the size of the latter; contrary to rule, the nasal, broadening outside, lies right down on the top of the maxillary (fig. 3, *n.*, *mx.*), leaving only a little chink in front. The nasals are (relatively) thick, convex shells; they are not entirely free from the subjacent endoskeletal bone (*eth.*). They are round in front, a good space apart, and leave the *Anatine* snout uncovered.

The premaxillaries (*px.*) run across under the snout, and meet at a large angle; they have pointed palatine processes (fig. 2), and short, tilted nasal processes (fig. 3).

The maxillaries (*mx.*) are of the average size, and only become pointed near their end; their sharp edge has a narrow palatine plate growing from it (fig. 2).

The quadrato-jugals (*q.j.*) are small, short, curved bones, slightly connected with the suspensorium (*q.*), and only touching the end of the maxillary. The squamosals (*sq.*) are in this species very instructive; they might be described as " preoperculars," *bent the wrong way:* a thing not impossible, for the outer bony plates are brought under the power of the endocranium to a very remarkable extent, and "without hands," are moulded upon it, cunningly.

In this species the supratemporal part of the bone is very narrow, and only clamps the edge of the unossified "tegmen," (*t.ty.*); Moreover, the axis of the supratemporal region is coincident with that of the stunted postorbital; the short descending part bends itself backwards at a very obtuse angle; this is the rounded space against, and partly on which, the small "annulus" (*a.ty.*) lies.

The parasphenoid (fig. 2, *pa.s.*) is a very remarkable bone, it is almost all " wings;" measured to the mid-line they are larger than the scoop-like fore part; they are also broader, and are obliquely truncate outside. Between these wings there is a fore-looking, triangular spur, as large as the hinder projection of the bone. The bone falls far short of the middle part of the ethmoid, in front.

The latter bone (*eth.*), in its subnasal extension, shows two sigmoid crests that bound the widely severed internal nostrils (*i.n.*); these are the anchylosed vomers (*v.*); the *right* is much larger than the *left*.

I shall compare the skull of the three species of this genus with that of the " norma," together; they differ remarkably *inter se;* but agree with each other in being almost the most abnormal of all the *tongue-bearing* Anura.

69. *Phryniscus nigricans.*—Adult female; 1¾ inch long. Costa Rica.

The last species had the average relation of length to greatest breadth; in this we have, suddenly, one of the longest skulls in the group, the breadth being little more than three-quarters of the length; the general outline is that of a parabola (Plate 41, figs. 6-9). In some things the asymmetry is like that of the last kind, in others the largest bone of a pair is on the opposite side. The ossification is more intense and *Salamandrian* than in *P. cruciger*. The skull is still more depressed, and the development of some parts has suffered severer arrests. The fitting on of the head to the spine is different, in that there was an almost directly *posterior* aspect of the occipital condyles; here (figs. 6-8, *oc.c.*) they are *postero-inferior*; the foramen magnum (*f.m.*) is still more oblique (superior in aspect) than in the last kind. Recently metamorphosed young of normal Anura show nearly as great an extension of the parotic region as this species; for here the horizontal canal runs into the substance of the thick unossified "tegmen" (*t.ty.*). That selvedge, a band above the huge optic fenestra (II.), the rim of the *duck-billed* snout, and the edge of the fenestra ovalis, are unossified; very little else of the cranium proper remains soft. Measured along the axis, there is but little difference in the relative extent of the fore, middle, and hind skull; the fore skull is somewhat longer than the rest, on account of the projection of the snout.

The semicircular canals project well from the bony hind skull; the supraoccipital region is short, the two oval and the large heart-shaped fontanelles (*fo., fo'.*) are very large; there is just an overlapping margin as a rudiment of the tegmen cranii running along the sides, and enclosing the fontanelles. The mid skull is wide, making it look short; the temporo-postorbital region is very wide. The mid skull bulges in the middle (fig. 8), but shrinks in front; the floor of the nasal region is like a highly arched "hard palate" in a human skull. The optic fenestræ (II.) occupy nearly a third of the orbital tract, and are almost as large as their counterparts for the setting in of the auditory capsules; the "serial homology" of these spaces can be well seen in such a skull as this. The girdle-bone extends from this space to nearly the verge of the snout, and thus occupies three cranial regions, besides taking in the ethmo-palatines (*e.pa.*), and uniting with the palatine ectostosis (*pa.*). The pro-rhinals (Plate 40, fig. 10, *p.rh.*) are small, and the fore palate is narrower than in the last. There is a narrow tract of cartilage on the edge of the upper surface of the pterygoid (fig. 6, *pg.*), and the facet of the pedicle (*pd.*) and most of the quadrate (*q.*) are unossified. The condyles (*q.c.*) are reniform, long, and very oblique, the *left* more than the *right*.

The mandibles (fig. 9) show a long condyle also, and very large mento-Meckelians (*m.mk.*); the dentary (*d.*) is very long, and the articulare (*ar.*) a mere trough for the cartilage (*mk.*). This is neither annulus tympanicus, nor columella; the stapes (figs. 7 and 8, *st.*) is large, oval, and umbonate.

The stylo-hyal (figs. 7-9, *st.h.*) is adherent; it is a narrow band, very slightly enlarging in the cerato-hyal region (*c.hy.*), and then becoming narrow again, without a lobe, as the retral hypo-hyal (*h.hy.*). The basal plate is very similar to that of the last kind, but the thyro-hyals (*t.hy.*) are still longer, and more divergent; they are strongly curved upwards and outwards. The investing bones, also, are very similar in the two species, allowance being made for the difference of outline of the skull. The fronto-parietals (*f.p.*) are better developed in the occipital and temporal regions, and as in the last, the parietal territory is anchylosed, and even somewhat adherent, to the bone below. They end in the frontal region so as to leave an emargination which slightly exposes the fontanelle (*fo.*); the right frontal is the longest, and yet it only touches the nasal of that side, the shorter *left* bone overlaps the nasal in front of it. These latter bones (*n.*) are very unsymmetrical; they are half their own width apart at the nearest point, and leave a large narrow-waisted bony tract bare, between them. Each bone has a thick rib, or boss, over its descending or facial process (figs. 6, 8, *n.*), and the whole facial edge (fig. 8) lies well down on the top of the maxillary (*mx.*); it is gently concave there, and notched further forwards for the nasal aperture (*e.n.*). Of these convex, rounded bones, the left is one-fourth larger than the right, and is altogether further back. The premaxillaries (*px.*) lie completely under the cochleate snout; they form a more definite angle at this juncture than in the last kind; the nasal process is longer, and the labials (*n.pl.u.l².*) are larger. The maxillary (fig. 8, *mx.*) is deeper, the short, unciform quadrate-jugal (*q.j.*) is a little more attached to the end of the maxillary; it is but little united with the quadrate. The squamosal (*sq.*) is more normal, and the shaft is bent on the upper part at a right angle; at that part the temporal region (fig. 6) is a large lozenge of bone, lying over the tegmen and horizontal canal; it is three times as wide as in the last kind. The postorbital process is very small. The rostral part of the parasphenoid (fig. 7, *pa.s.*) is larger relatively to the wide wings; there is a definite hinder, triangular process, and a triangular apophysis looking in the other direction between the fore part of the wings. The whole bone, instead of being nearly as long as the skull, as in *Dactylethra*, is only *two-fifths* its length. Here, the *left* vomer (*v.*) is larger than the right; the two protect the widely separated inner nostrils (*i.n.*); and the *nasal palate* is extremely like the "hard palate" of Man; all its elements are confluent.

70. *Phryniscus lævis.*—Adult female; 1¾ inch long. Ecuador.

This species, as far as my specimens show, is the largest of the three.

This is a slightly *longer* skull than that of *P. cruciger*, the breadth and the length of it being equal. On the whole, it is as asymmetrical as the other two; it has a broader snout, and is less ossified; it agrees in several things with *P. cruciger* more than with *P. varius*, but like the latter species it has no columella, and no "annulus;" its less intense ossification makes it a key to the difficulties in the other two. In one respect it is the most generalised of all the skulls of the *Phaneroglossal* types examined by me as yet, *e.g.*, the jugal arch is incomplete, as in the *Aglossa*.

The foramen magnum (Plate 40, fig. 6, *f.m.*) is large and obliquely superior; the occipital condyles (*oc.c.*) are large and posterior. Measured along the axis, the orbital region is longest, the nasal next, and the auditory the shortest; this part, however, is very wide, and the inner canals are very large and prominent; the posterior (*p.s.c.*) has, over it, a spiny epiotic prominence (fig. 8, *ep.*). There is a narrow synchondrosis both above and below, dividing the continuous lateral bony tracts (*pr.o., e.o.*).

The large, square parotic tracts are unossified at the tegminal edge (*t.ty.*), and this is barely covered by the squamosal (*sq.*). There is a large cordiform, and two smaller, oval, fontanelles and the dividing tracts are unossified over the middle part, the bone retreating, laterally, in the prootic region. The lateral roof-edge of the endocranium is moderately wide, and the main fontanelle is nearly covered by the roof bones (*f.p.*).

The prootic encloses the preauditory nerves (III., V.) up to the moderate optic fenestra (II.); there the cartilage (*o.s.*) is as extensive as the orbital part of the girdle-bone (*eth.*); that bone, however, runs far into the true nasal region, ossifying two-thirds of it; the broad overhanging snout is unossified in its front third. The girdle-bone runs well into its own wings, but it has not become anchylosed either with the palatines or vomers (fig. 7, *pa., v.*), as it does in the other species. The outer nostrils (*e.n.*) are very wide apart, the inner (*i.n.*) are scarcely more so, the whole fore face being so very broad. The outer angles of the largely ossified subnasal laminae (fig. 7) are but little more projecting than their general margin; the pro-rhinals (*p.rh.*) are slender and uncinate.

As in the other species, the ethmo-palatine cartilage is rounded off in front, so that there is but little pre-palatine; the cartilage is much diminished in size, but it can be seen outside the palatines and pterygoids (fig. 7, *pt.pa., pa., py.*), where they meet, and it becomes large as it approaches the pedicle and quadrate (fig. 8).

The two bony bars (*pa., py.*) have the average Bufonine development, the first is falcate, and the second is arcuate and forked; it becomes very wide behind, but the part covering the pedicle, and ossifying it, is short. The part inside the quadrate reaches no further backwards, for the hinge for the jaw (*q.c.*) goes scarcely further back than the pedicle on the left side, but has retreated twice as far on the right; this arrest of the metamorphosis is a correlate of the unfinished form of the pre-palatine. The condyles (*q.c.*) are large, reniform, and oblique, the left is least transverse of the two. The height of the suspensorium (fig. 8) is great, although it forms but little more than a right angle, in front, with the axis of the skull. The part of the pterygoid which passes inside the quadrate is pressed against the inner fork, and here there is a small crescentic aperture (blind), with its convexity inwards—a rudiment of the Eustachian passage (*eu.*).

There is, as in *P. rarius*, neither annulus nor columella; the stapes (figs. 7, 8, *st.*) is very large, oval, hollow, umbonate, and unossified. The stylo-hyal end of the hyoid (*st.h.*) is confluent, the whole bar (fig. 9, *c.hy.*) is narrow, and here is what I failed to find in the other two kinds, viz.: a projecting corner to the hypo-hyal. The notch in front of the basal plate is deep and round; the plate itself is very long, has only anterior lateral lobes, which are small and auriform; the long, very narrow cartilage (*b.h.br.*) is subcarinate, hollow above, and ends, behind, in two long, curved, highly divaricated thyro-hyals (*t.hy.*).

The investing bones are freer from the less ossified endocranium in this species; the fronto-parietals (*f.p.*) are separate, throughout, and show but little want of symmetry; their temporal angle is but little produced, and their fore ends are narrowed a little and oblique, they scarcely cover the fontanelle. The orbital edge is sharp, and the descending lamina thick (fig. 8), especially behind. The left nasal (*n.*) is the larger of the two, and projects further forwards; these bones are ovoidal, very convex, and show something of the ornamentation seen in *Bufo ornatus* and *Otilophus margaritifer*.

The feeble premaxillaries (*px.*) run far across as a sub-arcuate band under the snout, and their nasal processes are tilted forwards; these carry the inner labials (*u.l*^{1}.), to which are attached the outer valves (*u.l*^{2}.).

As in the other two, there is no septo-maxillary; the maxillaries (*mx.*) are high, roughly rounded in front, and gradually end in a blunt point, *which is free*. Their sides are sculptured, considerably, like the lateral part of the large, hollow nasals on which they rest (fig. 8). The *left* is considerably shorter than the right, and all that side of the face is drawn forwards. On the left side the jugal arch is unfinished for an extent equal to one-fifteenth part of the length of the skull; on the right side only half as far. The small tooth-like quadrato-jugals (*q.j.*) have their broad base grafted on the quadrate (*q.*); the *left* is the smaller bone.

The squamosals (*sq.*) are intermediate between those of the other two kinds; the supratemporal part is narrow, as in *P. cruciger* (Plate 41, fig. 1), but this large bone is bent on itself—almost as much as in *P. varius* (Plate 41, fig. 8).

The parasphenoid (*pa.s.*) is similar to that of the other two, but the cochleate median part is wider than the extended wings; where they are given off the bone is elevated on each side into a crescentic ridge, but there is no median apophysis imitating the triangular hinder part; the whole bone is considerably less than half the length of the skull.

The vomers (*v.*) are instructively separate in this species, explaining the ridges that defend the inside of the inner nostrils in the other two kinds.

They lie inside these small passages (*i.n.*), which like the outer openings are very wide apart; their shape is normal, they are cochleate, have a pre- and a post-narial spur, and an anterior lobe bent outwards upon the main part ; they are very accurately formed of a thin layer of dense bone, and are merely deficient in size, and in being edentulous—as in other Toads.

These three *species* have skulls that differ from one another more than whole *Families* of genera would be seen to do in highly specialised groups of Vertebrata, such as the Teleostean "Acanthoptera," or many groups of Carinate Birds. They come closest to the Neotropical Toads, with ornate skulls, but lie down below them, being more generalised, and they are evidently abortively developed.

These skulls form a great contrast to the Ranine "norma;" the main differences are as follows:—

1. The skull is below the average size, bears no teeth, and its ossification is excessive, and generalised.
2. There is much asymmetry in both the suspensoria, and the investing bones.
3. These latter are in several cases anchylosed to the endocranium; and in two species the palatines and vomers also.
4. The snout is very wide, and has a generalised rostrum.
5. The lower jaw is hinged to the head in front of the hind skull.
6. The tympano-Eustachian cleft is merely a small blind slit.
7. There is no "columella" nor "annulus," in two out of the three kinds, and in the third they are very small and arrested.
8. There are no septo-maxillaries, and the maxillaries either only touch the very small quadrato-jugal, or run short of it by a considerable space—as in the Aglossa.
9. The basal plate is very long and narrow, and has no hinder lateral lobes.
10. The left maxillary and nasal are less than the right (and sometimes the right nasal is the larger bone), and the parasphenoid is less than half the length of the skull.

Altogether, these may be said to lie at the very outside of their own sub-division, from the higher types of which they differ almost as much as the Aglossa; their place is between the ornate Toads and *Hylaplesia*.

Fifth Family. "ENGYSTOMIDÆ."

Ear rather imperfect; sacral apophyses dilated; toes free; no parotoids.

Genus *Engystoma*.

71. *Engystoma carolinense*.—Adult male; 11 lines long. Florida.

This is another very instructive instance of a small arrested skull (Plate 43, figs. 7, 8), the length and greatest breadth of which are equal, and of the hinge of the jaw being in front of the foramen ovale (V). Here the face forms a triangle, truncated in front, the moderately broad snout being transverse. This form makes the relative size of the hind skull very large, and yet the parotic processes are small; the occipital ring, and the auditory capsules are unusually wide—more than in the newly metamorphosed young of typical kinds. The occipital condyles ($oc.c.$) are large and posterior, they are separated by a space one half larger than their own breadth. The synchondrosis is rather wide, above and below, and there is an endosteal rudiment of both basi- and supraoccipital ($b.o.$, $s.o.$).

The roof runs on for a third the length of the cranial cavity, but it has two lesser fontanelles (fig. 7) in it; the main fontanelle is very large and heart-shaped, the front and side growths of the tegmen cranii being slight. The generalised occipito-auditory bones leave the very limited and rounded tegmen tympani (*t.ty.*) soft; in front, the bony matter forms a good margin in front of the foramen ovale (fig. 8, V.). Then more than half of the orbital walls are cartilaginous (*o.s.*), and in the middle of each space is the rather small optic fenestra (II.).

The girdle-bone (fig. 8) is in rudiments right and left, and there is no mesethmoidal bone. Each bony tract is composed of a large cochleate cranial part, separated by half its width from its fellow, and perforated by the orbito-nasal nerve. To this part there is a handle growing out at more than a right angle, and ending in an adze-shaped dilatation (fig. 8, *pa.*).

The fact is that the arrest of the bone towards the middle—above and below—is accompanied with an overgrowth beyond the proper alæ of this region, and the whole palatine tract, which has no ectostosis of its own, has become ossified from the ecto-ethmoidal, so that its three regions, *ethmo-*, *pre-*, and *post-*palatine, are all used up to form the handle to this curious "lateral ethmoid." The well-developed roof and floor of the nasal region (figs. 7 and 8) are entirely unossified ; there is a short prenasal (*p.n.*), the nostrils (*e.n.*) are almost tubular, and are defended by the normal valves (*u.l¹.u.l².*). Below, the internal nostrils (*i.n.*) are very wide apart, because of the breadth of the sub-nasal laminæ (fig. 8); these end in broad ear-shaped angles, outside, and near the middle have small apiculate pro-rhinals (*p.rh.*).

The pterygo-quadrate region is feeble, and the well-shaped, and distinctly-jointed pedicles (*pd.*), are a long distance apart. The pterygoid-bone (*py.*) has not ossified all the cartilaginous band (fig. 7); its forks are very short. The quadrate is partly ossified by the quadrato-jugal, and the condyles (*q.c.*) are oblique, and well-formed, with the hinder lobe of the trochlea much the larger. The Eustachian passage (*eu.*) is oval, and half the normal size. The annulus (fig. 11, *a.ty.*) is normal; its horns do not meet.

The stapes (figs. 7, 8, 11, and 11*a*, *st.*) is, relatively, the largest known to me, and it is also the hollowest, being like an oval Limpet-shell ; on its top is an oblong boss, and its substance is sub-osseous, except in front for a small space: the calcification is passing into true ossification over most of it. The columella is in one piece ; a considerable cartilaginous lobe passes within the stapes, and then it forms an arched rod, the medio-stapedial (*m.st.*); the extra-stapedial (*e.st.*) is peltate, and has a small rudiment of the supra-stapedial band. The lower hyoid bar has not quite lost its larval solidity; the stylo-hyal end (fig. 8, *st.h.*), is massive, and is only partially confluent above. The rest of the bar (fig. 10, *c.hy.*) is narrow, and only dilates a little in returning to the basal plate ; over the bend, an "extra-hyal" band is separated (*e.e.hy.*). The notch of the basal plate is large, the form wide, the lobes only moderately free ; a thick crest occupies the middle of its lower face, this expands in front, and is calcified; behind, there is a small bony basi-branchial ($b.br^1$.). The thyro-hyals ($t.hy$.) are normal, they are long, straight, and moderately divergent.

The mandible (fig. 9) is normal, but there is some endosteal deposit in the hinge.

The investing bones are very thin, but are dense, smooth laminæ; the fronto-parietals ($f.p$.) are separate; the two form a roughly pentagonal tract, they cover the hind skull largely, partly overlapping the canals, and well overlie the wide roof-space; their nasal suture is nearly transverse. The nasals (n.) are broad elegant shells of bone, showing the vascular *rete* through their thin clear substance; they meet along the middle, and send a curved horn over the sub-tubular nasal passage, in front, and another downwards, behind, to join the maxillary.

The promaxillaries, maxillaries, and quadrato-jugals ($px., mx., qj.$), are narrow, and, relatively, feeble bones; the squamosal (sq.) is very small, has an uncinate postorbital process, and binds merely on the anterior part of the small tegmen tympani ($t.ty$.).

The only other Batrachian known to me with a parasphenoid equal to this is *Pipa*; it is more like that of the "Urodeles" (Phil. Trans., 1877, Plates 24–26). It stretches from near the foramen magnum, to opposite the ethmoidal axillæ, and the width is such as nearly to reach the foramina ovalia (V.); it then expands in lozenge-shaped processes under the ear-capsules; the narrowing of the fore part takes place gently, and there the bone is like a (relatively) large spoon. The vomers (fig. 8, v.), on the contrary, are mere crescentic films of bone, bounding the inner edge of the internal nostrils ($i.n.$).

As compared with the "norma," this skull is—

1. Very small and arrested, the face being feeble, and the cranial cavity very large, relatively.

2. There are endosteal rudiments of the keystone and threshold bones—basi- and supraoccipitals.

3. The bones of the hind skull are generalised, or continuous on the same side.

4. The moieties of the girdle-bone are wide apart, and run into the palatine region, ossifying the cartilage, and suppressing the normal bony plate.

5. There is a distinct prenasal rostrum.

6. The parts of the middle ear are only feebly developed, and the stapes is a hollow semi-osseous shell.

7. The hyo-branchial apparatus is solid at its attachment, has an extra-hyal cartilage, is sub-crescentic, and possesses a distinct basi-branchial bone.

8. The roof- and floor-bones are very wide, the marginal bones narrow, and the vomers extremely minute; there are no septo-maxillaries.

B. b. *Toads with digital disks.*

First Family. "HYLÆDACTYLIDÆ."

Ear perfect; toes webbed; sacral apophyses dilated; and no parotoids.

Genus *Callula.*

72. *Callula pulchra.*—Adult female; $2\frac{3}{8}$ inches long. Pegu.

This is one of the shortest of the Batrachian skulls; the longest of which are short compared with those of other groups. The length, including the projecting occipital condyles, is scarcely more than three-fourths the breadth, but measured along the base, less than three-fourths (Plate 44, figs. 8, 9). The condyles of the quadrate end a little behind the fore edge of the parasphenoidal wings. If the outline of the face and cheeks were produced into a whole ellipse, the short diameter of that ellipse would bear, to the long, the proportion of 9 to 10.

The retreat of the hinge of the jaw is about equal to that of an average skull in the "Urodeles," and to what is seen in the type form at the time of transformation.

So far as to *form*; as to texture and substance, this skull is like that of any ordinary "Caducibranch" among the tailed Amphibia; and quite unlike the normal Anurous type of skull, for the bony tracts run past all the normal landmarks.

As to proportion of the parts, in detail, I may instance the usually small size of the "epi-hyal" element—transformed into the columella—which makes it seem, in most kinds, so poor a representative of the hyomandibular of Fishes; that is not a difficulty here (Plate 44, fig. 10, *m.st. e.st.*), for with its *new* form and functions it keeps its *old* size and proportions. The skull is well roofed, but much in the manner of the *Oriental Anura*, generally; the outer bony tracts are moderately thick, and keep their sutures well.

The occipital condyles (*oc.c.*) are large, bold, oval, and postero-inferior in position; they are separated by a gently emarginate tract two-thirds their own width.

The foramen magnum (fig. 8) is oblique, and looks upwards; the floor, behind, is quite ossified (fig. 9, *b.o.*); the roof has an exposed pentagonal tract of cartilage (fig. 8, *s.o.*), one side of which borders the foramen magnum; the passage for the 9th and 10th nerves *is not divided*. The auditory capsules show their canals strongly outside (fig. 8, *a.s.c., p.s.c., h.s.c.*), but the upper face is very much narrower than the lower, so that the stapedial series is well seen from above (fig. 8, *m.st.*), and when these are removed, the floor of the tympanum and the fenestra ovalis are seen. Beyond the horizontal canal the prootic region of bone projects like a handle, and this narrow prootic tract ends in a short and narrow unossified tegmen tympani (*t.ty.*), which is hooked in front, and from this hook, inwards, the face of the capsule is obliquely bevelled. Below (fig. 9), there is one continuous occipito-otic tract of bone from side to side, cartilage only remaining (*of basal origin*) to form a subconcave facet for the pedicle (*pd.*), and just where, behind this part, the stylo-hyal (*st.h.*) is confluent.

In this lower aspect we see the bone reaching round the foramen ovale (V.), right and left; and there, inside the bony bar that divides that passage from the large optic fenestra (II.) a little cartilage remains.

The orbital part of the skull is almost square, but it is wider in front than in the temporal region, which is pinched and bevelled. In front, the endocranium widens up to the ethmo-palatine wings, but the roof-bones are scant, externally (fig. 8).

The single fontanelle is heart-shaped, and occupies the middle third of the roof, or thereabouts; the hinder tegmen is longer than the tract in front of the fontanelle. The girdle-bone (*eth.*) is half the size of the cranial "large," for it is of great width, through the widening forwards of the skull, and it reaches from the optic fenestra to the verge of the nasal roofs (*al.n.*), above; only a small selvedge is left there, but the foremost third of the subnasal tract (*s.n.*) is unossified below; behind and above, the girdle-bone and prootics are confluent.

Both under and over views (figs. 8 and 9) give the idea of a *flat* skull; but the side view (fig. 10) corrects this error. Indeed, it is a very remarkable skull for a Batrachian, and more like that of a young Lizard with its high swelling brain, for the temporal region is very convex, and all along the height is considerable, and in remarkable contrast to the skulls of some of the large Tree-frogs (*e.g.*, *Rhacophorus*, *Phyllomedusa*) whose skulls are as flat as those of the *Rey tribe*. The orbital rim is developed along the fore half of the orbit, but not like a distinct flap; it is emarginate rather than lobate.

The nasal region is broad, and emarginate in front, there being no prenasal projection, and the bulging of the down-turned nasal roofs (*al.n.*) gives a somewhat bilobate form to the broad snout. Yet the snout is not wider than in the larger Oriental "Polypedatidæ," and not so broad as in *Rhacophorus*.

The nasals (*n.*) above, and the vomers (*v.*) below, leave the snout bare up to the rather small premaxillaries (*px.*), between whose processes the narrow pro-rhinals (fig. 9, *p.rh.*) are impacted. The submasal laminæ (*s.n.*) do not go far into the maxillaries (*mx.*), and these diverge rapidly from the nasal floor.

Nevertheless, the ethmo-palatines have the edge of their adze-shaped pre-palatine blades (*pr.pa.*) set between the wall and floor of these bones, and where the post-palatines end, there the cheeks are becoming very wide. The osseous counterparts of these tracts (*pa.*) are double on each side; the inner is the larger, and is not far from its fellow of the opposite side, and these two *internal* palatines are thin lanceolate, sharply crested bones. These sharp crests look as if they had once (*in a secular sense*) carried teeth; the saw-like crest on the same bones in *Bufo agua* suggest the same thing. The *left* outer bone is three times the size of the *right*, which is a little oval scale; these bones are inverted in the figure (fig. 9).

The well curved falcate fore region of the pterygoid (fig. 8, *pg.*) reaches half way over the prepalatine blade; one of its *forks* (fig. 9, *pal.*) is very short.

The core of the palato-suspensorial arch is persistent throughout, and can be seen, above (fig. 8), in the groove on the outer edge of the pterygoid. Also it forms a large oval condyle which fits on a shallow facet on the front of the ear-capsule (fig. 9, *pl.*)

The outer fork, or quadrate "pier," reaches but little further back (fig. 9), but is of great length when seen from the side (fig. 10); its condyle (figs. 9 and 10, *q.c.*) are large, convexo-concave, and reniform.

In descending, the quadrate pier is curved like a half-bent knee; its splint, the squamosal (*sq.*) fits to this bend; the hinder (inner) lobe of the condyle is higher than the other. The pedicle (*pd.*) grows inwards exactly half as far as it did in the Tadpole; the fixed band, from the foramen ovale to the enlarged part which becomes the condyle, has all been absorbed.

The quadrate (fig. 9) above the condyle is only slightly ossified by the quadrato-jugal. The annulus (*a.ty.*) is an open crescent; it is broad, and its front horn is higher than the other; its size is normal. The Eustachian (*eu.*) tube is small—half the normal size, and circular; the cavity altogether is very limited. The stapes (figs. 8, 10, 12) is very large, long-oval, with two anterior emarginations, and but little bossed. The columella, altogether, is relatively as large as its morphological counterpart in the *Skate*, viz.: the "hyomandibular."

The inter-stapedial (fig. 12, *i.st.*) is a short sub-oval segment of cartilage; the medio-stapedial (*m.st.*) is a long rod of bone, thick and clubbed, proximally; the supra-stapedial (*s.st.*) is a bud of cartilage growing from the upper edge of the short-stalked, peltate extra-stapedial (*e.st.*).

The mandible (fig. 10) is curved more than usual; its mento-Meckelian element (*m.mk.*) is unusually large, showing that metamorphosis did not lessen the actual size of the lower labials.

The stylo-hyal (fig. 9, *st.h.*) is continuous with the small tract of unossified cartilage at the outer part of the auditory region. The cerato-hyal (fig. 11, *c.hy.*) is rather broad, and turning back, as a hypo-hyal (*h.hy.*), without a lobe, it partly splits into two bands, a hyoid and "extra-hyal," as in some other kinds.

The notch of the basal plate is deep and the plate itself (*b.h.br.*) rather shorter than usual, and swollen. The fore lateral lobe is very large, and crenate in front; the hind lobe is normal, but outstanding. The thryo-hyals (*t.hy.*) are long and rather flat; between these roots there is a thick pentagonal wedge of bone—a "basi-branchial" (*b.br¹.*).

The investing bones are similar, as to thickness and strength, to those of the larger "Ranidæ" and "Polypedatidæ" of the same region.

The fronto-parietals (figs. 8, 10, *f.p.*) are very short and broad; their interorbital region forms, together, a square, with a small projection on each side, near the front; they are bevelled behind, as they pass into the expanded temporal region. Here, and over the fore half of the hind skull, they are modelled most accurately on to the endo-cranium, have rounded ends, and also are wrought into a rounded, low, transverse crest on each side.

In front these bones are scant where the ethmoidal wings grow out; they touch the nasals by their broad front edge, and the four bones, where they approach, leave a lozenge-shaped tract of the girdle-bone uncovered.

The nasals (figs. 8, 10, *n*.) are very large, very convex, and cover nearly as much ground as in the *Chelonia*. They are wide over the short facial process (fig. 10), meet at the middle, and leave, at their round, serrate, fore margin, the external nostrils, and front of the muzzle, bare. Notwithstanding the breadth of the muzzle, the premaxillaries (*px*.) are not large in the transverse direction, nor are the maxillaries (*mx*.) over large and high; the jugal process is curved downwards to meet the small quadrato-jugal (*q.j*.), which is slightly grafted on to the quadrate.

The squamosal (*sq*.) is but a modified "preopercular," with two short "horns," above, a postorbital, and a supra-temporal, spur; most of the bone is the descending bent splint to the suspensorium or quadrate; under the nostril, and its protecting labials (fig. 10, *e.n.*, $u.l^1.u.l^2$.) there is a noticeable septo-maxillary (*s.mx*.).

The parasphenoid (figs. 9, 10, *pa.s*) is very large and broad, and a good depth under the optic nerves (II.); it has the outline, on a smaller scale, of the whole cranial "boat," of which it forms the outer coating, save that it does not dilate in front; it is, however, very broad to the end, and is strongly clamped by *four* bones, viz.: the two *inner palatines* and the two vomers. These latter bones (*v*.) are flat, falcate, notched, in front, where they turn outwards, and touch the maxillaries; they are some distance apart; have no thick hind lobe, for they bear no teeth. They are quite as much *in front* of, as inside, the internal nostrils (*i.n*.), which are transverse, and reniform with the "hilus" behind.

The more important modifications of this skull, as compared with the *type*, are as follows:—

1. The general form, which is very short, wide, and deep.
2. A single, not large, fontanelle.
3. The more extensive ossification of the endocranium, the occipito-otics being continuous, and the right and left masses anchylosed, below; also the great extent of the girdle-bone.
4. The extent and narrowness of the "parotic wings."
5. The great breadth of the muzzle and the wide space between both outer and inner nares.
6. The absence of the almost universal dividing bar across the passage for the 9th and 10th nerves which arise, in this "Order," from a single ganglion.
7. The short pedicle, and long, but little retreated, quadrate region.
8. The double palatine bones.
9. The obliquity of the annulus tympanicus, and the very small cavity and opening.
10. The large stapes.
11. The great length of the columella, and its orbicular extra-stapedial.
12. The longitudinal subdivision of the lower part of the hyoid bar.

13. The solid basi-branchial bone.
14. The variation in some of the investing bones—
   a. Breadth of fronto-parietals.
   b. Great size of nasals.
   c. Feeble upper part of the long squamosals.
   d. The great breadth of the parasphenoid.
   e. The thin, shell-like, toothless vomers.

All these things seem to prove that *Cultula* is far removed from the typical Batrachian.

Second Family. HYLAPLESID.E.

Ear perfect; toes free; sacral apophyses cylindrical; no parotoids.

Genus *Hylaplesia*.

73. *Hylaplesia tinctoria.*—Adult female; 1¼ inch long. South America.

This is a long skull (Plate 44, figs. 1–7), the breadth being, contrary to rule, less than the length, or as 8 to 9. Its cheeks are very feeble, slightly bowed, but the cranium itself is very broad, and altogether forms an irregular oblong.

The quadrate condyles have only got as far back as the space between the Eustachian openings and the stylo-hyals (figs. 2 and 3, *q.c.*, *eu.*, *st.h.*). The breadth across the ethmo-palatines is three-fourths that across the quadrate hinges. The breadth across the ethmoidal wings, excluding the ethmo-palatines proper, is but little less than the breadth of the roof at the temporal angles; the roof is lessened very little, forwards; its orbital edge is gently sigmoid, as it narrows from the temples, then widens, and narrows again in front. The muzzle is by far the broadest I have seen; yet the nasal region is of normal *length*—or antero-posterior extent; but the *breadth* is just twice as great, so that the terms should be reversed. As in several of the "Caducibranchiate Urodeles," the endocranium is one continuous bony trough, for cartilage only remains as the occipital condyles (*oc.c.*); as the tegmen tympani (fig. 1, *t.ty.*); the facet for the pedicle (*pd.*); as a circle round the very small optic fenestra (II.); and in the fore and under parts of the transverse muzzle (*p.c.*, *s.n.*). The strength of the cranium is in great contrast with the weakness of the face; in the former, anchylosis has obliterated most of the landmarks, yet the fronto-sagittal suture is only lost behind (fig. 1, *f.p.**).

I can only do justice to this abnormal skull by comparing it with those of the two "Aglossal" types, *Pipa* and *Dactylethra* (see Phil. Trans., 1876, Plates 56–62).

As to the skull, there is no type amongst the tongue-bearing Frogs and Toads equal to this for softening down the hard distinction between them and the two tongueless "waifs" that make up the sub-order "Aglossa."

* This suture is too strongly marked, behind, in the figure, and so also is the slight ridge along the ethmoid.

The occipital condyles (*oc.c.*) are small and postero-external; *not so much outside as in Pipa*, but very much more than usual. They look as much upwards as downwards. The space between them is equal to the width of both, and is almost straight; the arch over the foramen magnum lies some way forwards, is very wide and somewhat angular. The condyles, and the projection backwards, first by the posterior and then by the horizontal canals (fig. 1, *ep.* by mistake), form a series of three irregular rounded steps on each side, the outer steps passing, point by point, further forwards; the outermost are a little in front of the superoccipital margin.

The auditory capsules are of the normal size, but they are very far apart, owing to the great width of the hind skull, which is altogether bony, not only by the thorough ossification of the cartilage right and left, above and below, but also by anchylosis of the inner with the outer elements, viz.: the fronto-parietals and parasphenoid (*f.p., pa.s.*).

The tegmen tympani (fig. 1, *t.ty.*) is broad in front and narrow behind, it is unossified; but the "canals" are all enclosed in bone. Below (fig. 2) there is the cartilaginous facet for the small, very external "pedicle" (*pd.*), and this tract just serves for union with the stylo-hyal (*st.h.*). But the fenestra ovalis (fig. 3, *eh.*) is well rimmed with bone; and the floor of the hind skull is sinuously flat, with very little scooping in the exoccipital region, right and left; the 9th and 10th nerves (IX., X.) have each their own passage, and this twin-hole is behind, rather than beneath, the arch. The facial and trifacial nerves (fig. 2, *pr.o.*, V.) also pass through a twin passage; from that point to the optic opening (II.) and its cartilaginous ring the skull is pinched. From the optic hole, forwards, the skull widens steadily up to the "axils" of the ethmoid, where the bone is gently scooped. Above, the inner bone can be seen there, in front of the fronto-parietals, and behind and between the nasals (*n.*), whose outlines can just be traced. The naked ethmo-septal bone is less than a third of the width of the roof; it projects in front a little, at the middle, and then the fore part, for about a third of the true nasal region, is unossified.

Below (figs. 2 and 5), the bone reaches as far forwards, but it is triangular with ragged sides; the right and left angles pass into the palatines, which are thrown across, inwards and backwards, as strong buttresses from the skull to the cheek (*eth., mx.*). The broad hatchet-shaped palatine (*pa.*) has ossified all its own overlying cartilage, the arcuate blade of which passes outside, in the maxillary, along the edge of the internal nostril (*i.n.*); this passage is large, oval, and looks forwards and inwards. The broad muzzle is hollow at the end of the septum nasi, and swells, sinuously, to the outside. The bones of the upper jaw (*px., mx., q.j.*) form a most elegant semi-oval—a bent bow—which at its arch lies under the two-leaved tract of subnasal cartilage (*s.n.l.*), and then binds upon the endocranial bars at three places on each side, namely, on the palatines, pterygoids, and quadrates (*pa., pg., q.c.*).

The outer nostrils (*e.n.*) are on the sides of the wide muzzle, and are twice the normal distance apart; they are well protected by the nasal roof (*al.n.*) above, by the first upper labial in front ($u.l^1.$), and by the second upper labial ($u.l^2.$) outside (see figs. 1, 2, 3, 5 and 6).

A little way outside the septum nasi the holes for the branches of the orbito-nasal nerves (fig. 5, *n.n.*) are very large and round; outside and in front of these outlets is the place for the "pro-rhinals," which are, at most, extremely feeble.*

The palatines (*pa.*) are confluent with the ethmoid; the pterygoids (*pg.*) are in great contrast with them, they are very feeble, the smallest, relatively, even with their cartilaginous model, that I have, as yet, seen.

Each bone is a gently curved needle, with a groove in which the cartilago lies outside and above, the point nearly touches the palatine, the "eye" is not finished by bone, it is the small, oval, outwardly-turned Eustachian opening (*eu.*). There the bone becomes forked, and the inner fork is a small foot, with a sub-convex cartilaginous sole—the pedicle (*pd.*). The outer fork is large, it is the unossified quadrate (fig. 7, *sp.*) which passes downwards and a little backwards, and ends in the large reniform condyle whose direction is unusually transverse, the fore lobe being but little in advance of the other.

The annulus (fig. 3, *a.ty.*) is of the normal diameter, its breadth moderate, its horns sharp and open, so that the whole is but three-fifths of a circle. This very open crescent is always a correlate of a small tympano-Eustachian cavity; and these may be combined, as they are here, with a very large columella; for an over large columella is *ichthyic*; when the metamorphosis is most complete the "epi-hyal" element is arrested whilst very small, and wrought into the elegant tympanic "key."

The stapes (figs. 3 and 7, *st.*) is large and oval, but with a concave deficiency in its antero-superior margin; it has no boss.

The columella is as large, proximally, and larger, distally, than the stapes; the inter-stapedial segment (*i.st.*) is as long as the stapes but has a concavity below; its distal third is ossified and united by suture with the obliquely clubbed upper end of the medio-stapedial (*m.st.*); the staff is no longer than the knob, and becoming cartilaginous obliquely, it passes into the orbicular extra-stapedial (*e.st.*). This part is evenly circular, and its diameter is equal to the length of the stapes; it has a supra-stapedial bud (*s.st.*), which is short, thick, and mammillate.

The mandible (fig. 4) is stronger than the cheek, its mento-Meckelians (*m.mk.*) are large, its coronoid region low, and its condyle long and cylindroidal; the dentary (*d.*) is short, and neither it nor the articular (*ar.*) hides more than half the pith (*mk.*).

* This little skull was worked out two or three years ago, and became accidentally dried when my work was nearly finished, so that I could not trace the pro-rhinals into the premaxillaries. I have no doubt of their existence; in skulls most akin to this, such as the species of *Phryniscus*, they arise outside and in front of the nerve-passage, and lie on each side as a very fine thread of hyaline cartilage between the laminæ of each premaxillary, in a mass of connective tissue, with the point inwards as in most of the toothless types. When so small as this, they always come away with the bone when it is detached in search for them; and *it* has to be stained and mounted for high powers—$\frac{1}{8}$ inch object glass—before the pro-rhinals can be demonstrated.

The stylo-hyals are articulated (or partly confluent) with the ear-mass; they are narrow, but the cerato-hyal (fig. 4, *c.hy.*) broadens up to the base, and the hypo-hyal (*h.hy.*) has its lobe behind, and not in front, as is the rule.

The basal plate is relatively rather small and long, the front lobe smallish and rounded, the hind lobe normal; the thyro-hyals (*t.hy.*) are large, and not so divergent as in most kinds.

The investing bones are as remarkable as the endoskeletal, and the two kinds are largely anchylosed together. The fronto-parietals (*f.p.*) form, together, an oblong tract, largely confluent with the underlying bone, and not anchylosed together, except behind, the fronto-sagittal suture being straight and nearly perfect. The two bones are longest at the middle; they form, in front, a low-angled projection; they largely overlap the endocranium, especially over the optic passages, where they are unusually thick (fig. 3). The moderate temporal angle is ribbed, and where it rides over the ear-mass, and coalesces with it, it is crenate, along a concave edge.

From one double canal-arch to the other (fig. 1, *f.p.*) the posterior parietal edge is straight; over the posterior canal the outlines are lost. I suspect that this part does not cover any secondary fontanelles.

The large, broad, conchoidal nasals (*n.*) are wide apart, and send down a short, blunt facial process (fig. 3, *f.p.* by mistake); they cover a third of the cartilaginous snout; as seen laterally, their edge is twice crescentically emarginate.

The twin bones that finish the inferior and external facial arch (*p.c.*) are wider than the rest of the arch, the maxillaries (*mx.*) soon narrowing in; and the small quadrato-jugals (*q.j.*), which simply articulate with the quadrate, are narrower still; there are no teeth in the jaws.

Both the nasal and palatine processes, as well as the body, of the premaxillaries, are small and feeble; the contiguous part of the maxillary, on each side, is rather high but does not reach the nasal at any point. The external nostrils, thrust out to the sides of the wide muzzle, are, however, well protected (figs. 3 and 6). The inner, upper labial (*u.l*¹.) is larger than the premaxillary; it is semi-oval, with a dilated base; it partly rests on the maxillary. The outer labial (*u.l*².) is pedate below and rounded above; it lies outside and below the nostril, and equals in size the projecting part of the roof (*al.n.*).

Between the two labials there is a lozenge-shaped septo-maxillary (fig. 6, *s.mx.*), and inside the outer labial there is a second larger bone—a pre-orbital (*p.ob.*). This bone is two-thirds the size of the cartilage it is attached to, and of the same shape, but reversed; it has its counterpart in *Pipa* (Phil. Trans., 1876, Plate 62).

The squamosal (figs. 1, 3, and 7, *sq.*) is better developed than in *Callula*; it is narrow, sigmoid above (fig. 1); is bent upon itself at less than a right angle; and the postorbital region of the upper part is hollowed out for the annulus and extra-stapedial, and helps to increase the size of the tympanic cavity.

This structure, which exists in some degree in most Anura, comes in this case very

DEVELOPMENT OF THE SKULL IN THE BATRACHIA.   251

near the remarkable modifications seen in *Pipa* (ibid., Plate 62, figs. 2 and 9, &c.); in that strange type the outer and middle ear meet in, and cover, a curious spoon-shaped process of the squamosal.

The parasphenoid (fig. 2, *pa.s.*) is evidently large, but its boundaries are very indistinct through anchylosis; it has between its three main rays a rather rare "apophysis;" this is a transverse, rounded projection, looking forwards, and downwards; it has a hollow in front of it, and a mammillate elevation behind.

There is not a trace of the right and left vomer; in this, again, we see a character in which this type agrees with *Pipa*.

But *Hylaplesia* agrees in some important characters with the other Aglossal type, viz.: *Dactylethra* (Phil. Trans., 1876, Plate 59). This similarity is to be seen in the *Salamandrian* extension of bone in the endocranium; in the *superior aspect* of the occipital condyles—not to the same extent, but as much seen above as below; in the huge size of the "middle ear," and especially its inter- and extra-stapedial elements. I look upon these similarities rather as an expression of the generalised nature of both kinds than as suggesting genetic relationships.

With the adopted and natural "norma" the skull of *Hylaplesia* presents remarkable contrasts; they may be summed up as follows:—

1. The general form of the skull is as much longer than what is typical, as that of *Callula* is shorter.
2. The great breadth and strength of the cranium as compared with the feebleness of the face.
3. The arrested retreat of the quadrates and their condyles.
4. The extreme breadth of the muzzle.
5. The intense ossification of the endocranium, and the anchylosis with it of the investing bones.
6. The absence of vomers.
7. The apophysis on the parasphenoid.
8. The superior as well as inferior aspect of the occipital condyles.
9. The small size of the tympano-Eustachian cavity, and the increase of room in that "cleft" by the hollowing out of the squamosal, above.
10. The very open annulus and large stapes.
11. The huge size, relatively, of the columella, and especially of its proximal and distal elements.
12. The large size of the three pairs of labials—the lower as mento-Meckelians.
13. The presence of a considerable pre-orbital besides the septo-maxillary.
14. The reversed position of the hypo-hyal lobe, the smallness of the basal plate, and large size of the thyro-hyals.

These are some of the more outstanding peculiarities of this strong little skull, whose generalised nature is shown, also, in its unlooked-for agreement with the archaic and non-typical skulls of the "Aglossa."

This archaic, generalised skull helps us to see how severe, "in number, weight, and measure," the morphological law is, that has, *at last*, reduced the Anurous type of skull to the *elected simplicity* of that of the Common Frog.[*]

SUMMARY.

A.—*Primitive form of chondrocranium.*

In a Common Frog or Toad, soon after hatching, whilst the *true outer* (or cutaneous) gills are present, cartilage appears in the cephalic region.

This first endoskeletal framework consists of three sub-parallel bands on each side, that converge a little forwards, are some distance apart, and are almost entirely in front of the notochord (Phil. Trans., 1876, Plate 55, figs. 1, 2).

Other cartilages form about the same time, immediately under the skin, as labials in front, and as branchial pouches behind; but these I shall not now describe; only the true endocranial elements.

The innermost pair of bands, together, form a lyriform structure; they are the largest; by their hind part they embrace the notochord at its apex; they diverge suddenly, enclose a large pyriform space under the fore-brain, converge nearly to touching, and then diverge again, as short, broad, decurved horns.

These are the "trabeculæ cranii;" they are *para*-chordal, behind, and the rest of each bar is *pro*-chordal: there is no other parachordal cartilage, as yet; the huge notochord, which only gradually lessens to its rounded end, has, right and left, two pairs of "muscle-plates" enclosing it in the cranial region.

Where the trabeculæ are most bent, behind and in front of the eye-ball, there the second band is continuous with the trabeculæ; by this double conjugation it encloses an oval (sub-ocular) space.

In front, each of these outer bands turns inwards towards the horns of the trabeculæ, and develops an oval, short, segment, which also turns inwards. This twice-conjugated, second bar, is the "suspensorium" of the mandible; the short segment is the mandible itself, or the articulo-Meckelian rod. The hinder conjugating part is the "pedicle," and the fore band is the pterygo-palatine rudiment.

A little behind that rudiment, on its under face, the suspensorium has a broad lozenge-shaped cartilage articulated with it—the third band; this is only half as long, but twice as broad as the second, and is the lower half of the hyoid arch—the cerato-hyal: the upper half, or "epi-hyal," is not developed until two or three months after the transformation of the Tadpole. Opposite the junction of each first and second muscle-plate, there is a hollow ball of cartilage, unfinished (membranous) above; these

[*] For the description of the skulls of the "Aglossa"—Nos. 74 and 75—the reader is referred to my former paper (Phil. Trans., 1876, Plates 56–62, pp. 625–665); and "Summary," *infra*, p. 255.

are the auditory capsules; large masses of cells, the rudiments of the ganglia of the 5th and 7th nerves, separate these globes from the pedicles of the suspensoria.

The eye-balls are not taken into account; the nasal sacs are entirely membranous, at present.

The parts displayed in the dissected head, and in sections through all the parts of the head, lend colour to the suggestion that all the three pairs of cartilages are serially homologous: I believe this to be quite opposed to the true interpretation of the parts; that the inner bands (trabeculæ) are prematurely developed, paired, axial parts, growing beyond the notochord, but *parachordal* in their hinder part: moreover, in other types, as the "Urodeles" and "Marsipobranchs," the part embracing the notochord is, from the first, much larger than in the "Anura."

B. a.—*Perfect chondrocranium:—before the formation of bony centres—in the "Phaneroglossa."*

The simple cartilaginous bars that were seen at first are soon developed into a perfect chondrocranium of the *Petromyzine* type.

The cranial notochord, besides its own mesoblastic sheath, which is now and then chondrified even in the "Anura," becomes enclosed, right and left, between two solid bars of cartilage—the extension, backwards, of the apices of the trabeculæ (and not as separate plates, as in the "Urodeles"); and these two basal plates are fused together, for a short distance, in front of the notochord (see Phil. Trans., 1876, Plate 55, and the figures of larval skulls in the present paper).

In the nasal region the trabeculæ coalesce, and then send their elongated horns forwards, and downwards; in the interorbital region, each bar sends upwards a crest, which becomes thick and bulbous near the ethmoidal region, or closing-in part of the skull. After a while, behind the ear-capsules, a wall, and then a roof, is formed—the occipital arch.

The large, bowed, twice-conjugated suspensoria develop a crest along their outer edge, and this grows into a large leafy plate in the ethmoidal region—the "orbitar process," which may (exceptionally), as in *Bufo vulgaris*, coalesce with the ethmoid.

The free mandibles grow larger, form a condyle and an "olecranon," and carry the (suctorial) lower labials between them, as the cornua trabeculæ carry the upper or overlapping labials.

The third bar (hyoid—second visceral arch) grows more perfect in form, as well as gets larger in size, but retains its primitive place under the antorbital region.

The auditory (parachordal), orbital, and nasal regions are nearly equal in length, and the auditory sacs become gradually completely invested with their own cartilaginous coat, which takes the form of the swelling and arching cavities within.

The azygous tubular tract of skeletal tissue surrounding the notochord—membranous or cartilaginous, as the case may be—is stopped in front by the coalescence of the basal plates: they, then diverging, leave a large *pituitary* space—for some days membranous. In two places an "inter-trabecular" tract of cartilage appears; behind, filling in this large inter-orbital pituitary space, flat and thin; and in front, over the coalesced parts, as a rising wall between the nasal sacs: this is the first rudiment of the perpendicular ethmoid and septum nasi.

Afterwards, but not in larval life, a *third* inter-trabecular tract fills in the re-entering angle between the cornua trabeculæ.

The upgrowing wall turns over above, and forms more or less roof; this is deficient along the middle of the interorbital part of the skull, also often, in one, or mostly two, places, between the exit of the 5th nerves—these are the main and secondary "fontanelles."

The ear-sacs get, now, a *secondary floor* from the basal plate; parachordal tracts reach to the cleft now forming in their fundus, which is becoming the fenestra ovalis, right and left, the membrane closing which chondrifies as the stapes.

The nasal sacs now acquire a distinct roof of cartilage, which soon coalesces with the middle wall or perpendicular plate; the ethmoidal end of the skull, also, besides closing in the skull-cavity, grows out as wings ("aliethmoids"), behind and round the nasal sacs.

The quadrate end of the suspensorium sends in a spike towards the cornu trabeculæ in front of the inner nasal opening; this is generally attached by a ligament, but sometimes touches the cornu; this is the rudiment of the pre-palatine. A ridge appears on the pterygo-palatine conjugation, which often grows into the orbital fenestra as a flap; this is the post-palatine rudiment.

One or two pairs of cartilaginous plates are now well-formed on the upper lip and answer to the anterior dorsal plate and angulo-labials of the Lamprey: and the thick divided, semilunar mass of cartilage, between the free mandibles, forms the sucking disk (or inferior labials).

The *four* subcutaneous cartilages margining the *three* branchial clefts have grown, the first and fourth into thin, baggy pouches, and the second and third into broad bars. These are covered with a free growth of branchial tufts, that break out between the clefts, but are hidden under the great membranous operculum—closed entirely on the right side.

The hyo-branchial cleft is still open, but the mandibulo-hyoid never opens *outside*; over it the opercular membrane becomes cartilaginous, the cartilage growing downwards and forwards from the upper edge of the auditory sac, or from the elbow of the suspensorium; it becomes an independent spiracular cartilage.

The massive cerato-hyals are conjugated by simple cartilage—the basi-hyal; behind it a pear-shaped plate appears, its base foremost; it represents two basi-branchials; embracing this, and running backwards there is a pair of broad, flat, hypo-branchials, and each of these gives off four rudimentary *true*, or intra-branchial, cartilages—small "cerato-branchials"—with, as is the case in the hyoid, at present, no upper element.

This inferior, arrested, *intra*-branchial framework is covered with a "lower velum," and the projection of the four pairs of rudimentary cerato-branchials makes it have a crenate margin (Plate 1, fig. 4); it is the broad rudiment of the membrane which runs under the pharynx in the Lamprey at the entrance of the "branchial canal;" the copious growths of gill-tufts on each side, divided by the three clefts, lie under it, partially covered.

All these larval structures may exist in the chondrocranium before the *Petromyzine* skull is modified either by *outer* or *inner* bones, and the skull of the larva explains, and is explained by, the skull of *Petromyzon* and its congeners.

To the Morphologist there is, here, a pause; but, in fact, in the growth of the Tadpole's skull, there is none; changes begin soon, and the work of transformation then goes on steadily.

B. *b.*—*Perfect chondrocranium—before the formation of bony centres—in the " Aglossa."*

The chondrocranium of the larval *Dactylethra* (Phil. Trans., 1876, Plate 56) is extremely flat and outspread; it has all the essential parts of a Batrachian larval skull, nevertheless. Its form is almost triangular, with the base in front; and although the condyles of the quadrate are carried forwards almost as far as the trabecular horns, they are a great distance apart. Even behind, the skull is greatly widened by lateral outgrowths of cartilage, a wide "tegmen tympani" already growing from each auditory capsule; and from this, and confluent with it, a very large ear-shaped "spiracular" cartilage, which spreads wider than, and almost touches the base of, the small orbitar process, and the hinge for the cerato-hyal. The suspensoria, also, are wide bands running, without an outbend, forwards and gently outwards, to the front of the orbitar process. They then bend inwards and are confluent by a very wide band with the trabeculæ, each band being pterygo-palatine and pre-palatine all in one plate. As in Skates, the trabeculæ are enormous; they are separated from the suspensoria by a very narrow subocular fenestra, and rise as convexo-concave shells, converging at the *hinder third* of the interorbital region to finish the cranial cavity. As in Skates, however, there is a long hollow from it into the nasal trough, which soon widens again. The cornua trabeculæ are stretched out, as wide arms, in front, and their free end is confluent with the pre-palatine spur. They are narrowish bands, and curve round the front of nasal passages which are near the frontal margin, but very wide apart. Their fore margin shows no notch, for they are completely fused together by a large, but indistinct, inter-trabecular tract. The continuous cartilaginous labial, in front of the cornua, is partly confluent with them, and runs wider out; it is the same as the "anterior dorsal cartilage" of the Lamprey; the "angular cartilage" of that Fish is represented by the suddenly narrowed part of this band, which runs back as the pith of the long oval palpus, right and left—a "feeler" or *barbel* which reaches to the end of the abdomen. The lower labials are small, rounded rods, placed *across* the space between the long, slender mandibles, and not arranged as a semicircle to support a sucking disk. The hyoids are very massive; and the branchial pouches are all fused together into a case, with more slits outside—through which the branchial tufts *do not* protrude; and with wider openings inside.

The larval chondrocranium of *Pipa* is similar to that of *Dactylethra*, but differs in several particulars (ibid., Plates 60, 61). The auditory capsules and cranial notochord reach almost half way to the frontal wall. The trabeculæ are completely fused up to their *retral* cornua. The "sub-ocular fenestra" is a mere crescentic slit with its convexity outwards. The pterygo-palatine band is narrow and very small in front of that slit, and the pre-palatine projects forwards under the dilated end of each hook-shaped trabecular cornu. The nasal openings are nearer together; the quadrate condyles are wide apart but not so far forwards in position; the pedicle is merely the inner and posterior horn of a four-winged, dilated, suspensorium; the otic process is, already, nearly as wide as the pedicle, and longer.

The condyle for the cerato-hyal is a large outwardly-projecting, oblique, pyriform process; over it there is a small orbitar process. The labials are scarcely chondrified as yet, and there are no palpi; as in *Man*, and the *Pig*, the Meckelian rods meet and coalesce, directly; and I find no inferior labials. The cerato-hyals are large, oblong, square above, and with a hinder submesial process; *they are entirely absorbed before hatching.*

Scarcely a trace of branchiæ appear; the intra-branchial arches are merely represented by a basi-hypobranchial band which, at the mid line, unites the extra-branchials. These are continuous above and below, and form a *flat* plate on each side with three lanceolate slits in it, and a dentated upper margin: this growth has its upper part largely absorbed on either side before hatching; at which time a very remarkable hypo-basibranchial tract has developed, which in the early larva was merely a conjugational band uniting the extra-branchials.*

C.—*Order of appearance of parts in the skull during metamorphosis in the*
"Phaneroglossa."

The first splint-bone to appear in the Tadpole's skull is the first we find in the gradation of the types, viz.: the parasphenoid, *first seen,* outside the merely "Cartilaginous" Fishes, in the Acipenseridæ and the Dipnoi.

The former have no ex-occipital, but the latter have; this endoskeletal bone comes

* For the transformation of these two extraordinary types of larval crania, I must refer to the former paper (Part II.); I still have to show how the adult skulls of the *Aglossa* differ from those of typical *Phaneroglossa.*

next in the Tadpole, after it the prootic, right and left. Later on the fronto-parietals, and after them the premaxillaries appear; much later the maxillaries, nasals and quadrato-jugals, and later still, as a rule, the squamosals: last of all the *generalised* annular ethmoid or "girdle-bone;" it is mostly contemporary with the shaft of the columella.

But the appearance of the bony tracts, outer or inner, is but a small part of the matter; the changes in the fundamental chondrocranium are the most marvellous. The organic concussion of the metamorphosis, so to speak, causes the removal of some things that are disturbed by it, and the transformation of others, which, in new forms, are put to new uses.

Whilst the large upper labials are still at their highest development two new small pairs appear, fitting, the inner to the end of the nasal process of each premaxillary, and the outer round each outer nostril, as a valve; this is at the time that the opercular (or "spiracular") cartilage becomes free.

The band from the junction of the pedicle with the skull in front of the ear to the end of the suspensorium with its condyle for the mandible, is, in the larva, a long arcuate tract two-thirds the length of the head: the cross band uniting this with the ethmoid, on the contrary, is very short. These relative lengths are, during metamorphosis, entirely reversed; the suspensorium becomes short and the pterygo-palatine band two-thirds the length of the skull. The pedicle, and the longitudinal part of the suspensorium (without the orbitar process), together, answering to the suborbital band of the Lamprey; the upper and outer *regions* are absorbed.

As to position, also, the condyle for the short mandible gets, in many cases, to the end of, or even behind, the skull, whereas it did reach nearly to the front; and then carries a long mandible which passes to the front. The outer or hyoid bar is carried with this pier, and gets under the fore part of the ear-capsules instead of, as once, in front of the eye-ball. Thus a small *terminal* suctorial, is exchanged for a widely-gaping *inferior*, mouth; a mouth like that of a Lamprey is exchanged for one like that of a Crocodile.

The cornua trabeculæ become merely the angulated subnasal tract or floor, united in front and above by the intertrabecular tract. Each cornu near its inner edge, below, develops a small hook-shaped secondary cornu, the "pro-rhinal," which is imbedded between the laminæ of the premaxillary.

Whilst the pterygo-palatines are becoming, from a short band, a long arch, the palatine bones appear in front, transversely, and the pterygoid bones, behind, as a forked tract—forked to invest the lower face of the *new* and *short* pedicle and the inner face of the quadrate region.

In some forms—species of *Rana*—the pedicle has its own tract of bone—a distinct, "metapterygoid." Whilst the bones are appearing the sub-apical part of the suspensorium grows into a large leaf of cartilage, ready to become pedate, with a terminal, sub-convex, oval condyle. This condyle, so far from its original root, articulates with the *sub-auditory wing* of the basal plate, so also does the upper hyoid bar (columella), and the stylo-hyal end of the lower bar, except, when, as in *Phyllomedusa bicolor* and *Acris Pickeringii*, it fuses with the hinder edge of the pedicle.

D.—*What larval structures are retained (in a modified form), and what parts are moulted or absorbed.*

Some of the structures are but little changed, but the new things of the permanent creature are rapidly forming; we can see what is cast off, what is kept, and what things are superadded.

As to several of the larval structures, the morphological force makes clean work of them; some are shed, completely, or moulted off, as the horny jaws, dentated rugæ, and papillose oral appendages. Whilst the huge upper labials are absorbed, the lower labials are used for making the "chin" of the adult; the spiracular cartilage—coming in slowly—is kept to form the annulus tympanicus, but the four pairs of "extra-branchials" are absorbed, and remnants only of them and of the true *inner* visceral arches are retained. A summer *Bee-hive* is not a more lively place than a Tadpole's organism at this stage; every cell is busy, every element is moving, working, and changing; the size, shape, and direction of organs and parts are all being lessened, or made larger; moulded into new shapes, and turned fore or aft, outwards or inwards, as the governing force listeth. As the result of all this working towards a new zoological end, that which began life as a *Fish* but little higher than a Lamprey, fulfills its after-life as a *Reptile* but little lower than a Turtle.

After the horny jaws and dentated horny rugæ have been moulted, before the upper labials have been absorbed, a new pair appear on each side, attached to the nasal processes of the premaxillaries and to the outer nostrils. The lower labials become relatively less, get into a line continuous with that of the mandible of the same side, coalesce with those bars and become ossified by the grafting of the dentaries upon them.

The "orbitar processes," after their transportation to a post-oral position, become, in most cases, quite absorbed; so also, in most cases, does the upper part of the pedicle; it is retained in some, as in *Bufo vulgaris*, where the pterygoid bone binds down upon the "lobe," or the sub-apical, enlarged part of the pedicle.

The "spiracular cartilage" may chondrify continuously, either with the "elbow" of the pedicle (otic process), or with the tegmen tympani. It may become lessened in size, considerably, before it takes on its adult form, as the annulus tympanicus.

As I have said, the four other outer cartilages belonging to the 3rd, 4th, 5th, and 6th arches, viz.: the extra-branchials, become almost entirely absorbed; in some cases the proper hyoid bar does, also.

I believe that I am right in calling the larval rudiment of the "annulus tympanicus" the *spiracular cartilage*; I put it into the category of the "inter-branchial rays"—as in the Sharks; the "tegmen tympani" is the "extra-visceral" tract from which it grows, as a rule.

Certainly, in many of the Anura, more or less cartilage appears in the outer edge of the hyoid bar, but quite distinct from it, corresponding exactly to the basal part of the pectinated inter-branchials of the hyoid bar in the Shark and *Chimæra*.

The hyoid bars, the soft basi-hyal, the rudimentary cerato-branchials, the paired hypo-branchials, and the median basi-branchial all fuse together to become the hyobranchial apparatus of the adult. The narrowed and ossified hypo-branchials are true "thyro-hyals," like those of the Mammal.

Among the newly appearing parts are the stapes (*early*), and the columella (*late*). The stapes appears about the time of the first bony tracts; but the columella, as a rule, not until some months after metamorphosis. It appears whilst the tail is still large in *Pseudis*; and long before hatching in *Pipa*; in *Dactylethra* it is as late as in the ordinary " Phaneroglossa."

E.—*The normal (or Ranine) adult skull compared with sub-typical and aberrant forms in the " Anura," generally, where larval structures are in some degree retained, or where generalised (ichthyic) characters turn up; lastly, the residuum of characters which are universal.*

The prootics (or *spheno-prootics*) and ex-occipitals are not separately ossified, as in the " norma," in *Pseudis, Dactylethra*, and others. In some kinds, as *Acris, Pseudophryne, Diplopelma*, there is a small basioccipital; in some of these small kinds there is an evident endosteal super-occipital.

The girdle-bone is absent in some dwarf kinds, as in some small West African *Rana*, in *Gomphobates*, and in some species of *Diplopelma*.

The girdle-bone is in two pieces in *Diplopelma ornatum*, in *Engystoma*, and in *Pseudophryne*; there is a large T-shaped, persistent membrane bone over the ethmoidal region in *Dactylethra*, a crescentic bar in *Rappia bicolor*, and a similar bone, not persistent, in *Rana temporaria*.

As a rule, there are two lesser fontanelles over the hind skull besides the main space, but in many kinds there is only the large one; in *Alytes* there is one lesser, hinder fontanelle.

In some of the Frogs (*Rana*, sp.), still better in the Hylidæ, there is a superorbital "eave" of cartilage; in *Phyllomedusa bicolor, Hyla rubra*, and especially in *Alytes obstetricans*, there is a separate superorbital cartilage, besides.

The nasal region is often largely ossified from the girdle-bone; this tract may be a true " sphenethmoid," as in *Dactylethra*, where it does not finish the ethmoidal region in front, but runs back to the auditory capsules.

The nasal roofs are only partially covered by their own proper cartilage in the genus *Bufo*, in *Dactylethra*, and in some others; the pro-rhinals are distinct in *Rana esculenta* and in *Dactylethra*; the two nasal fenestræ of *Bufo* answer to the series of slits in *Myxine*.

In *Bufo vulgaris* and some other species of that genus (not in *B. calamita*) the ethmo-palatine becomes segmented from the ethmoid, in front, and from the pterygoid cartilage, behind; in *B. vulgaris*, also, the pedicle does not lose its apex.

In *Bufo ornatus* and in *Otilophus margaritifer*, as also in a *Ruppia* from Lagos, and in *Diplopelma berdmorei*, the "orbitar process" is permanent, but small.

In many kinds the quadrato-jugal bone sets up ossification in the quadrate region, often to a considerable extent. In *Dactylethra*, *Pelodytes*, and *Bombinator* there is no palatine bone; it is often subdivided into two, often has a sharp crest, and is composed of several bones in *Rana pipiens*, and may be entirely endosteal as in some kinds of *Diplopelma*. There is a mesopterygoid bone in *Rana pipiens*, and a metapterygoid in several *Ranæ*.

Except, probably, in some arrested forms, the "mento-Meckelian" bone is always formed in the "Phaneroglossa," but is not distinct in the "Aglossa;" in these latter types the endosteal and ectosteal "articulare" are completely fused, and ossify the articulo-Meckelian cartilage very largely; this takes place, to a less extent, in many of the "Phaneroglossa."

The *upper hyoid element* or columella is absent in *Pseudophryne Bibronii*, *Phryniscus lævis*, *Phryniscus varius*, and *Bombinator igneus*; there is a small bony rudiment in *Pelobates fuscus*, and a small unossified rudiment in *Rhinoderma Darwinii*.

I find no trace of an *annulus tympanicus* in *Pelobates*, *Bombinator*, nor in two out of three of the species of *Phryniscus*, viz.: *P. lævis* and *P. varius*; it exists without a columella in *Pseudophryne Bibronii*, and with a rudiment in *Rhinoderma*.

In several high kinds, viz.: *Rana halecina*, *R. palustris*, *Phyllomedusa bicolor*, &c., the proximal part of the columella (shaft or "medio-stapedial") becomes segmented off from the distal ("extra-stapedial"). This is not in conformity with the segmentation of a normal branchial arch, or even of the hyoid of *Chimæra*, where there is a well-formed "pharyngo-hyal" above the "epi-hyal."

In the Sturgeon this kind of division does take place; the parts are termed in it "hyo-mandibular" and "symplectic."

But in the last-mentioned type the pharyngo-branchials themselves are in two segments.[*]

I find that in about half of the Anura the upper hyoid element is subdivided as in a normal branchial arch: here the terms for the segments are "inter-stapedial" for the upper, and "medio-stapedial" for the lower, hyal piece.

The lower band of the hyoid, which was coeval in development with the mandible, and *antorbital* in position, becomes, as a rule, confluent with the floor of the tympanum—a growth from the basal plate or "parachordal."

It may articulate with that part as in the Common Frog, or be suspended by a longish ligament as in *Dactylethra*, or the bar may be partially absorbed (often on one side) as in *Rana hexadactyla*, or the middle part may become fibrous as in *Rana tigrina*, or the proximal and distal parts, as in *Bufo agua*, or the whole may vanish as

[*] As pointed out to me by Mr. Howes.

in *Xenophrys monticola* and in *Pipa*, or its apex may coalesce with the pedicle, as in *Phyllomedusa* and *Acris*; but as a rule it is confluent above.

In *Cailula*, *Diplopelma*, and *Engystoma*, the basi-hyobranchial plate is sub-carinate below, and a basi-branchial bone appears between the thyro-hyals; in *Alytes* and *Pelodytes*, a V-shaped splint is applied to this part below.

In the genus *Phryniscus* the basal plate is almost as narrow as in Fishes; its postero-lateral processes are ossified in *Bombinator*, and the cerato-hyals have two small centres in them on each side, in *Pelobates*; in the two last, and in some of the "Hylidæ," the front notch is so large as to make the basal plate like two parts united by a conjugational band, behind.

In *Hyla rubra* and *H. Ewingii* the basal plate extends, behind, as the rudiment of an additional basi-branchial piece.

The "investing bones" show some curious variations in this group. The *parasphenoid* has a distinct centre, in front, in *Rana pipiens* and in *R. halecina*.

The *vomers* are absent in *Pipa* and *Hylaplesia*; single in *Dactylethra*; and confluent with the ossified nasal floor in some *Phrynisci* (e.g., *P. cruciger* and *P. varius*): they remain distinct in *P. lævis*.

The *fronto-parietals* are found right and left, as symmetrical bones, and may *divide* into frontal and parietal for a time; these soon coalesce again, and in many cases the median suture is lost; in some cases (*Camariolius tasmaniensis*, *Acris Pickeringii*, and *Rappia bicolor*) they are extremely small.

The *nasals* are always present; they are extremely feeble in *Dactylethra*.

The *septo-maxillaries* are very inconstant; in some kinds, as *Pipa*, *Hylaplesia*, *Bufo ornatus*, and *Rana pipiens*, there is a second præ-orbital behind on each side of them; in the latter a "lacrymal" is also present: in *Rana pipiens* there is no proper septo-maxillary, but several irregular palato-maxillaries.

The *premaxillaries* are always double, and the *maxillaries* nearly always articulate with the *quadrato-jugal* to form a perfect cheek bar; this bar is deficient in the "Aglossa" and in *Phryniscus lævis*; in *Phryniscus* the investing bones are very unequal, right and left; and in *P. varius* and *P. cruciger* the cheek is scarcely finished; the quadrato-jugal is suppressed in *Pipa*.

The *squamosal* is remarkably modified in the Aglossa, being hollowed out to help to form the drum-cavity in *Pipa* and having its lower part aborted in *Dactylethra*. The hollowing out of the squamosal under the "annulus," is seen again in *Hylaplesia*, and more or less in many kinds. Its postorbital process is, at times, very short, in others very long.

In *Dactylethra*, only, have I found any rudiment of the bony "annulus tympanicus," it has two ossicles of this nature on each side; and it only has a single vomer.

The *dentary* in the "Phaneroglossa" constantly grafts itself on the inferior labial after that has become fused with the distal end of the Meckelian rod.

In *Rana pipiens* there are two or three splint bones (*inter-suspensorials*) between the quadrate cartilage and the hind crus of the pterygoid bone.

In some types (e.g., *Ceratophrys*, *Calyptocephalus*, *Pelobates*, and *Nototrema*) the surface of the investing bones is ornate, and almost Ganoid.

The following endoskeletal and exoskeletal bones are constant throughout the "Anura" in the adult skull :—

1. The ex-occipitals.*
2. The prootics.
3. The pterygoids.
4. The articulars.
5. The thyro-hyals.
6. The parasphenoid.
7. The nasals.
8. The fronto-parietals.
9. The premaxillaries.
10. The maxillaries.
11. The squamosals.
12. The dentaries.

The skull is always finished with cartilage, below (contrary to the Urodeles), and never finished with cartilage, above (as in the Sharks).

The occipital condyles are always double, and quite distinct, often wide apart; but the occipital arch is not always restricted to a right and left bone; there may be a rudiment, above and below, of a median centre.

The skull is always closed in front by the ethmoid (cartilage or bone): it is at times very unfinished and membranous in the orbital region (e.g., *Rappia bicolor*, *Camariolius tasmaniensis*, and *Acris Pickeringii*).

The whole (true) ethmoidal region is formed by the trabeculæ and intertrabecula; the nasal roofs are distinct cartilages at first (like the eye-balls and auditory capsules); their floor is formed by the trabeculæ.

F.—*On the likeness and unlikeness of the skulls of the "Urodela" and "Anura."*

The difference between the skull of a Tadpole and that of a larval Urodele is very great from the first (see Phil. Trans., 1877, Plates 22–24). In the latter the trabeculæ largely embrace the huge cranial notochord, and are some time before they close in in front of the membranous space below the fore brain. They finish their cornua, in the internasal region, afterwards, and they do not finish the occipital floor; *distinct* parachordals appear there.

The suspensorium is in them, at first, quite distinct from the trabecula, when it does coalesce it unites, first, with the wall of the skull, above the orbito-nasal nerve; in the Tadpole it is from the first *continuous*, as cartilage, below that nerve. Their ethmo-

* The prootics and ex-occipitals do not always arise from independent centres.

palatine is always a distinct cartilage and rather late in appearance; the pterygoid cartilage is in the Urodele an outgrowth from the suspensorium—a little later in appearing; the post-palatine is a separate cartilage, also.

The hinge for the jaw is in them, at first, just where it gets to be in the transforming Tadpole, when the tail is nearly absorbed; the cerato-hyal has in that stage the same position it has, at first, in the embryo of the Urodele. There are no labials, no extra-branchials, and no spiracular cartilage in the Urodeles,* and they develop *three* or *four* true *intra-*branchial arches.

The *floor* is never finished with cartilage any more than the *roof;* the former is often divided into two compartments by the persistence, for a long period, of the apices of the trabeculæ, which keep separate from the *posterior parachordal* tracts.

In one species (viz.: *Ranodon sibiricus*, see WIEDERSHEIM, "Das Kopfskelet der Urodelen," plate 5, figs. 69, 70), the palato-quadrate arch *becomes* continuous; an exception similar to that of *Bufo vulgaris*, where it *becomes* segmented.

The frontals and parietals are always long narrow bones; the parasphenoid is a broad generalised plate; the vomers and palatines are both dentigerous; the latter become strangely transposed, during metamorphosis, in the "Caducibranchiata."

The quadrate ossifies of itself, and there is neither a quadrato-jugal nor a jugal; the bony arch of the cheek is always (as in the Anura Aglossa and Teleostei) unfinished.

The squamosal has no postorbital process, but it has, at times, a very distinct *lower* supratemporal process. The premaxillaries are generally double; and sometimes there is only one. The maxillaries are large in the higher, but small, or even suppressed, in the lower, types.

Besides the cerato-hyal, and its lower hypo-hyal segment or segments, united by a distinct basi-hyal, there is in the larger and some of the smaller kinds, an epi-hyal element, not infrequently subdivided, so as to show a "pharyngo-hyal" also. This last is found in the "suspensorio-stapedial" ligament, and the former in the "hyo-suspensorial;" the upper piece answers to the pharyngo-hyal of the *Chimæra*, and to the inter-stapedial of the Frog; the lower piece to the epi-hyal of the one, and to the medio-stapedial of the other.†

Even this very imperfect comparison of the skull of the Urodeles, with what we have just seen in the Anura, shows how far these groups are apart, notwithstanding their many points of similarity; a thorough comparison of the larval skull of the latter, with that of the Lamprey, will be given in my next communication, which will treat of the cranio-facial skeleton of that Fish.

* In my paper on the Urodeles (Phil. Trans., 1877, p. 587) I expressed an opinion, now found to be wrong, viz.: that the cartilage which in some Urodeles passes from the suspensorium to the stapes was the same as the "spiracular cartilage" of the Tadpole.

† In *Rana halecina* and *R. palustris* (Plate 5, figs. 5 and 10) I have shown the *Acipenserine* subdivision of the Frog's columella; of course the "inter-stapedial" is due to the normal subdivision of the proximal piece.

List of Abbreviations.

The Roman figures indicate nerves or their foramina.

a.	Articulation.	*fo.*	Fontanelle.
al.e.	Aliethmoid.	*f.o.*	Fenestra ovalis.
al.n.	Alinasal.	*f.p.*	Fronto-parietal.
al.sp.	Alisphenoid.	*f.ty.*	Floor of tympanum.
al.s.	Aliseptal.	*h.*	Heart.
ar.	Articulare.	*h.br.*	Hyo-branchial.
ar.c.	Articular condyle.	*h.hy.*	Hypo-hyal.
a.s.c.	Anterior semicircular canal.	*h.s.c.*	Horizontal semicircular canal.
a.ty.	Annulus tympanicus.	*h.s.l.*	Hyo-suspensorial ligament.
au.	Auditory capsule.	*hy.c.*	Hyoid condyle.
a v.	Anterior velum.	*hy.f.*	Facet for hyoid.
b.br.	Basi-branchial.	*i.n.*	Internal nostril.
b.br.l.	Basi-branchial lobe.	*i.sp.*	Inter-suspensorial.
b.h br.	Basi-hyobranchial.	*i.st.*	Inter-stapedial.
b.hy.	Basi-hyal.	*i.tr.*	Intertrabecula.
b.o.	Basioccipital.	*iv.*	Investing mass.
br.p.	Branchial pouch.	*l.*	Lacrymal.
br.r.	Branchial rays.	*l.l.*	Lower labial.
c.br.	Cerato-branchial.	*l.f.*	Labial fontanelle.
c.hy.	Cerato-hyal.	*l.v.*	Lower velum.
cl.	Visceral cleft.	*lr.*	Larynx.
co.	Columella.	*mk.*	Meckel's cartilage.
c.pd.	Condyle of the pedicle.	*m.mk.*	Mento-Meckelian.
c.tr.	Cornu trabeculae.	*ms.pg.*	Mesopterygoid.
d.	Dentary.	*m.st.*	Mesostapedial.
e.hy.	Epi-hyal.	*mt.pg.*	Metapterygoid.
e.n.	External nostril.	*m.ty.*	Membrana tympani.
e.o.	Ex-occipital.	*mx.*	Maxillary.
ep.	Epiotic.	*n.*	Nasal.
e.pa.	Ethmo-palatine.	*na.*	Nasal capsule.
e.st.	Extra-stapedial.	*ne.*	Notochord.
eth.	Ethmoid.	*n.f.*	Nasal floor.
eu.	Eustachian pouch.	*n.n.*	Nasal nerve.
ex.br.	Extra-branchial.	*n.p.*	Nasal passage.
ex.hy.	Extra-hyal.	*n.px.*	Nasal process of premaxillary.
f.	Frontal.	*n.r.*	Nasal roof.
f.m.	Foramen magnum.	*n.w.*	Nasal wall.

oc.c.	Occipital condyle.	*px.*	Premaxillary.
o.fo.	Orbital fontanelle.	*py.*	Pituitary region.
o.p.	Oral papillae.	*q.*	Quadrate.
o.r.	Oval rugae.	*q.c.*	Quadrate condyle.
or.p.	Orbitar process.	*q.j.*	Quadrato-jugal.
o.s.	Orbito-sphenoid.	*s.mx.*	Septo-maxillary.
o.s.f.	Orbito-sphenoidal fontanelle.	*s.n.*	Septum nasi.
ot.p.	Otic process.	*s.n.l.*	Subnasal lamina.
p.	Parietal.	*s.o.*	Supraoccipital.
pa.	Palatine.	*s.ob.*	Superorbital.
pa.s.	Parasphenoid.	*s.o.f.*	Suborbital fenestra.
pd.	Pedicle.	*sp.*	Suspensorium.
pd.f.	Facet for pedicle.	*sp.c.*	Spiracular cartilage.
pd.m.	Meniscus of pedicle.	*sq.*	Squamosal.
p.e.	Perpendicular ethmoid.	*s.st.*	Supra-stapedial.
pg.	Pterygoid.	*st.*	Stapes.
pg.c.	Pterygoid cartilage.	*st.h.*	Stylo-hyal.
p.mx.	Palatine portion of maxillary.	*t.cr.*	Tegmen cranii.
p.n.	Prenasal.	*ty.*	Tongue.
p.ob.	Preorbital.	*t.hy.*	Thyro-hyal.
p.p.	Palatine papilla.	*tr.*	Trabecula.
p.pg.	Palato-pterygoid.	*t.ty.*	Tegmen tympani.
pr.o.	Prootic.	*ty.f.*	Tympanic floor.
pr.pa.	Pre-palatine.	*u.l^{1,2}.*	Permanent upper labials.
p.rh.	Pro-rhinal.	*u.l^{1,2}.*	Temporary upper labials.
p.s.c.	Posterior semicircular canal.	*v.*	Vomer.
pt.o.	Pterotic.	*vb.*	Vestibule.
pt.pa.	Post-palatine.	*vs.*	Blood vessel.
p.v.	Posterior velum.		

List of Families and Genera, with References to the Pages.

Families.	Genera.	Page.
Ranidæ	Rana	19
	Tomopterna	62
	Pyxicephalus	66
Cystignathidæ	Pseudis	69
	Gomphobates	86
	Cystignathus	89
	Pleurodema	97
	Leptodactylus	99
	Cacotus	102
	Cyclorhamphus	107
Discoglossidæ	Discoglossus	112
	Pelodytes	114
	Xenophrys	117
	Calyptocephalus	122
Alytidæ	Alytes	131
Hyperoliidæ	Hyperolius	134
Bombinatoridæ	Bombinator	136
Pelobatidæ	Pelobates	139
Polypedatidæ	Polypedates	144
	Rhacophorus	147
	Ixalus	149
	Hylarana	150
	Rappia	155
Hylodidæ	Hylodes	160
	Acris	164
Hylidæ	Hyla	169
	Litoria	182
	Notodromas	184
Pelodryadidæ	Pelodryas	190
	Phyllomedusa	194
Bufonidæ	Bufo	198
	Otilophus	219
Rhinodermatidæ	Rhinoderma	222
	Diplopelma	225
Brachycephalidæ	Pseudophryne	230
Phryniscidæ	Phryniscus	233
Engystomidæ	Engystoma	240
Hyladactylidæ	Callula	243
Hylaplesidæ	Hylaplesia	247
Dactylethridæ	Dactylethra	255
Pipidæ	Pipa	256

PLATE I.

Figures.		Number of times magnified.
1 (p. 69)	Youngest Tadpole of *Pseudis paradoxa* (A).—First larva; length, $10\frac{1}{3}$ inches; head and body, $3\frac{1}{3}$ inches; tail, 7 inches; greatest width of ditto, 4 inches; depth of body, $2\frac{5}{6}$ inches; hind legs, $\frac{1}{2}$ inch. South America .	Nat. size.
2	Mouth of same	2
3 (p. 22)	*Rana pipiens* (A).—Youngest larva; length, 5 inches; tail, 3 inches; hind legs, $\frac{1}{3}$ inch. North America. (Undissected palate)	$3\frac{3}{4}$
4	The same. Floor of mouth undissected	$3\frac{3}{4}$
5	The same. Section of head (*vertical*) with brain removed	$3\frac{3}{4}$

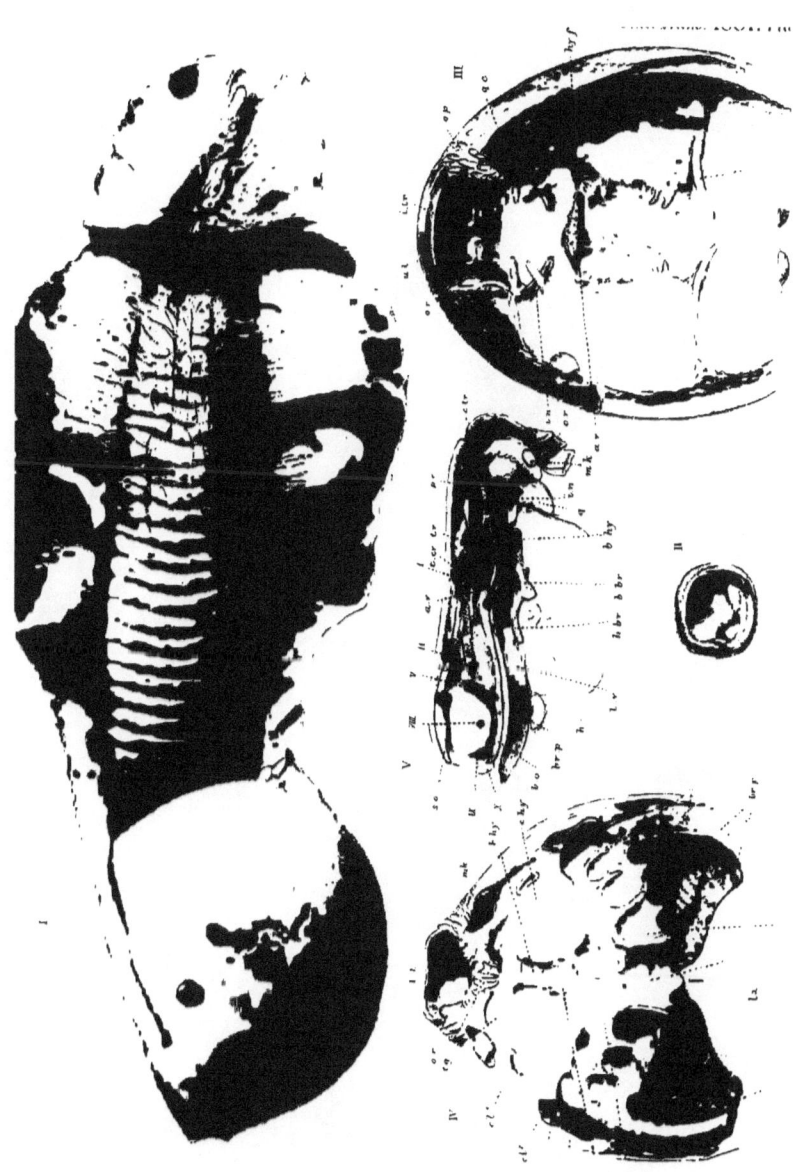

PLATE 2.

Figures.		Number of times magnified.
1 (p. 69)	Skull of Tadpole of *Pseudis paradoxa* (A).—Upper view	$2\frac{3}{3}$
2	The same. Lower view	$2\frac{2}{3}$
3	The same. End view	2
4	Section of occipital arch, below	8
5 (p. 19)	Skull of Tadpole of *Rana clamata* (A).—First larva; length, 3¼ inches; tail, 2 inches; hind legs, 5 lines. Cambridge, Mass., U.S. Upper view	5
6	The same. Lower view	5
7	The same. Part of lower arches	5
8 (p. 22)	Hyoid and branchial arches of larva of *Rana pipiens* (B).—Length, 3⅔ inches; tail, 2¼ inches; hind legs, ¾ inch. Upper view	5

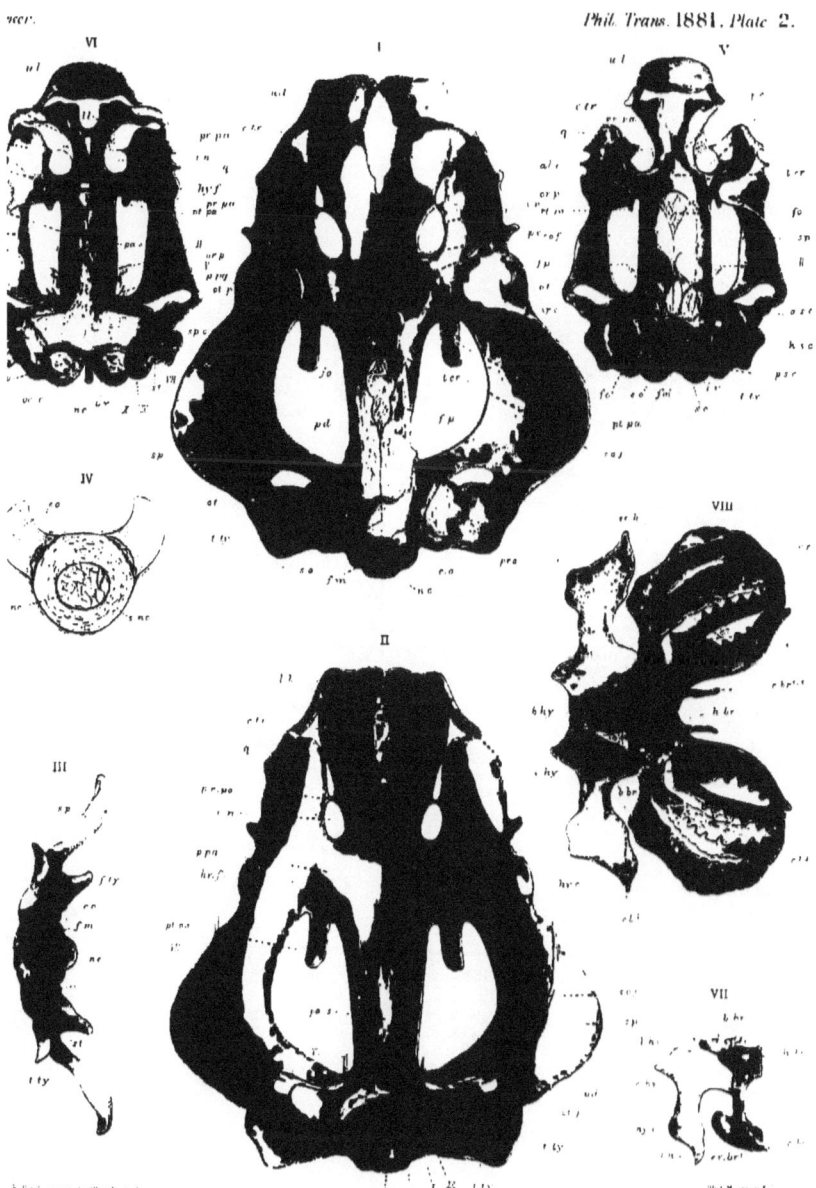

PLATE 3.

Figures.		Number of times magnified.
1 (p. 22)	Skull of Tadpole of *Rana pipiens* (B).—Second larva. Upper view	$5\frac{1}{4}$
2	The same. Lower view	$5\frac{1}{4}$
3	The same. Side view	$5\frac{1}{4}$
4 (p. 30)	Skull of Tadpole of *Rana pipiens* (C).—Third larva; all the legs free, and tail lessening rapidly. Upper view	$4\frac{1}{2}$
5	The same. Lower view	$4\frac{1}{2}$
6	The same. Side view	$4\frac{1}{2}$
7	1st of a series of transversely vertical sections of the same	9
8	2nd section	9
9	3rd section	9
10	4th section	9
11	5th section	9
12	6th section	9
13	7th section	9

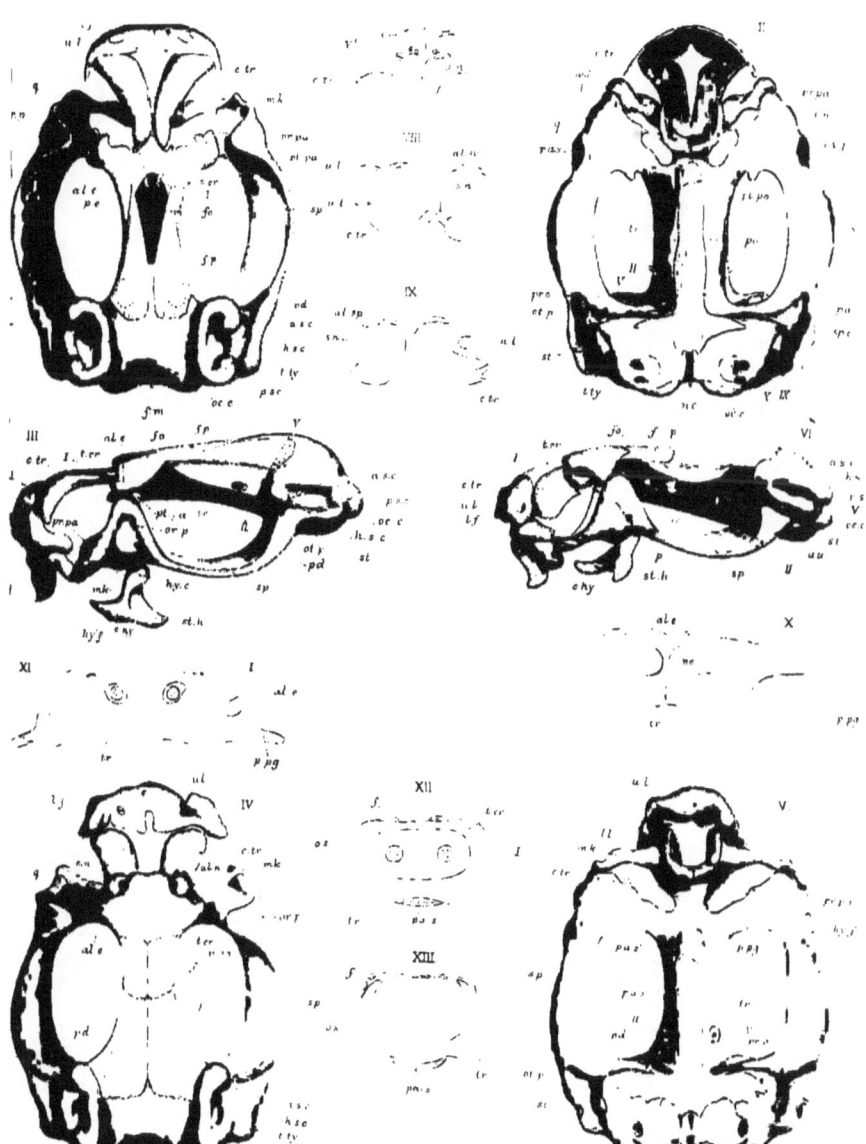

PLATE 4.

Figures.		Number of times magnified.
1 (p. 29)	Skull of *Rana clamata* (B).—Second larva; 3 inches 5 lines long; tail, 2¼ inches; hind legs, 1⅓ inch; fore legs hidden. Upper view	5
2	The same. Lower view	5
3	Hyo-branchials of the same. Upper view	5
4	The same; showing junction of 1st *extra-* with 1st intra-branchial.	10
5 (p. 33)	Skull of *Rana clamata* (C).—Third larva; length, 3¼ inches; tail, 1½½ inch; all the legs free. Upper view.	5
6	The same. Lower view	5
7	The same. Lower jaws and lower labials. Upper view.	5
8 (p. 34)	Skull of larva of Bull-frog (? species).—Body with clouded, ornate patches; entire length, 1¾ inch; tail, ⅓ inch; all the legs large and free. India. Upper view.	5
9	The same. Lower view	5
10	Lower arches of same. Upper view	5

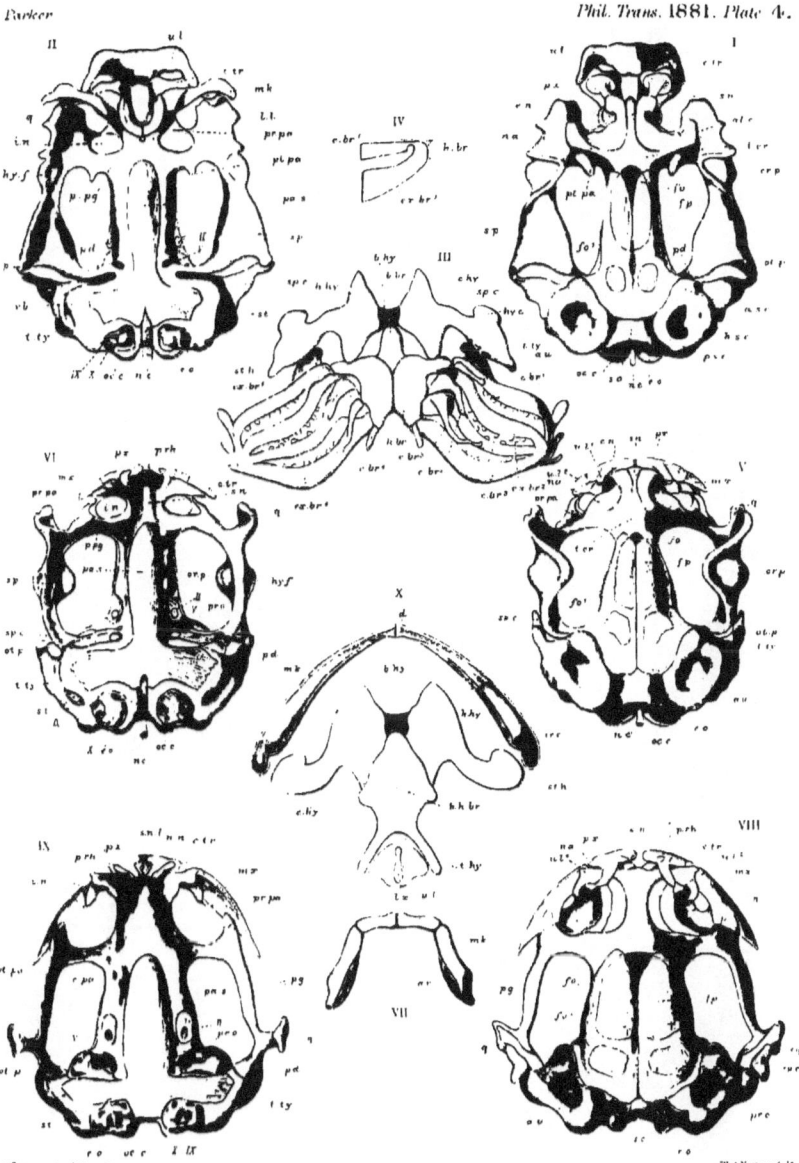

PLATE 5.

Figures.		Number of times magnified.
1 (p. 37)	Skull of *Rana halecina*.—Immature male; length 1¾ inch. North America. Upper view, with investing bones removed from right side.	5
2	The same. Lower view.	5
3	Mandible of same. Side view	5
4	Half of hyo-branchial plate of same.	5
5	The same. Columella and stapes. Outer view.	10
6 (p. 35)	Skull of *Rana palustris*, Cambridge, Mass., U.S.—Newly metamorphosed; 11 lines long. Upper view, with investing bones removed from right side	7½
7	The same. Lower view	7½
8	The same. Mandible. Side view	7½
9	The same. Hyo-branchials. Upper view	7½
10	The same. Columella and stapes. Outer view.	15
11 (p. 43)	Skull of *Rana pygmæa*.—Adult male; ⅝ inch long. Anamallays Mountains, Malabar, S.W. India. Upper view.	10
12	The same. Lower view.	10
13	The same. Mandible. Outer view.	10
14	The same. Hyo-branchials (*half*).	10
15	The same. Columella and stapes. Outer view.	20

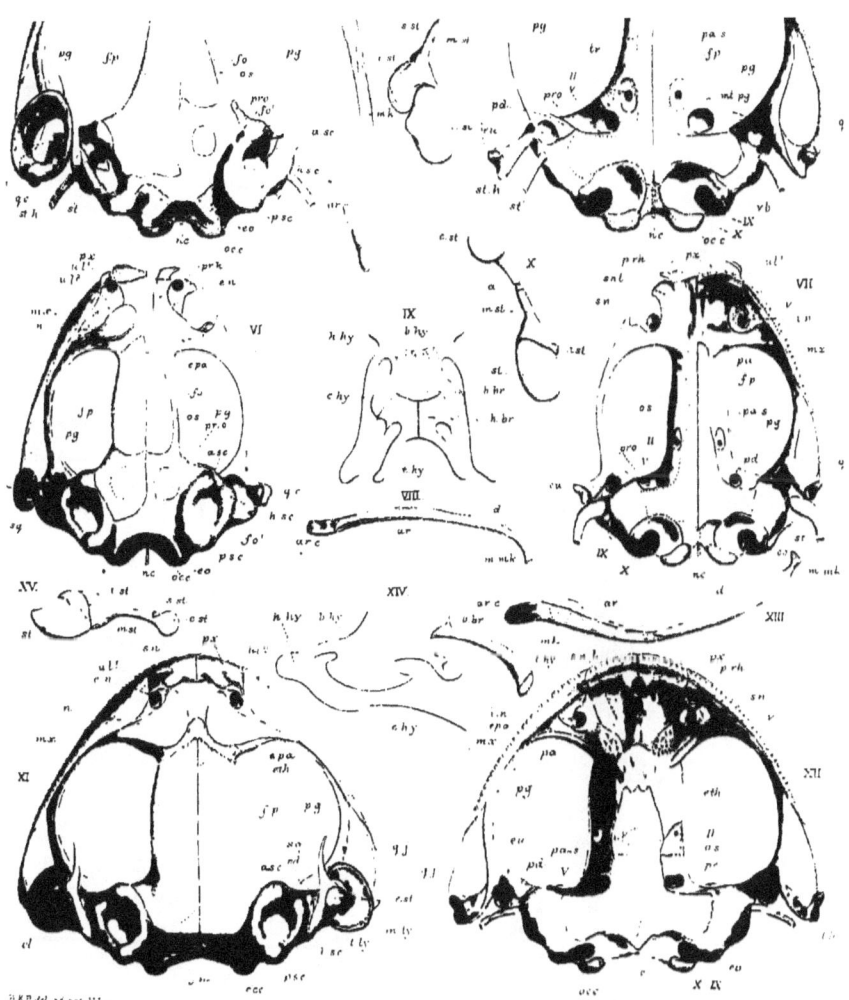

PLATE 6.

Figures.		Number of times magnified.
1 (p. 45)	Skull of *Rana tigrina*.—Adult female; $5\frac{1}{4}$ inches long. Ceylon. Upper view.	2
2	The same. Lower view.	2
3	The same. Side view	2
4	The same. Hyo-branchials. Upper view.	2
5	The same. Stapes and columella. Outer view.	4
6 (p. 40)	Skull of *Rana gracilis*.—Adult male; $1\frac{1}{2}$ inch long. Ceylon. Upper view	$6\frac{2}{3}$
7	The same. Lower view	$6\frac{2}{7}$
8	The same. Mandible. Outer view.	$6\frac{2}{3}$
9	The same. Hyo-branchials (*half*)	$6\frac{2}{3}$
10	The same. Stapes and columella. Outer view.	20
11 (p. 41)	Mandible of *Rana cyanophlyctis* (adult, see Plate 10). Outer view	$6\frac{2}{3}$
12	The same. Stapes and columella.	20

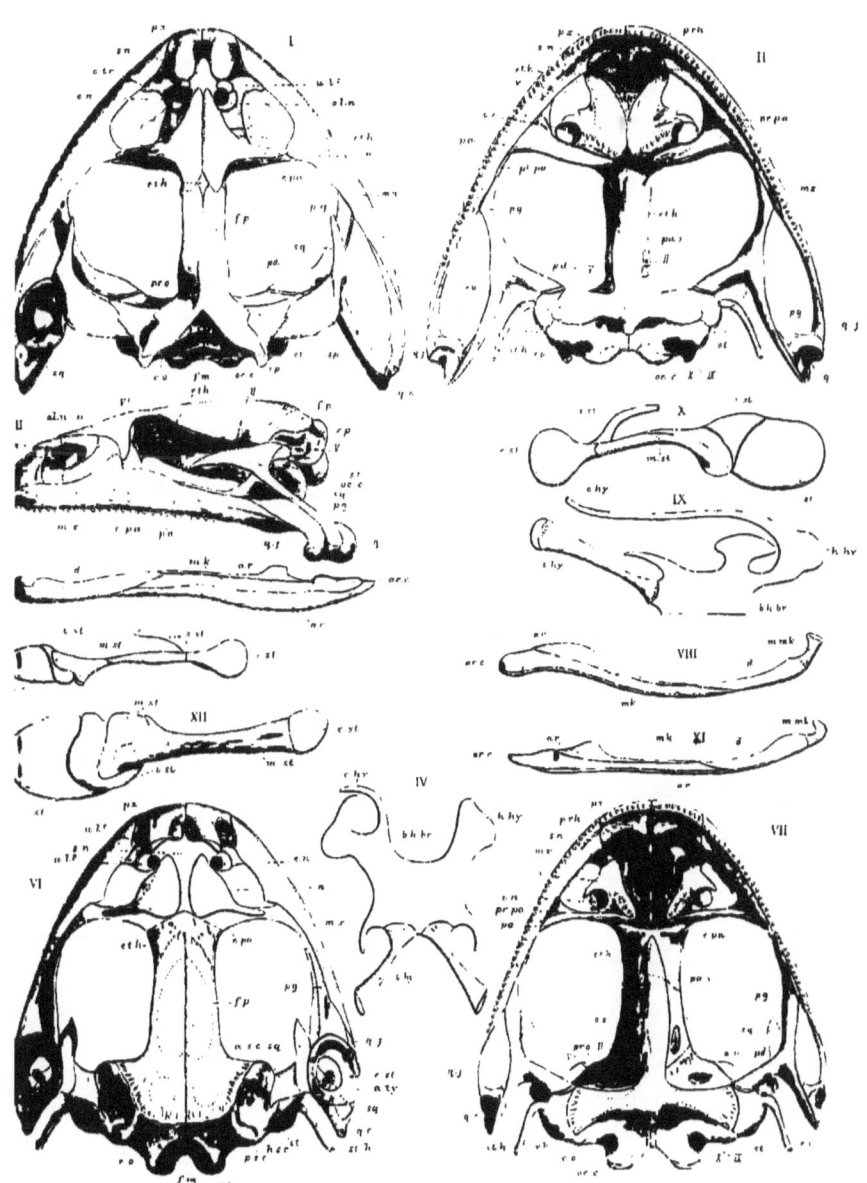

PLATE 7.

Figures.		Number of times magnified.
1 (p. 48)	Skull of *Rana hexadactyla.*—Adult female; $5\frac{1}{6}$ inches long. Ceylon. Upper view	2
2	The same. Lower view.	2
3	The same. Side view	2
4	The same. Hyo-branchials. Upper view	2
5	The same. Stapes and columella. Outer view.	4
6 (p. 50)	Skull of *Rana Kuhli.*—Almost adult male; $2\frac{1}{2}$ inches long. Ceylon. Upper view	4
7	The same. Lower view	4
8	The same. Side view	4
9	The same. Hyo-branchials (*half*)	4
10	The same. Stapes, columella, and *annulus tympanicus*. Outer view	8

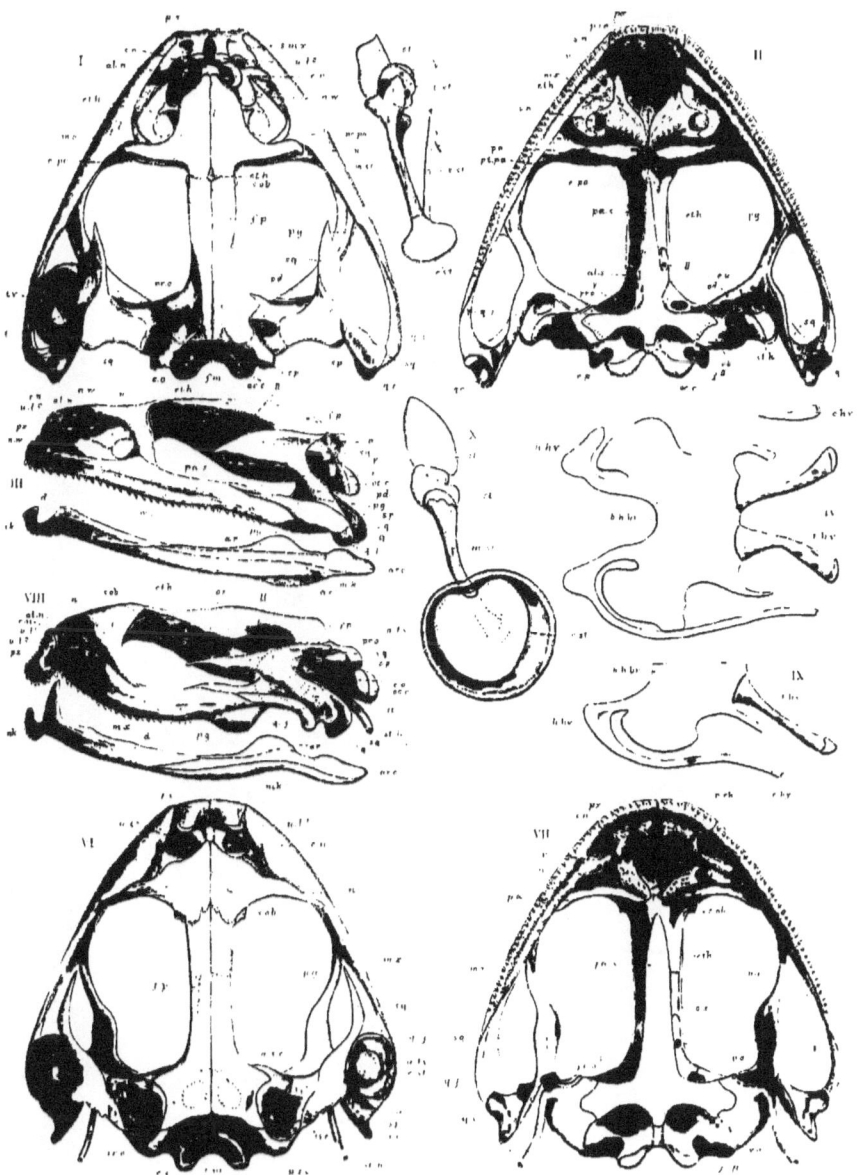

PLATE 8.

Figures.		Number of times magnified.
1 (p. 59)	Skull of *Rana pipiens.*—Adult; 6 inches long. North America. Upper view	2
2	The same. Lower view. . .	2
3	The same. Side view . .	2
4	The same. End view	2
5	The same. Hyo-branchials (*part*) . .	2

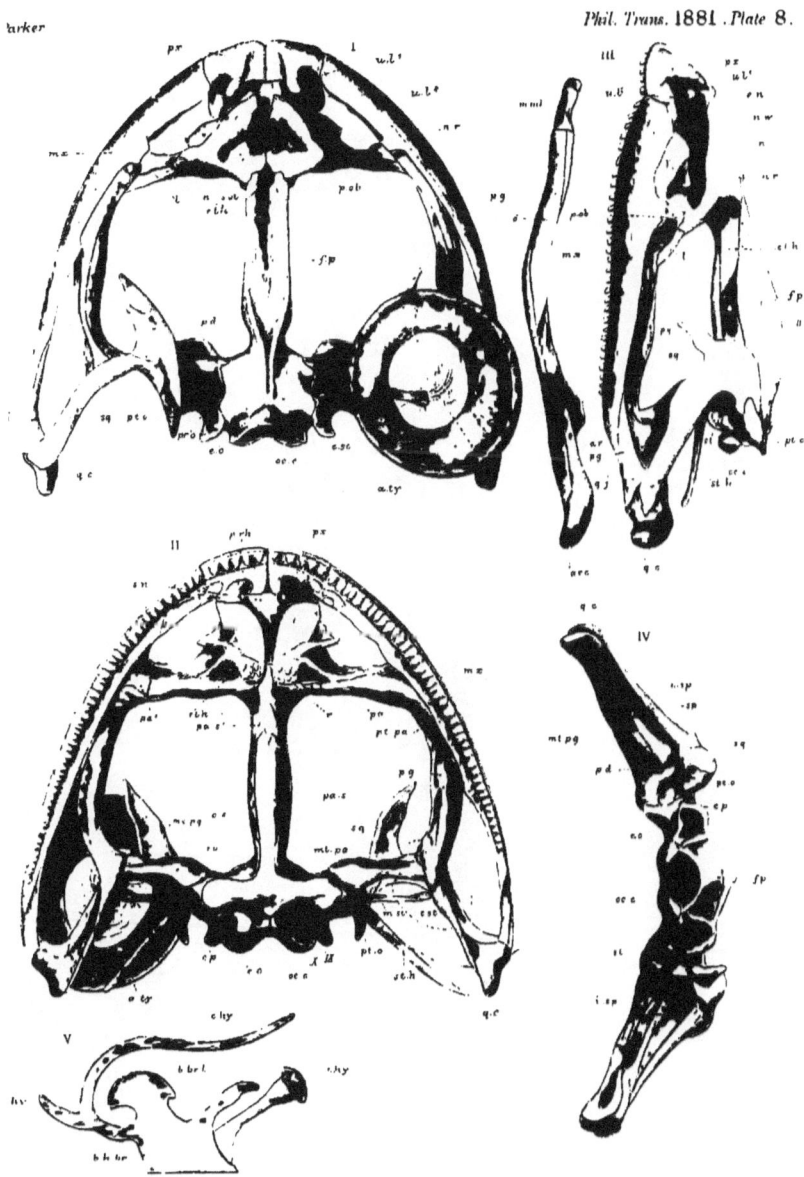

PLATE 9.

Figures.		Number of times magnified.
1 (p. 30)	Skull of Tadpole of *Rana pipiens* (C).—Third larva (continued from Plate 3). Lower view	4½
2	Part of same object	4½
3	Part of "extra-branchial" cartilages of the same . . .	4½
4	8th section of same stage of skull (continued from Plate 3)	9
5	9th section	9
6	10th section	9
7 (p. 53)	Skull of adult *Rana pipiens* (continued from Plate 8).—Part of lower view	4½
8	The same. Auditory region. Side view	3
9	The same. Stapes and columella. Outer view .	3
10	The same object. Inner view	3
11	The same skull. Front view of auditory region . . .	3
12	The same. Suspensorium. Lower view	3
13	The same object. Obliquely upper view	3
14	Part of same object. Inner view . . .	3

Parker.

Phil. Trans. 1881. *Plate* 9.

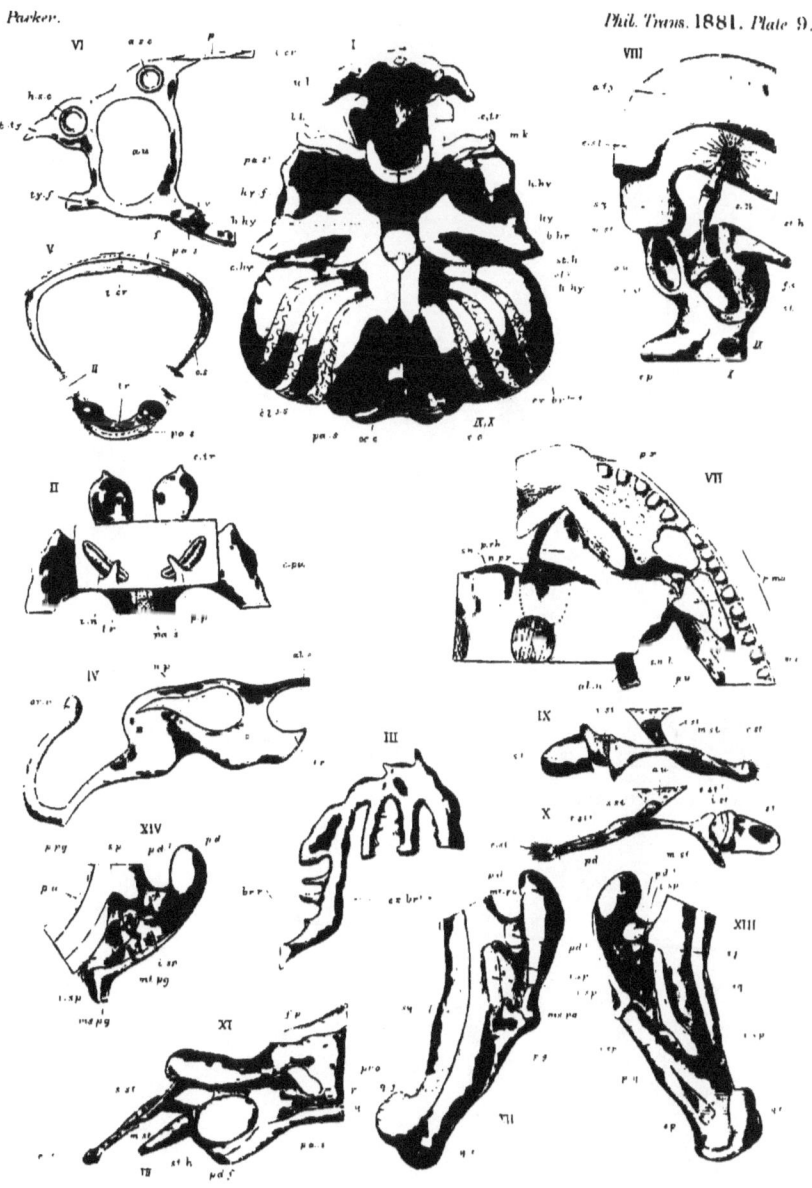

PLATE 10.

Figures.		Number of times magnified.
1 (p. 83)	Skull of *Pseudis paradoxa* (E).—Old male; $2\frac{1}{2}$ inches long. Surinam. (HYRTL's prepn. in Mus. Coll. Surg., England) Upper view	4
2	The same. Lower view	4
3	The same. Side view	4
4	The same. Stapes and columella. Outer view	8
5 (p. 69)	Skull of *Pseudis paradoxa* (A).—First larva. Side view	2
6	Hyo-branchial arches of the same (*half*). Upper view	2
7 (p. 41)	Skull of *Rana cyanophlyctis*.—Adult male; $1\frac{3}{4}$ inch long. Ceylon. Upper view .	$6\frac{2}{3}$
8	The same. Lower view . . .	$6\frac{2}{7}$
9	The same. Hyo-branchials (*half*)	$6\frac{2}{7}$
10	The same. Columella and stapes	$13\frac{1}{3}$

Parker.

Phil. Trans. 1881. Plate 10.

PLATE 11.

Figures.		Number of times magnified.
1 (p. 73)	Skull of *Pseudis paradoxa* (B).—Second larva; 7 inches long; tail, $4\frac{3}{4}$ inches; greatest width of tail, 2 inches; hind legs, 3 inches; fore legs hidden. Upper view. .	3
2	The same. Lower view	3
3	The same. Side view	3
4	The same. Hyoid, and part of branchial arches. Upper view	3
5	Upper part of branchial pouches of the same.	3
6 (p. 75)	Skull of *Pseudis paradoxa* (C).—Third larva; tail, 5 inches long; greatest width of tail, $1\frac{1}{2}$ inch; all the legs large and free. Upper view	6
7	The same. Lower view.	6
8	Slice of auditory region of the same. Outer view . . .	12
9	The same object. Inner view.	12

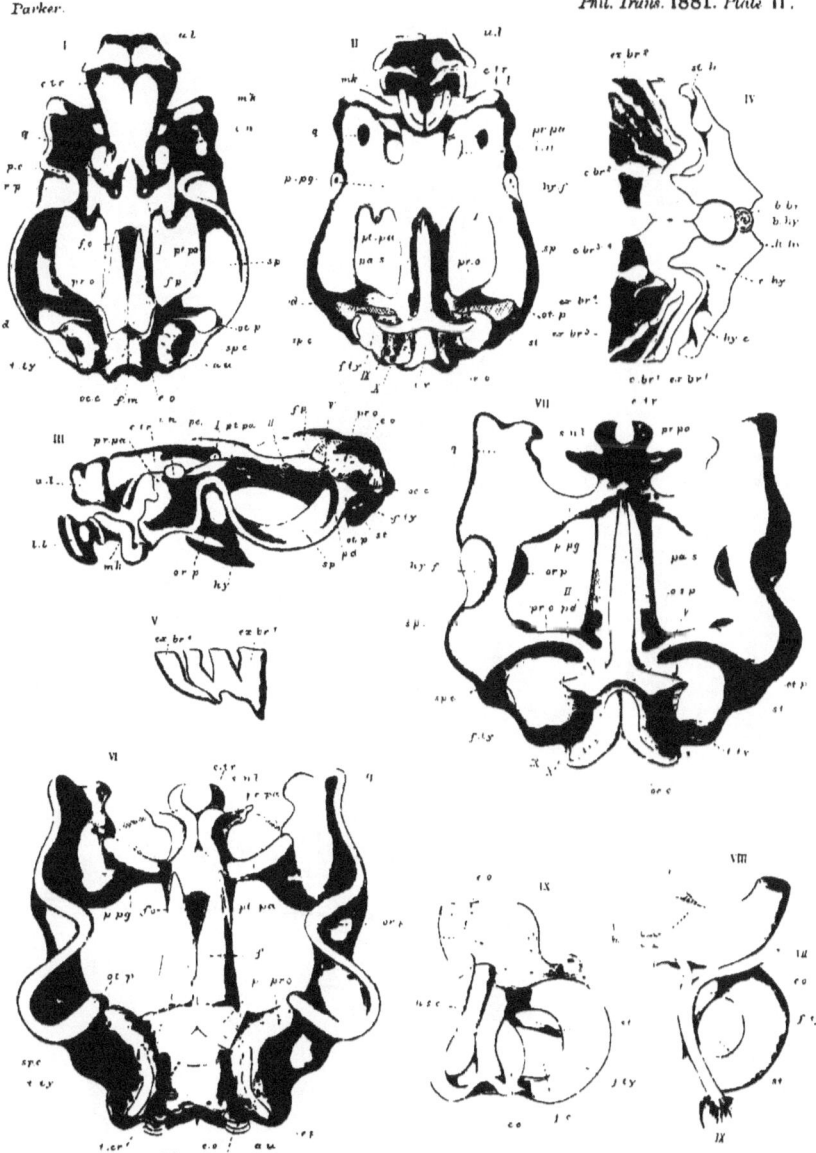

PLATE 12.

Figures.		Number of times magnified.
1 (p. 75)	Skull of the *Pseudis paradoxa* (C).—Third stage. Side view	6
2 (p. 78)	Skull of *Pseudis paradoxa* (D).—Fourth larva; tail, 3 inches long; greatest width of tail, 7 lines; all the legs large and free. Upper view . .	6
3	The same. Lower view .	6
4	The same. Side view	6
5	Hyo-branchials of the same skull. Upper view . . .	6
6	Slice from auditory region of the same skull. Outer view	12
7	The same object. Inner view.	12

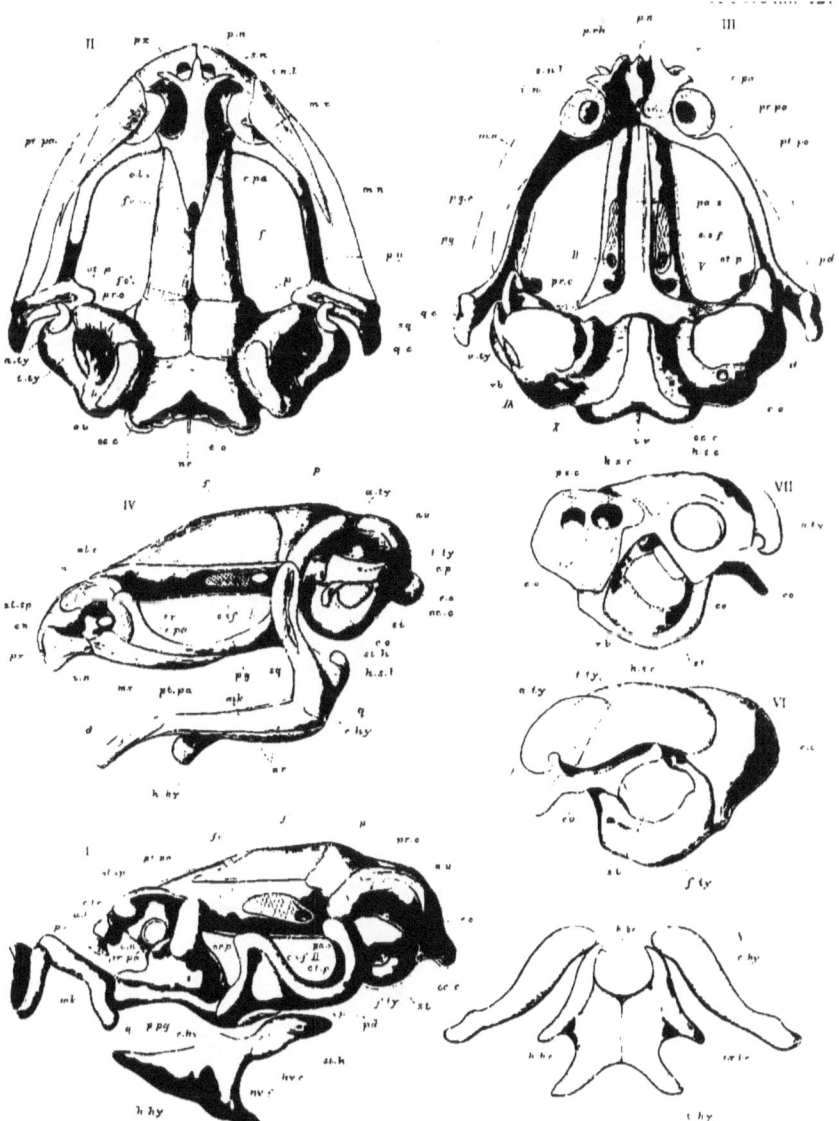

PLATE 13.

Figures.		Number of times magnified.
1 (p. 86)	Skull—with investing bones only partially drawn—of *Gomphobates* —— ? sp.—Adult (?); 10 lines long. River Plate. Upper view	9
2	The same. Lower view	9
3	Mandible of same. Lower view . . .	9
4	Hyo-branchials (*half*) of the same	9
5	Slice from auditory region of the same. Outer view . .	15
6	The same object. Inner view	15
7 (p. 60)	Skull of *Rana* —— ? sp.—Adult female; 1 inch long. Lagos. Upper view	8
8	The same. Lower view	8
9	The same. Mandible. Lower view .	8
10	The same. Hyo-branchials (*half*)	8
11	The same. Stapes and columella. Outer view	16
12	The same. Premaxillaries and permanent upper labials. Inner view	16

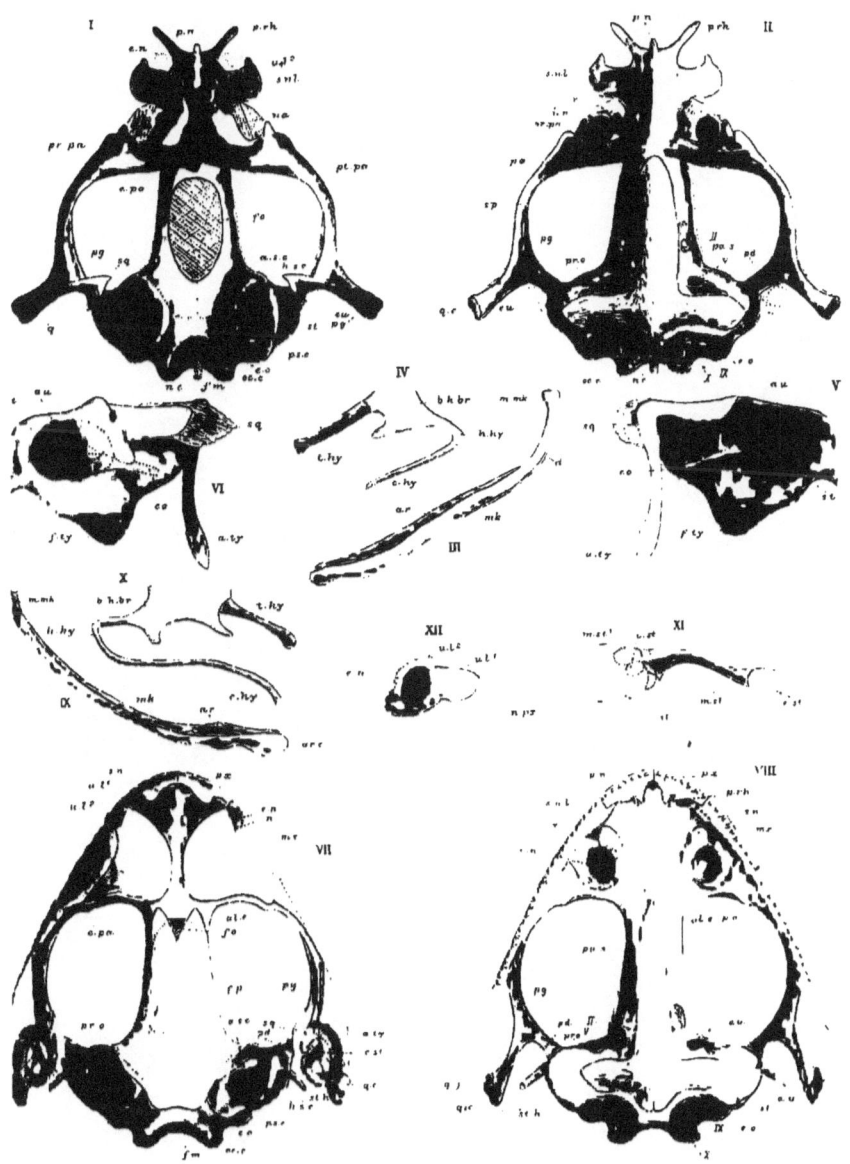

PLATE 14.

Figures.		Number of times magnified.
1 (p. 66)	Skull of *Pyxicephalus rufescens*.—Adult male; 1 inch 5 lines long. India. Upper view	$6\frac{2}{3}$
2	The same. Lower view	$6\frac{2}{3}$
3	Inferior arches of the same	$6\frac{2}{3}$
4	Annulus, columella, and stapes of the same	$13\frac{1}{3}$
5 (p. 62)	Skull of *Tomopterna breviceps* (A).—Female; half-grown; $1\frac{1}{4}$ inch long. South India. Upper view . . .	$6\frac{3}{4}$
6	The same. Lower view	$6\frac{2}{3}$
7	Lower jaw of same. Upper view	$6\frac{2}{3}$
8	Hyo-branchial plate (*half*)	$6\frac{3}{4}$
9	Fore part of skull of the same. Lower view	15
10	Part of auditory capsule of the same, with stylo-hyal, columella, and stapes	$13\frac{1}{3}$

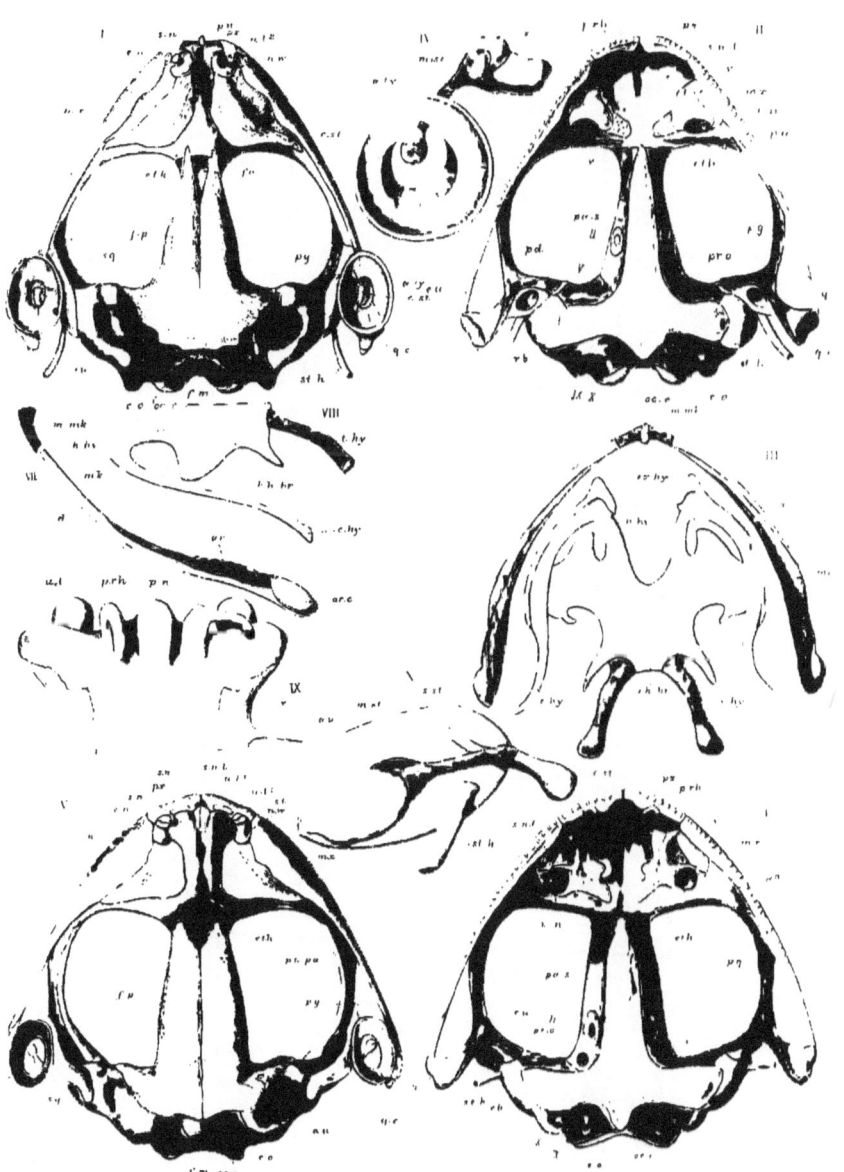

PLATE 15.

Figures.		Number of times magnified.
1 (p. 63)	Skull of *Tomopterna breviceps* (B).—Adult female; 2 inches long. Ceylon. Upper view	$4\frac{2}{3}$
2	The same. Lower view	$4\frac{2}{3}$
3	The same. Side view*	$4\frac{2}{3}$
4	Hyo-branchials of the same	$4\frac{2}{3}$
5	Fore part of skull of *Pyxicephalus rufescens*. Lower view	8
6 (p. 105)	Skull of Tadpole of *Camariolius tasmaniensis* (?) (B).— Tadpole; $\frac{3}{4}$ inch long; legs, $\frac{1}{30}$ of an inch. Australia. Upper view	$22\frac{1}{2}$
7	The same. Lower view	$22\frac{1}{2}$
8 (p. 190)	Hyo-branchials of *Pelodryas cerulæus* (see Plate 34) . .	4
9 (p. 194)	Hyo-branchials of *Phyllomedusa bicolor* (see Plate 34) .	$3\frac{1}{3}$

* Meckel's cartilage is lettered *d*, by mistake, in this figure.

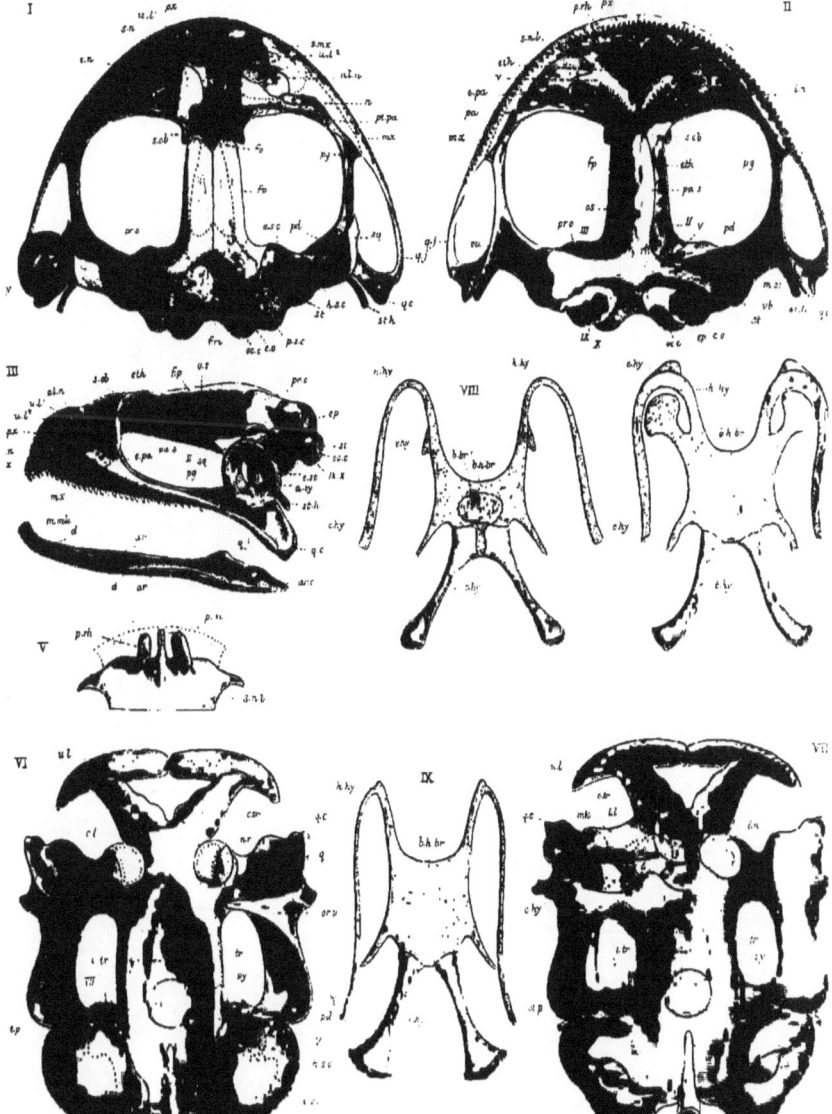

PLATE 16.

Figures.		Number of times magnified.
1 (p. 93)	Skull of *Cystignathus ocellatus* (var) (C).—Adult male; 5¼ inches long. Dominica. Upper view .	2
2	The same. Lower view .	2
3	The same. Side view . .	2
4	The same. Hyo-branchials . .	2
5	The same. Stapes and columella . . .	4
6 (p. 95)	Skull of *Cystignathus typhonius*.—Adult female; 1½ inch long. Porto Rico, West Indies. Upper view .	5⅓
7	The same. Lower view.	5⅓
8	The same. Lower jaw	5⅓
9	The same. Hyo-branchials . . .	5⅓
10	The same. Stapes and columella	10⅔

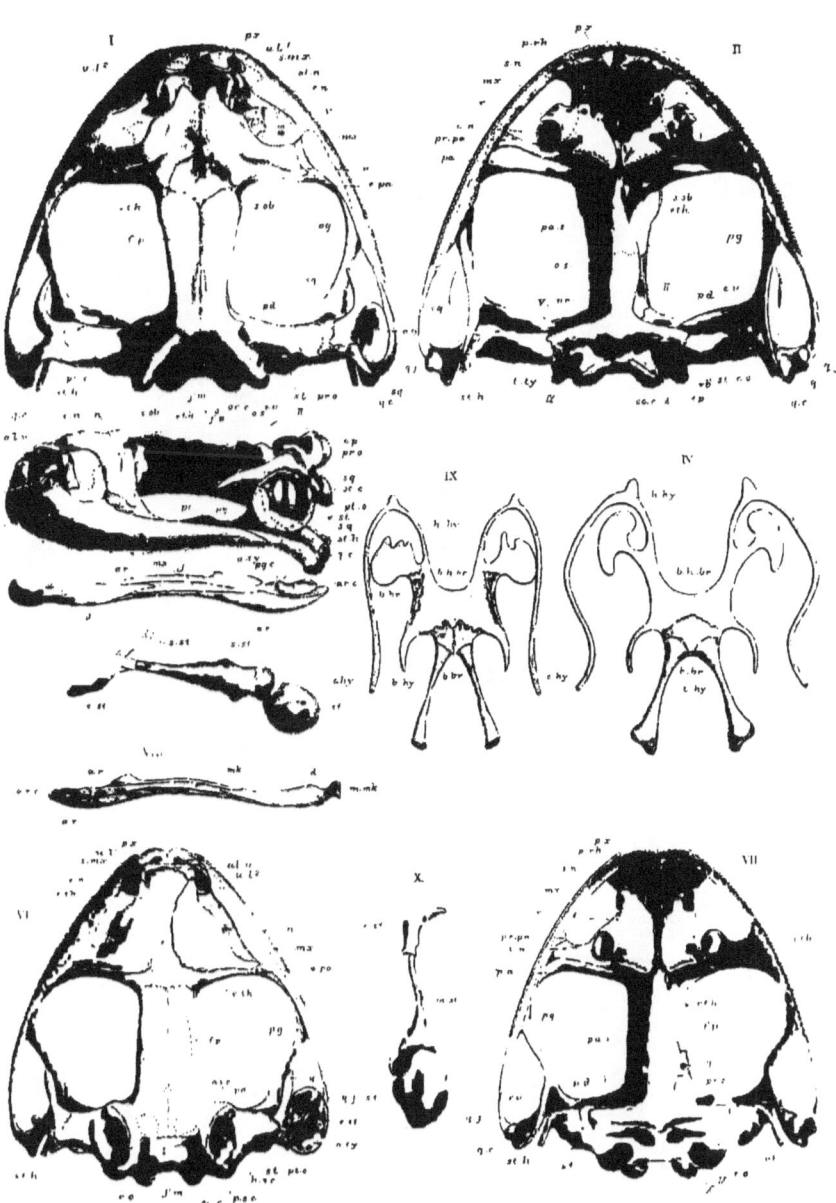

PLATE 17.

Figures.		Number of times magnified.
1 (p. 89)	Skull of Tadpole of *Cystignathus ocellatus* (?) (A).—$2\frac{1}{6}$ inches long; tail, $1\frac{1}{2}$ inch; hind legs not visible. Rodsio—a tributary of Rio dos Macacos—above the Falls, Brazils, May, 1865. (A. AGASSIZ.) Upper view	8
2	The same. Lower view	8
3	Mandibles and lower labials of the same. Upper view	8
4	Hyo-branchials of the same. Upper view	8
5 (p. 91)	Skull of larger larva of the same gathering (B); $3\frac{1}{4}$ inches long; tail, $2\frac{1}{4}$ inches; hind legs 1 line. Upper view	6
6	The same. Lower view	6
7	Mandibles and lower labials of the same	6
8 (p. 92)	Skull of Tadpole of *Cystignathus*——? sp.—Length, 1 inch; tail, $\frac{3}{4}$ inch; hind legs, 1 line. Lake Janarg, Manaoo, Brazils. Upper view	12
9	The same. Lower view	12

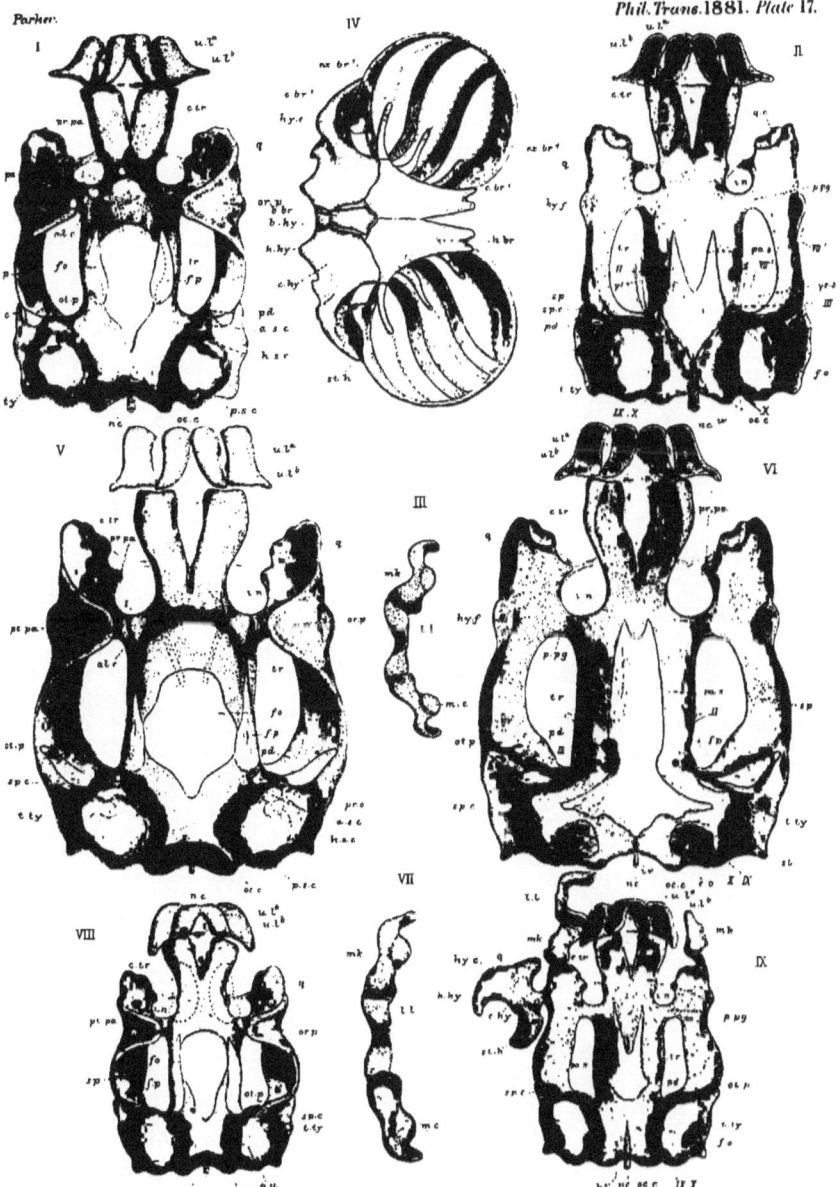

PLATE 18.

Figures.		Number of times magnified.
1 (p. 97)	Skull of *Pleurodema Bibronii*—Adult female; $1\frac{1}{3}$ inch long. Chili. Upper view	$6\frac{2}{3}$
2	The same. Lower view	$6\frac{2}{3}$
3	The same. Mandible and hyo-branchials (*half*). Upper view	$6\frac{2}{3}$
4	The same. Columella and stapes. Outer view	$13\frac{1}{3}$
5 (p. 99)	Skull of *Lymnodynastes tasmaniensis.*—Adult female; $1\frac{2}{3}$ inch long. Tasmania. Upper view	$6\frac{2}{3}$
6	The same. Lower view	$6\frac{2}{3}$
7	The same. Mandible and hyo-branchials (*half*). Upper view	$6\frac{2}{3}$
8	Auditory region of the same, with stapes and columella. Outer view	$13\frac{1}{3}$

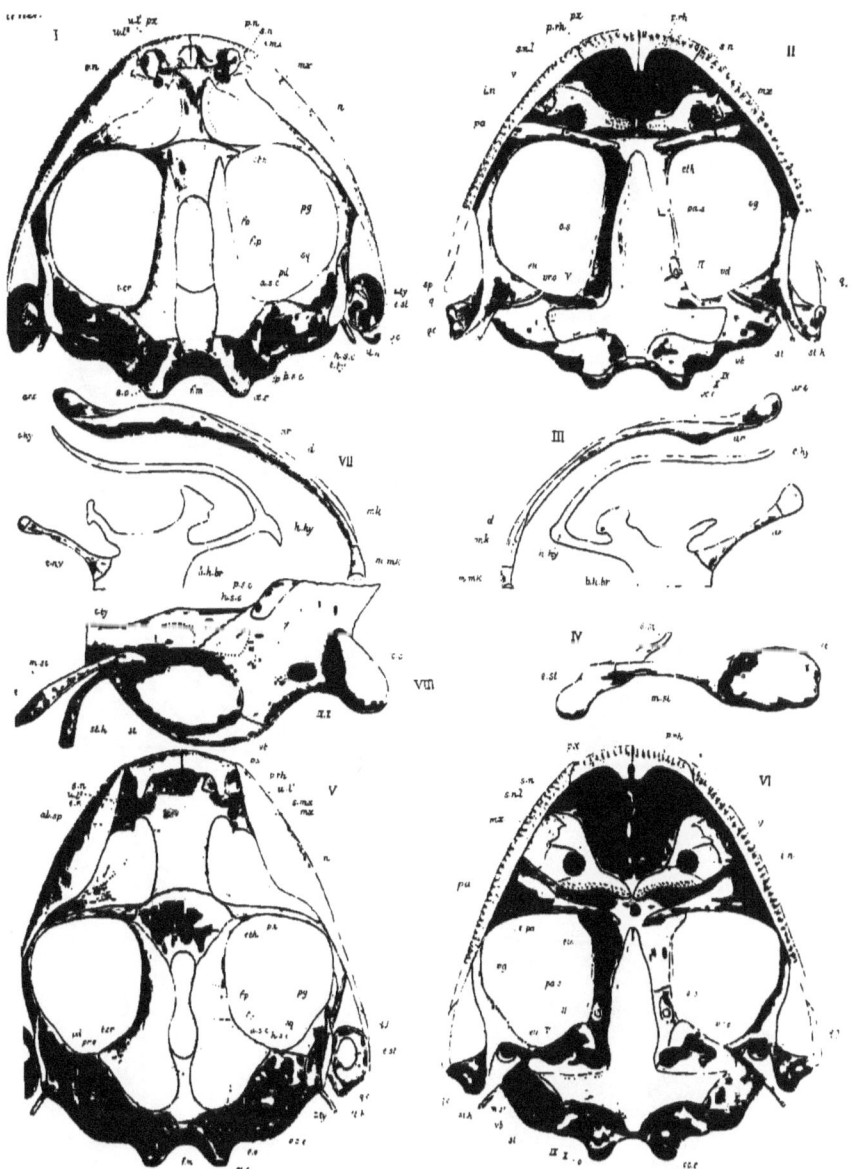

PLATE 19.

Figures.		Number of times magnified.
1 (p. 102)	Skull of *Camariolius tasmaniensis* (?) (A).—Adult female; ¾ inch long. Australia. Upper view	10
2	The same. Lower view	10
3	The same. Mandible. Upper view	10
4	The same. Hyo-branchials (*half*)	10
5	The same. Stapes and columella. Outer view	20
6 (p. 158)	Skull of *Rappia bicolor*.—Adult female; ¾ inch long. Dog-trap Road, Paramatta, Australia. Upper view	8
7	The same. Lower view	8
8	Hyo-branchials of same (*half*)	8
9	The same. Fore part of chondrocranium. Lower view	16
10	The same. Stapes and columella. Inner view	24
11 (p. 182)	Skull of *Litoria marmorata*.—Adult male; 1½ inch long. Australia. Upper view	5⅓
12	The same. Lower view	5⅓
13	The same. Mandible. Upper view	5⅓
14	The same. Hyo-branchials. Upper view	5⅓
15	The same. Stapes and columella. Outer view	10⅔
16	Part of same object. Side view	10⅔

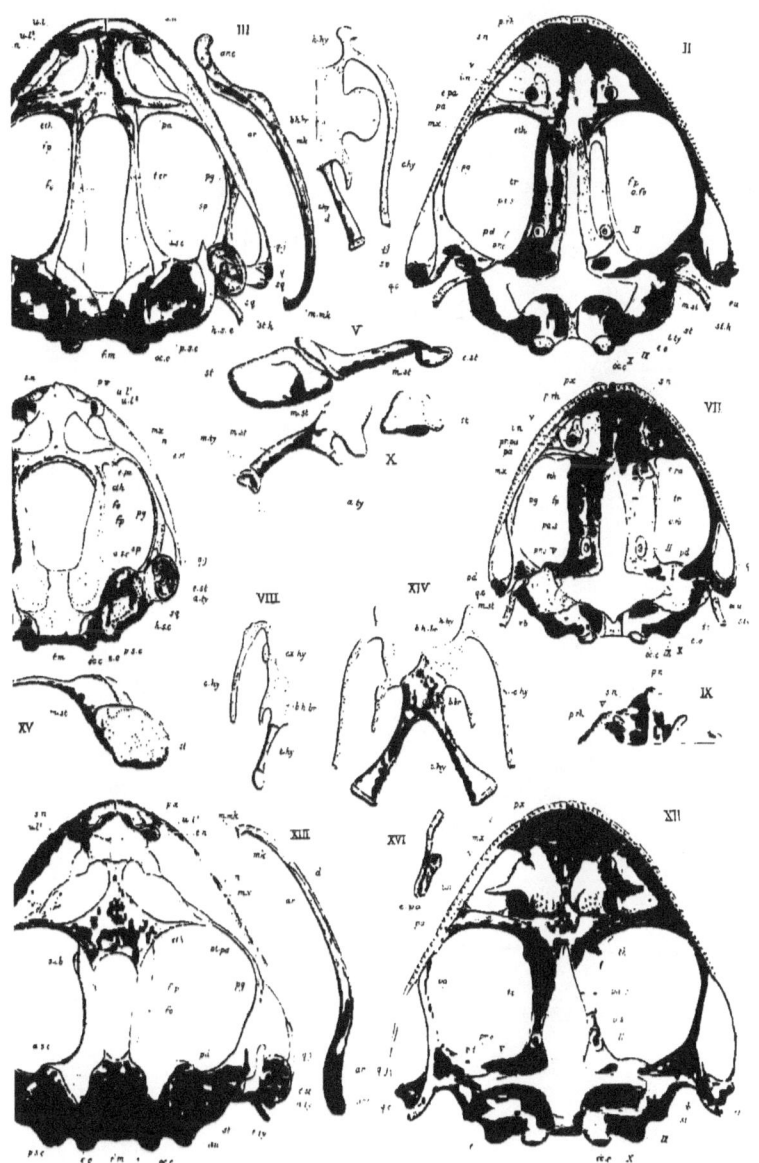

PLATE 20.

Figures.		Number of times magnified.
1 (p. 107)	Skull of *Cyclorhamphus marmoratus*.—Adult female: $1\frac{5}{6}$ inch long. Vinco Caya, Peruvian Andes, 16,000 feet high. Upper view	$4\frac{2}{3}$
2	The same. Lower view	$4\frac{2}{3}$
3	The same. Side view	$4\frac{2}{3}$
4	The same. Hyo-branchials. Upper view	$4\frac{2}{3}$
5	The same. Pedicle and part of skull; left side. Upper view	$9\frac{1}{3}$
5A	Part of same object—ossified *meniscus*	$9\frac{1}{3}$
6	The same skull. Stapes, columella, and annulus. Outer view	$9\frac{1}{3}$
7 (p. 112)	Skull of *Discoglossus pictus*.—Adult male; $2\frac{1}{6}$ inches long. South Europe. Upper view	$4\frac{2}{3}$
8	The same. Lower view	$4\frac{2}{3}$
9	The same. Mandible. Outer view	$4\frac{2}{3}$
10	The same. Hyo-branchials (*half*)	$4\frac{2}{3}$
11	The same. Stapes and columella. Outer view . . .	14

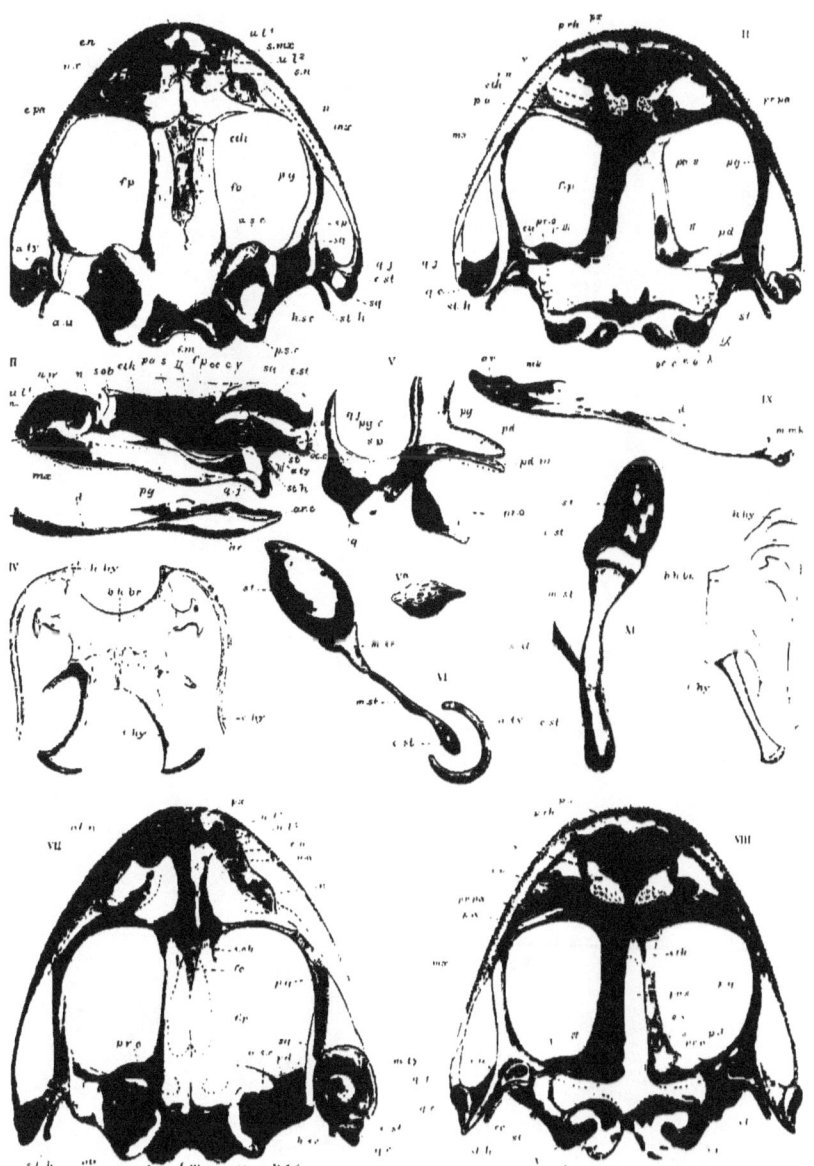

PLATE 21.

Figures.		Number of times magnified.
1 (p. 122)	Skull of *Calyptocephalus Gayi* (A). — Adult female; 5½ inches long. Chili. Upper view	2
2	The same. Lower view	2
3	The same. End view	2
4	The same. Hyo-branchials. Upper view	2
5	The same. Stapes and columella. Outer view . .	4
6	Part of columella. Side view	4

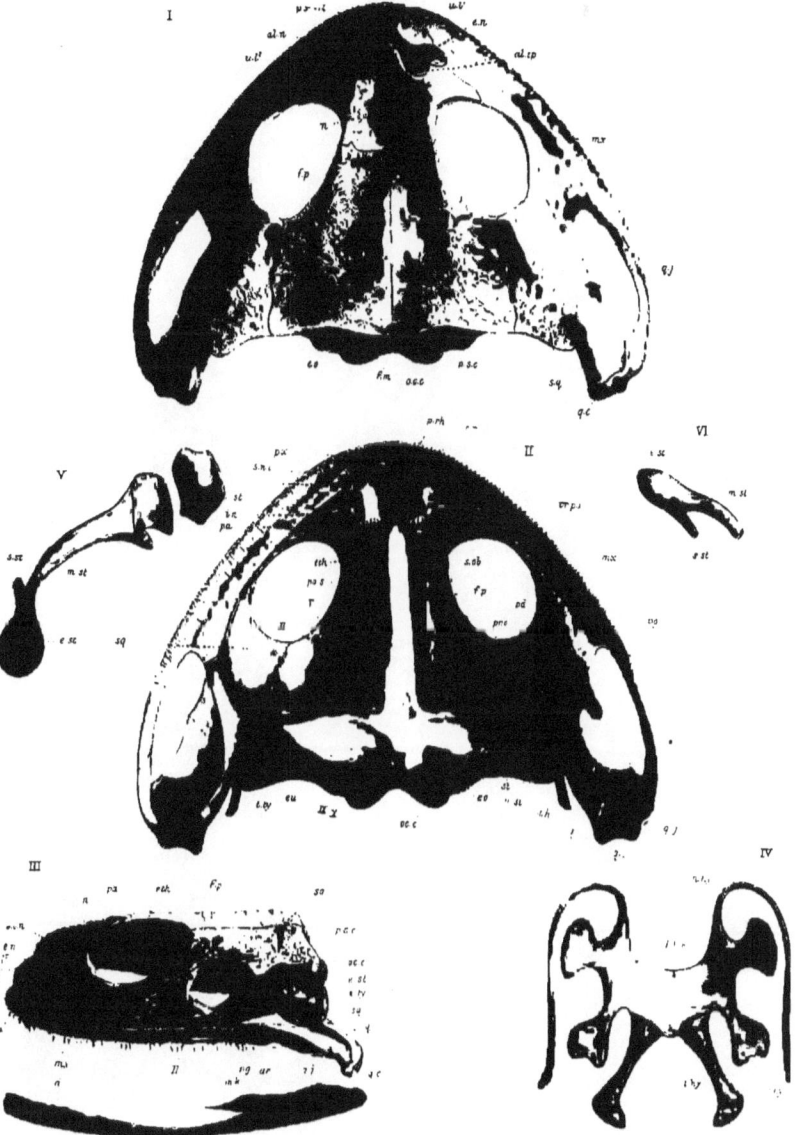

PLATE 22.

Figures.		Number of times magnified.
1 (p. 122)	Skull of adult *Calyptocephalus Gayi* (A).—End view . .	2
2 (p. 127)	Skull of Tadpole of the same species (B).—4¾ inches long; tail, 2¾ inches; hind legs, ⅛ inch. Upper view . . .	5
3	The same. Lower view	5
4	The same. Mandible and lower labials. Lower view . .	5
5	The same. Hyo-branchials of left side. Upper view . .	5
6 (p. 110)	Skull of Tadpole of *Cyclorhamphus culeus*.—3¼ inches long; tail, 2 inches; hind legs, 7 lines. Puno, Lake Titicaca, Peru. Upper view	5
7	The same. Lower view	5
8	The same. Mandibles and lower labials. Lower views .	5
9	The same. Hyo-branchials of left side. Upper view. .	5

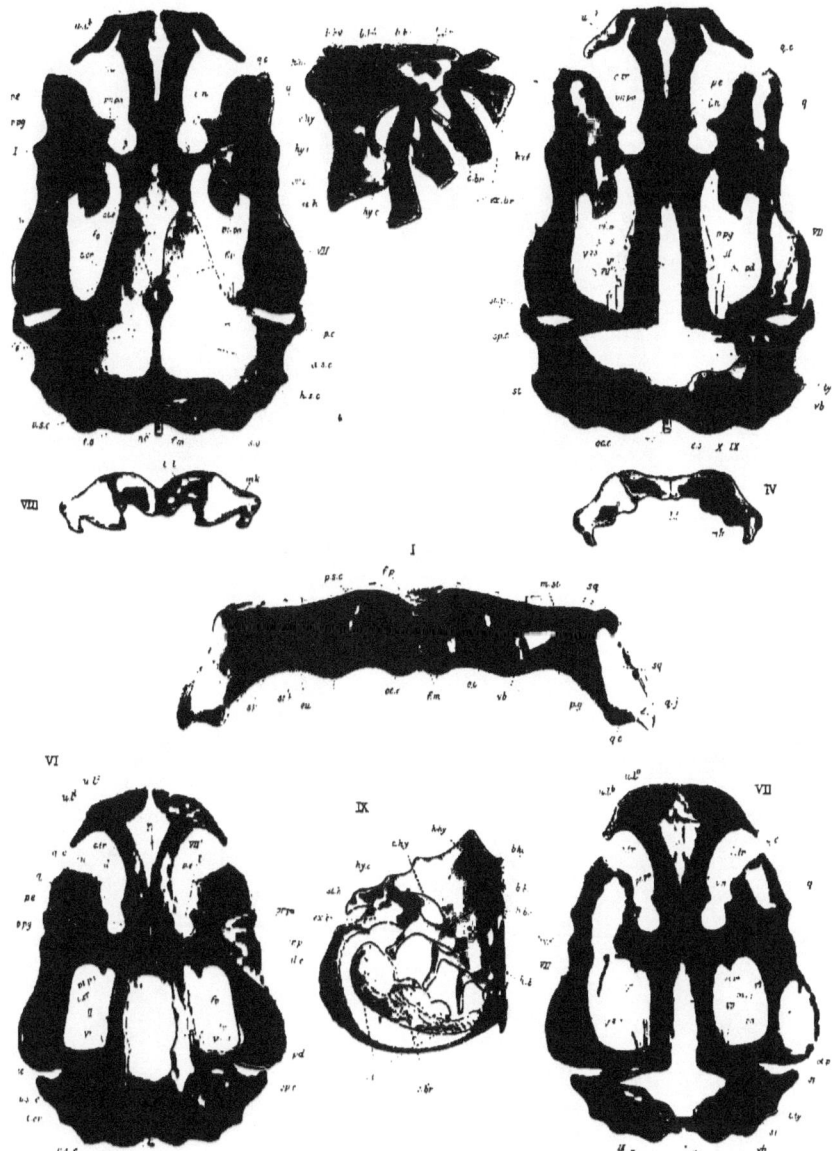

PLATE 23.

Figures.		Number of times magnified.
1 (p. 114)	Skull of *Pelodytes punctatus*.—Adult male; $1\frac{1}{2}$ inch long. Europe. Upper view	$6\frac{2}{3}$
2	The same. Lower view	$6\frac{2}{3}$
3	The same. Mandibles and hyo-branchials. Lower view .	$6\frac{2}{3}$
4	The same. Stapes and columella. Outer view	$13\frac{1}{3}$
5 (p. 117)	Skull of *Xenophrys monticola*.—Adult male; 3 inches long; length of hind leg, $4\frac{2}{3}$ inches. Darjeeling. Upper view	$3\frac{1}{3}$
6	The same. Lower view	$3\frac{1}{3}$
7	The same. Side view	$3\frac{1}{3}$
8	The same. Hyo-branchials (*half*)	$3\frac{1}{3}$
9	The same. Fore part of chondrocranium. Lower view .	$3\frac{1}{3}$
10	The same. Stapes and columella. Outer view	$6\frac{2}{3}$

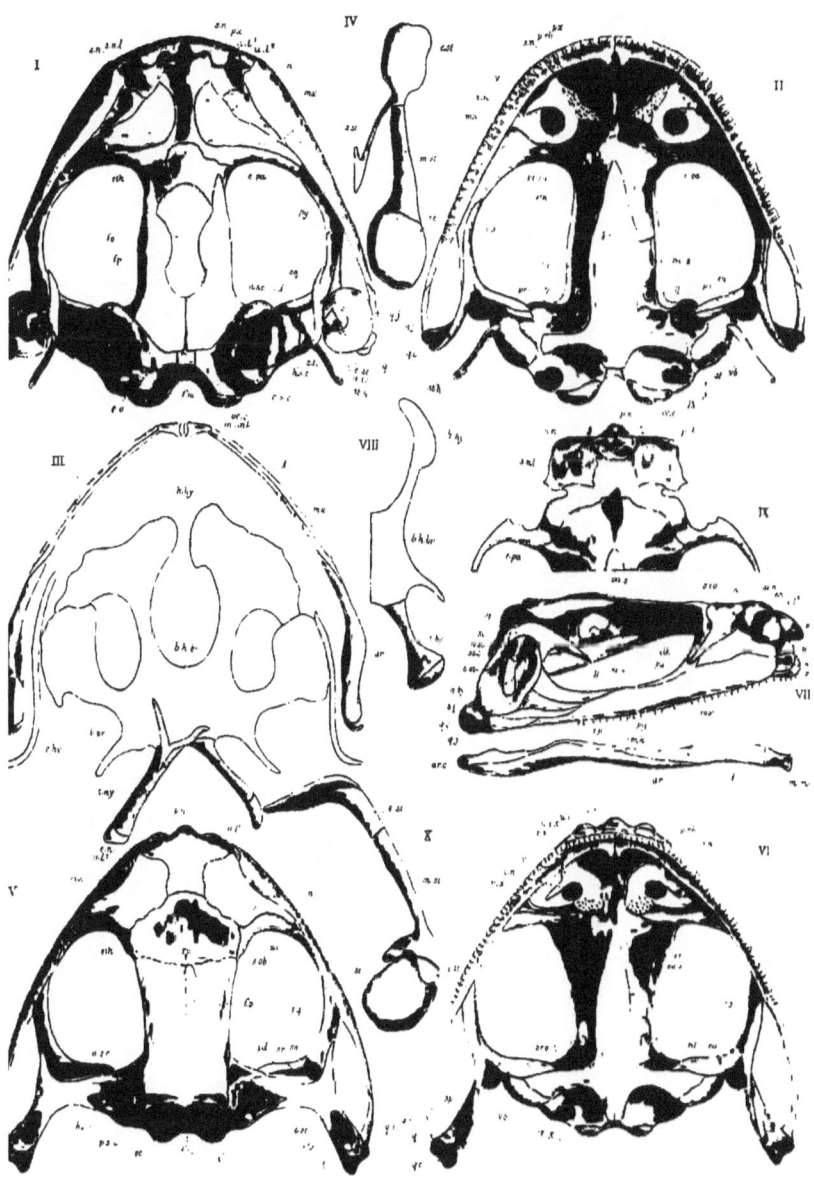

PLATE 24.

Figures.		Number of times magnified.
1 (p. 131)	Skull of *Alytes obstetricans.*—Adult female; 1 inch 10 lines long. Europe. Upper view	4½
2	The same. Lower view	4½
3	The same. Mandible. Outer view.	4½
4	The same. Hyo-branchials. Lower view.	4½
5	The same. Stapes and columella. Outer view . . .	9
6 (p. 134)	Skull of *Hyperolius marmoratus.*—Adult female; 1¼ inch long. Paramatta, Australia. Upper view. . . .	7½
7	The same. Lower view	7½
8	The same. Side view	7½
9	The same. Hyo-branchials (*half*)	7½
10	The same. Stapes and columella. Outer view . . .	15

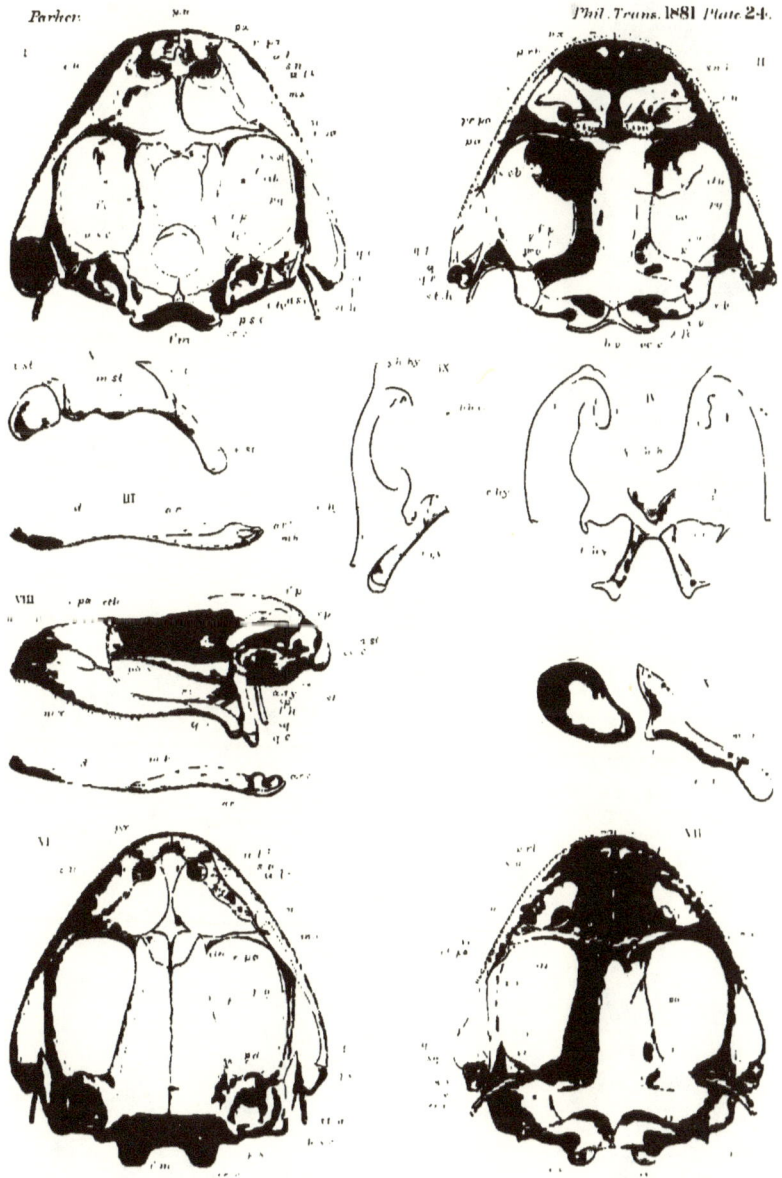

PLATE 25.

Figures.		Number of times magnified.
1 (p. 136)	Skull of *Bombinator igneus.*—Adult female; 1¾ inch long. Europe. Upper view	6
2	The same. Lower view	6
3	The same. Mandibles and hyo-branchials. Upper view .	6
4	The same. Auditory region. Outer view	12
5 (p. 139)	Skull of *Pelobates fuscus.*—Adult male; 2 inches 5 lines long. Europe. Upper view	4
6	The same. Lower view	4
7	The same. Side view	4
8	The same. End view	4
9	The same. Hyo-branchials. Upper view . .	4
10	The same. Mandible. Outer view. . .	4
11	The same. Auditory region. Outer view .	6

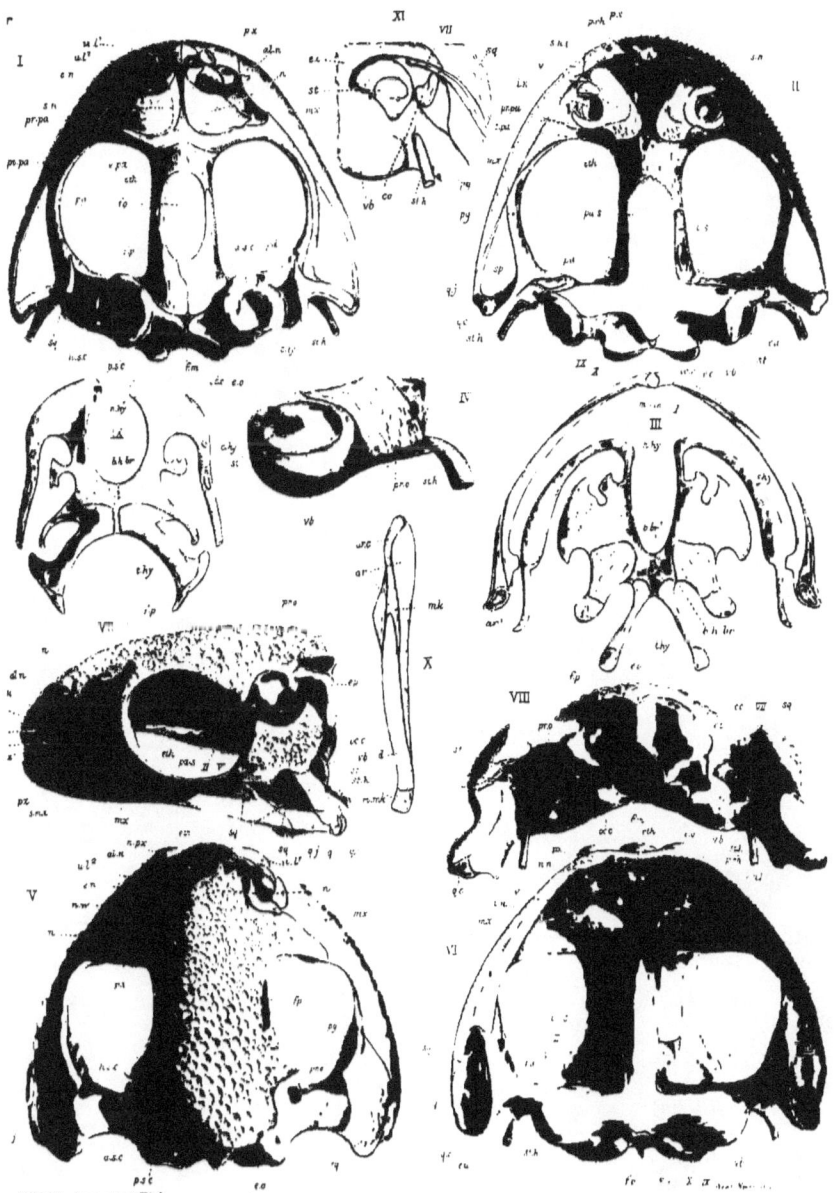

PLATE 26.

Figures.		Number of times magnified.
1 (p. 144)	Skull of *Polypedates chloronotus*.—Adult male; 2 inches long. India. Upper view.	5⅓
2	The same. Lower view	5⅓
3	The same. Mandible. Upper view	5⅓
4	The same. Hyo-branchials (*half*). Upper view . . .	5⅓
5 (p. 147)	Skull of *Rhacophorus maximus*.—Adult male; 3⅙ inches long. North India. Upper view	3⅓
6	The same. Lower view.	3⅓
7	The same. Side view	3⅓
8	The same. Hyo-branchials. Upper view . .	3⅓
9	The same. Stapes and columella. Outer view	6⅔

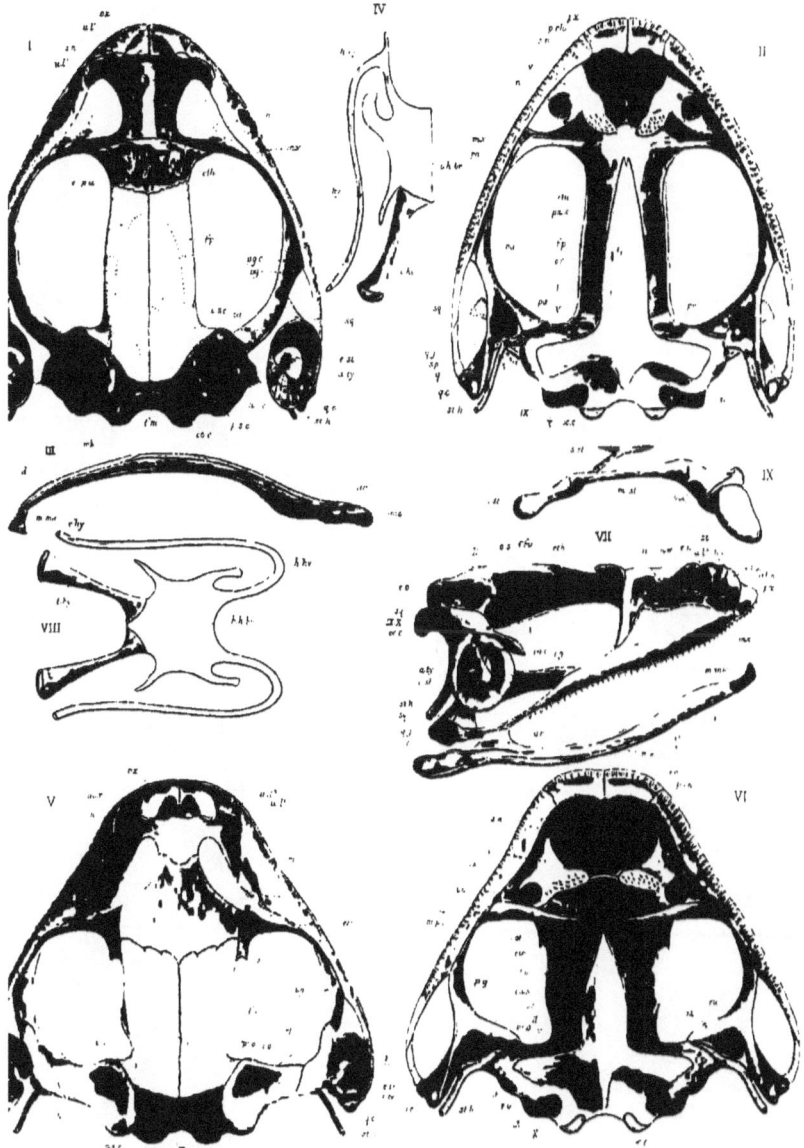

PLATE 27.

Figures.		Number of times magnified.
1 (p. 145)	Skull of *Polypedates maculatus.*—Adult male; 2 inches 1 line long. India. Upper view.	$5\frac{1}{3}$
2	The same. Lower view	$5\frac{1}{3}$
3	The same. Mandibles and hyo-branchials. Upper view.	$5\frac{1}{3}$
4	The same. Stapes and columella. Outer view.	$10\frac{2}{3}$
5 (p. 149)	Skull of *Ixalus variabilis.*—Adult female; 1 inch 1 line long. Ceylon. Upper view	$6\frac{2}{3}$
6	The same. Lower view	$6\frac{2}{3}$
7	The same. Mandibles and hyo-branchials. Upper view.	$6\frac{2}{3}$

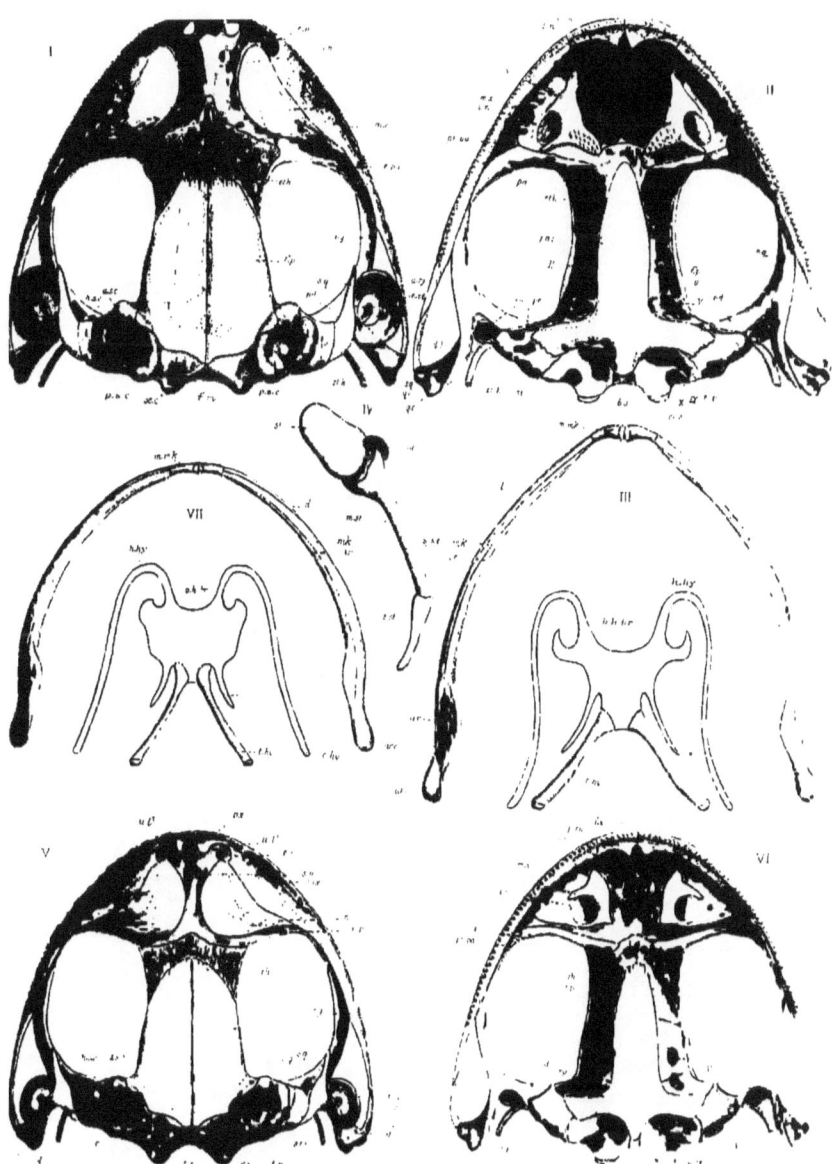

PLATE 28.

Figures.		Number of times magnified.
1 (p. 150)	Skull of *Hylarana malabarica.*—Young; ⅔ inch long. India	10
2	The same. Lower view.	10
3	The same. Mandibles and hyo-branchials. Upper view .	10
4	The same. Stapes and columella. Outer view . .	20
5	Hind part of fig. 1	20
6 (p. 155)	Skull of *Rappia* ——? sp.—Adult female; ⅘ inch long. Lagos. Upper view	10
7	The same. Lower view.	10
8	The same. Mandibles and hyo-branchials. Upper view .	10
9	The same skull. Fore part of chondrocranium. Upper view	20
10	The same skull. Stapes and columella. Outer view . .	20

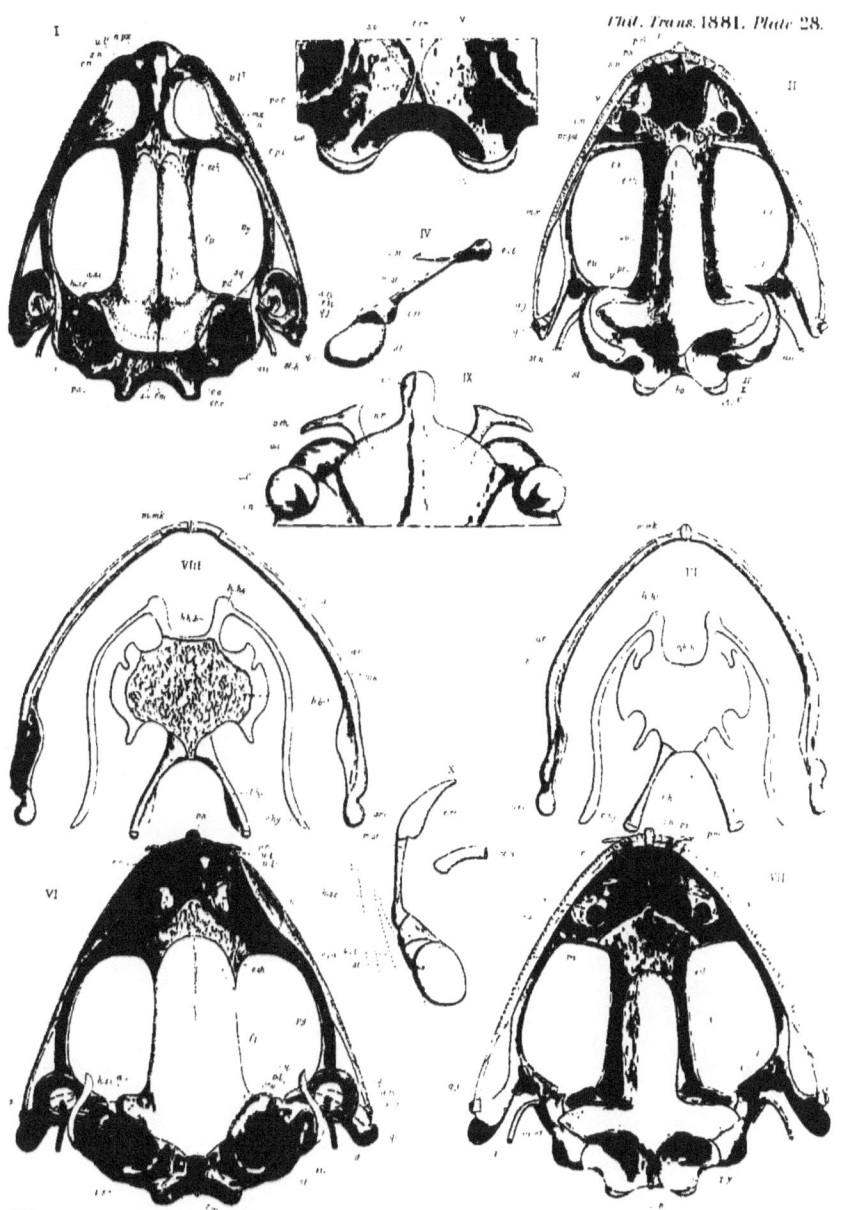

PLATE 29.

Figures.		Number of times magnified.
1 (p. 152)	Skull of *Hylarana temporalis.*—Adult male; $2\frac{1}{3}$ inches long. Ceylon. Upper view.	4
2	The same. Lower view	4
3	The same. Mandible. Outer view.	4
4	The same. Hyo-branchials. Upper view	4
5	The same. Stapes and columella. Outer view.	8
6 (p. 149)	*Ixalus variabilis.* Stapes and columella. Outer view	8
7 (p. 160)	Skull of *Hylodes martinicensis.*—Adult female; $1\frac{1}{2}$ inch long. Martinique. Upper view.	6
8	The same. Lower view	6
9	The same. Mandible. Outer view	6
10	The same. Hyo-branchials. Upper view.	6
11	The same. Stapes, columella, and annulus. Outer view.	12

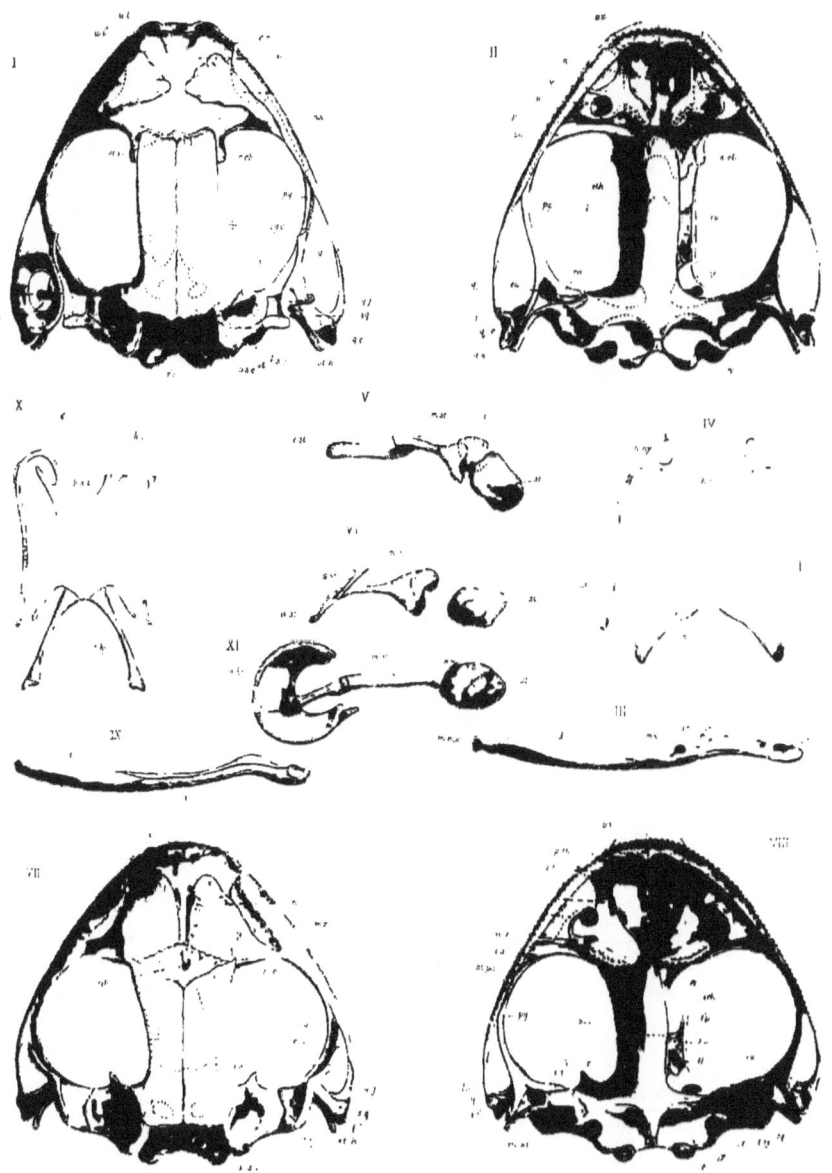

PLATE 30.

Figures.		Number of times magnified.
1 (p. 164)	Skull of *Acris Pickeringii* (A).—Adult female; 10 lines long. Cambridge, Mass., U.S. Upper view ...	10
2	The same. Lower view. . .	10
3	The same. Lower jaw	10
4	The same. Hyo-branchials. . .	10
5	The same. Stapes and columella	20
6 (p. 167)	Skull of *Acris Pickeringii* (B).—Larva, 1 inch 2 lines long; tail, $\frac{3}{4}$ inch; hind legs, $\frac{1}{4}$ inch. Same locality. Upper view	10
7	The same. Lower view.	10
8 (p. 180)	Skull of Tadpole of *Hyla* ——? sp.—1 inch long; hind legs, 5 lines. Rio Janeiro. Upper view	15
9	The same. Lower view	15
10 (p. 188)	Skull of Tadpole of *Nototrema marsupiatum* (B).—Larva, $2\frac{1}{4}$ inches long; tail, $1\frac{1}{3}$ inch; hind legs, 13 lines. South America. Upper view	$7\frac{1}{2}$
11	The same. Lower view	$7\frac{1}{2}$
12	The same. Mandibles and lower labials . .	$7\frac{1}{2}$
13	The same. Hyoid bar	$7\frac{1}{2}$

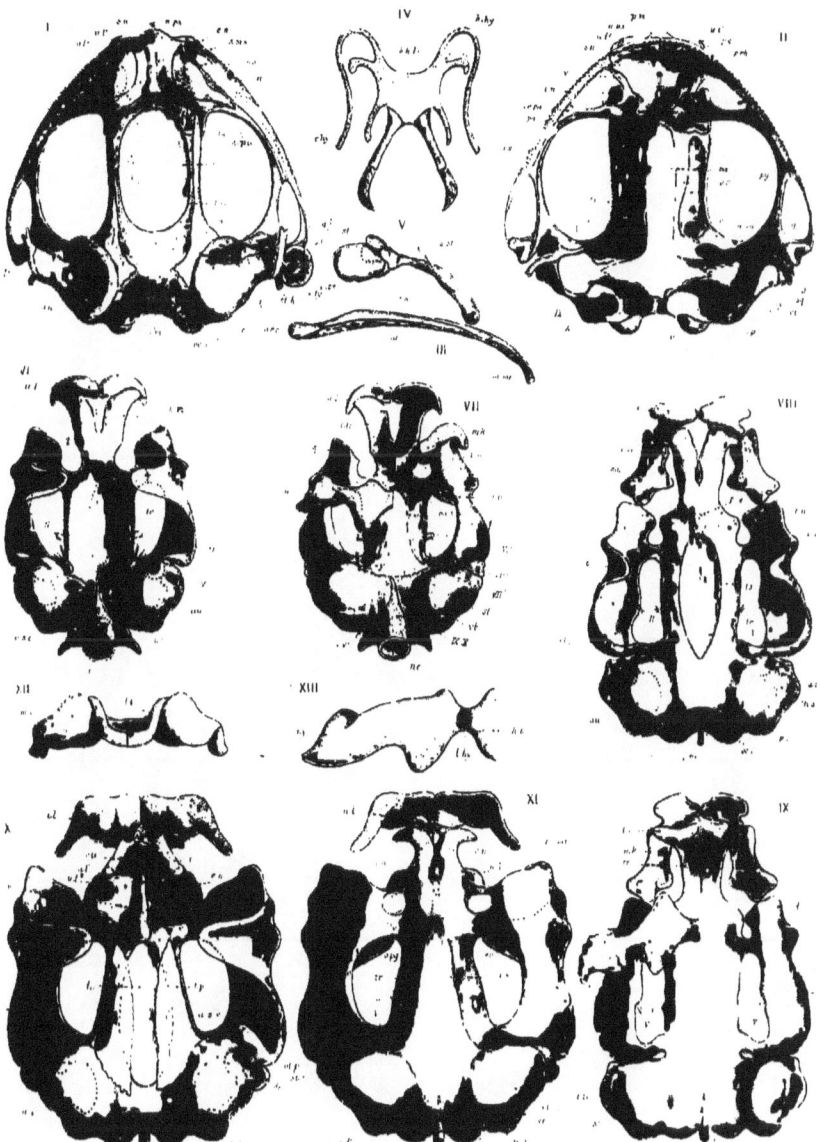

PLATE 31.

Figures.		Number of times magnified.
1 (p. 169)	Skull of *Hyla Ewingii*.—Adult female ; 1¼ inch long. Van Diemen's Land. Upper view	6
2	The same. Lower view	6
3	The same. Mandibles and hyo-branchials. Upper view .	6
4	The same. Front part of chondrocranium. Upper view.	12
5	The same. Stapes, columella, and annulus. Outer view.	12
6 (p. 171)	Skull of *Hyla phyllochroa*.—Adult female ; 1 inch 5 lines long. Cape York, Australia. Upper view . .	6
7	The same. Lower view	6
8	The same. Mandibles and hyo-branchials. Upper view.	6
9	The same. Stapes, columella, and stylo-hyal. Inner view	12

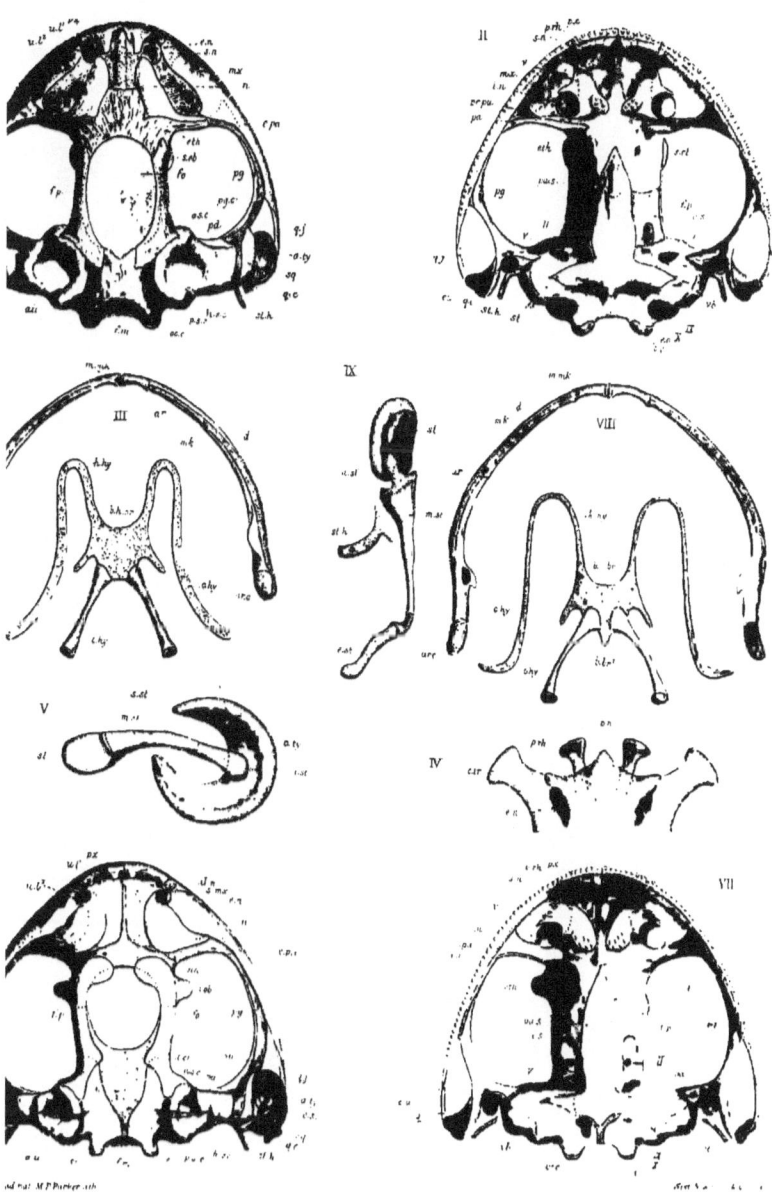

PLATE 32.

Figures.		Number of times magnified.
1 (p. 173)	Skull of *Hyla arborea*.—Adult male; $1\frac{1}{3}$ inch long. South Europe. Upper view	$7\frac{1}{5}$
2	The same. Lower view	$7\frac{1}{5}$
3	The same. Mandibles, and hyo-branchial arches. Upper view	$7\frac{1}{5}$
4	The same. Anterior part of chondrocranium. Lower view	$14\frac{2}{5}$
5	The same. Stapes and columella. Outer view	$14\frac{2}{5}$
6 (p. 175)	Skull of *Hyla albomarginata*.—Adult female; $2\frac{1}{3}$ inches long. Brazils. Upper view	$4\frac{1}{5}$
7	The same. Lower view	$4\frac{1}{5}$
8	The same. Mandibles, and hyo-branchial arches. Upper view	$4\frac{1}{5}$
9	The same. Stapes and columella. Outer view (*inverted*)	$8\frac{2}{5}$

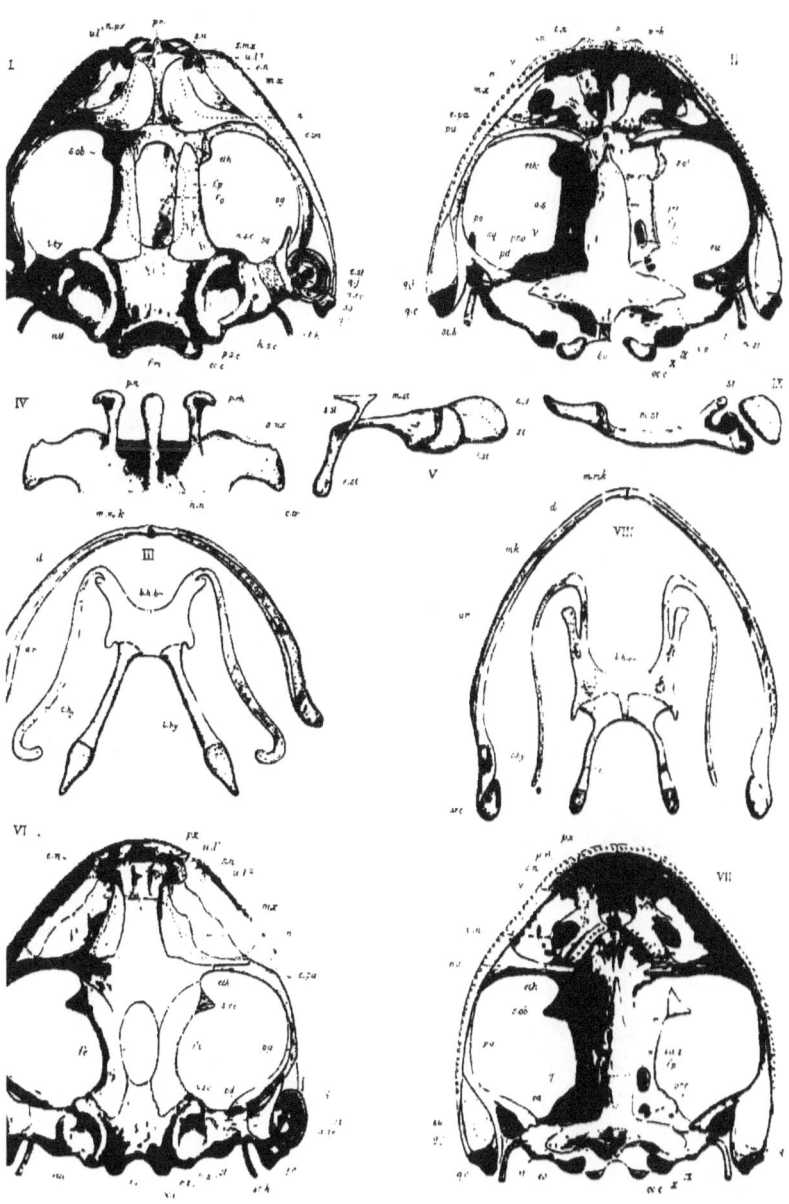

PLATE 33.

Figures.		Number of times magnified.
1 (p. 184)	Skull of *Nototrema marsupiatum* (A).—Adult male; 1¾ inch long. South America. Upper view.	5⅓
2	The same. Lower view.	5¼
3	The same. Side view	5¼
4	The same. Hyo-branchials. Upper view	5⅓
5	The same. Stapes and columella. Outer view.	16
6 (p. 178)	Skull of *Hyla rubra*.—Adult male; 1 inch 11 lines long. South America. Upper view.	5⅓
7	The same. Lower view	5½
8	The same. Mandible. Outer view.	5⅓
9	The same. Hyo-branchials. Upper view	5⅓
10	The same. Stapes and columella. Outer view.	16
11	The same objects. Inner view	16

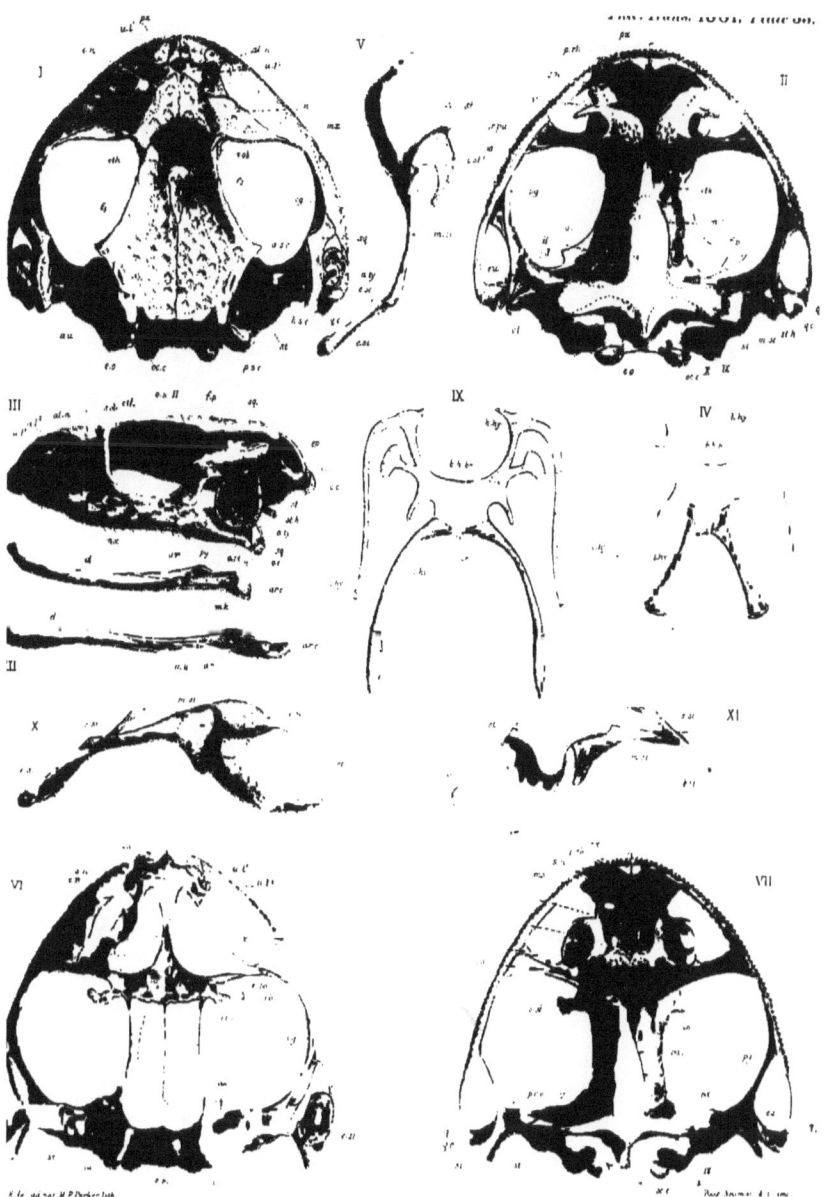

PLATE 34.

Figures.		Number of times magnified.
1 (p. 190)	Skull of *Pelodryas cæruleus.*—Adult male; 3 inches long. New South Wales. Upper view	$3\frac{3}{5}$
2	The same. Lower view	$3\frac{3}{5}$
3	The same. Side view	$3\frac{3}{5}$
4	The same. End view (more than *half*)	$3\frac{3}{5}$
5	The same. Stapes and columella. Outer view . .	$7\frac{1}{5}$
6	The same objects. Inner view	$7\frac{1}{5}$
7 (p. 194)	Skull of *Phyllomedusa bicolor.*—Adult female; $3\frac{1}{2}$ inches long. Santarem, River Amazon, South America. Upper view	3
8	The same. Lower view	3
9	The same. Side view	3
10	The same. Stapes and columella. Outer view	6
11	The same. Suspensorium and squamosal. Side view . .	3

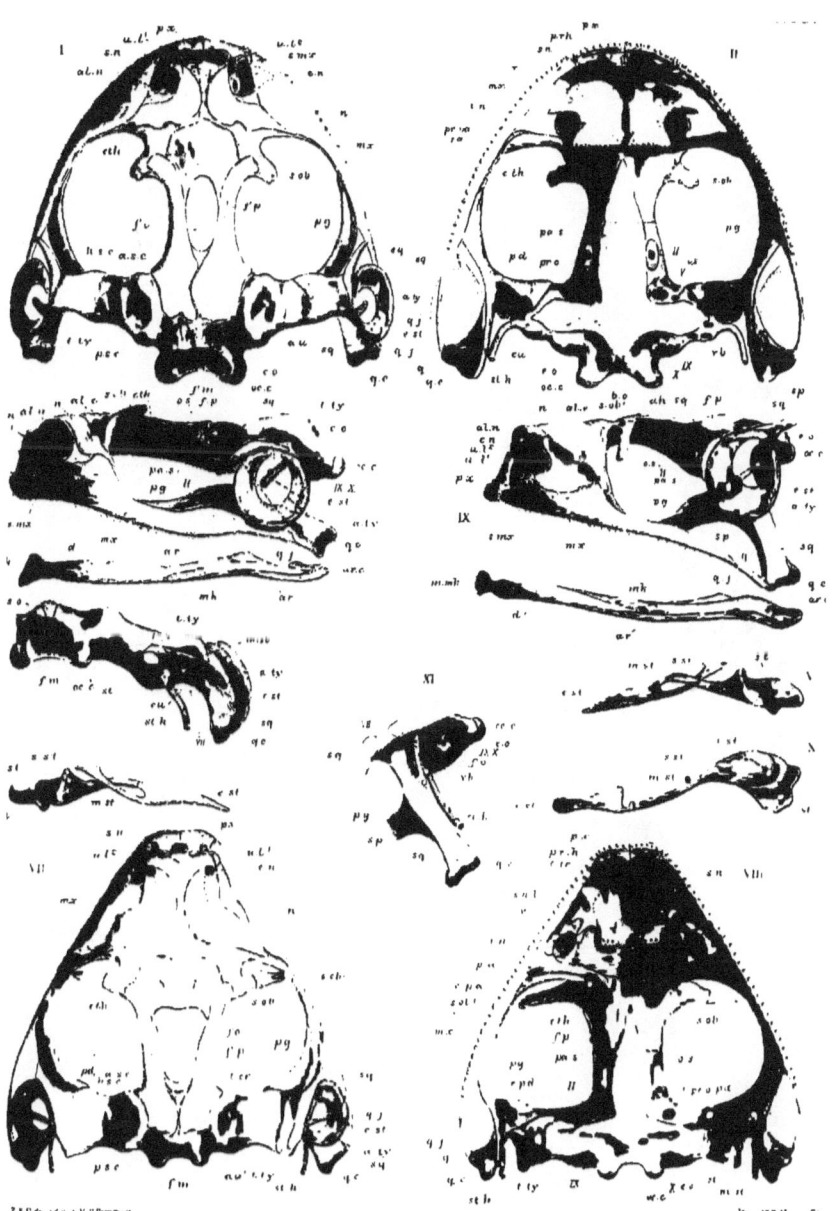

PLATE 35.

Figures.		Number of times magnified.
1 (p. 198)	Skull of *Bufo pantherinus*.—Adult female; 4¼ inches long. Africa. Upper view	2⅔
2	The same. Lower view	2⅔
3	The same. Side view	⅔
4	The same. Stapes and columella. Outer view	5⅓
5	Stapes and columella of *Bufo chilensis* (A).—Adult male; 3 inches long. Arequipa, Peru. Outer view . . .	5⅓
6	The same. Ethmoidal region. Lower view	2⅔
7 (p. 201)	Skull of *Bufo melanostictus*.—Half-grown male; 2½ inches long. India. Upper view	4
8	The same. Lower view	4
9	The same. Side view	4
10	The same. Stapes and columella. Outer view .	8
11	The same objects. Inner view	8

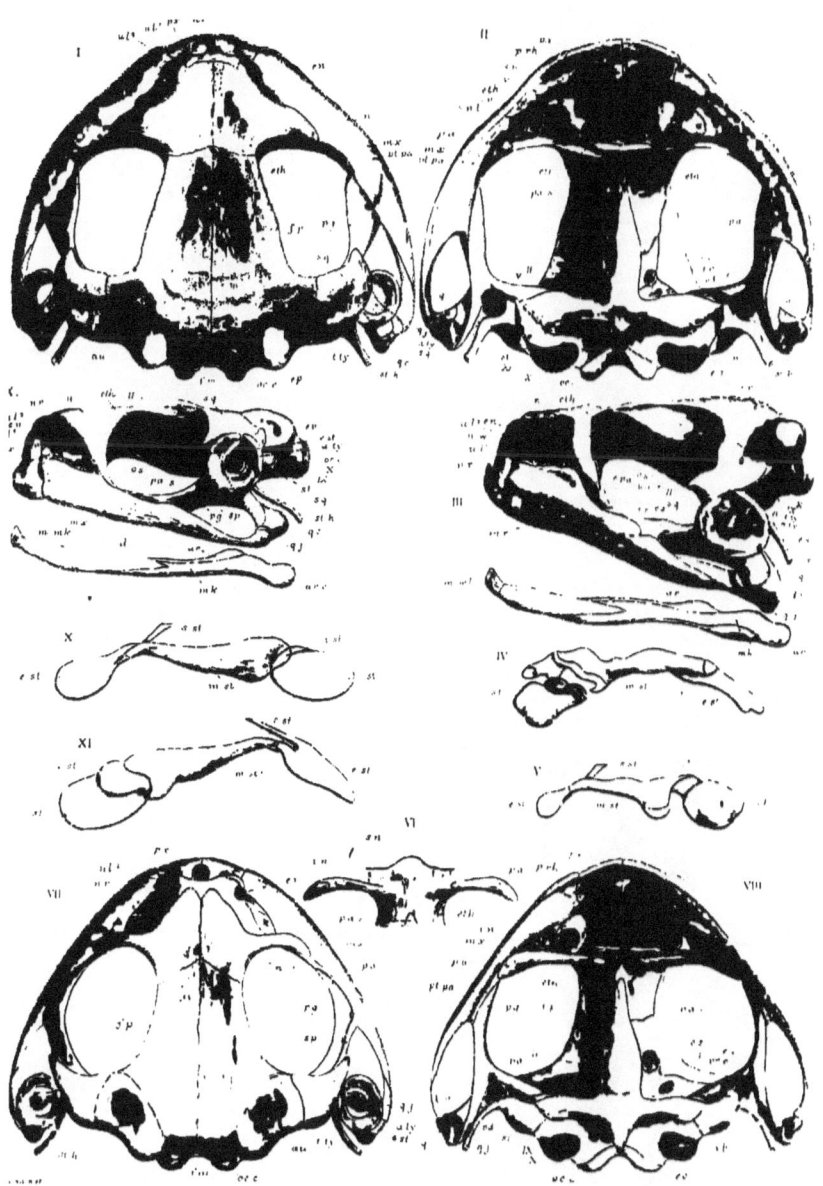

PLATE 36.

Figures.		Number of times magnified.
1 (p. 202)	Skull of *Bufo agua*.—Old female; 6¼ inches long. South America. Upper view	2
2	The same. Lower view. .	2
3	The same. Side view	2
4	Auditory region and suspensorium of a younger female: 5 inches long. Back view	2¾
5	Hyo-branchials of same species. Upper view . .	2
6	The same. Stapes and columella. Outer view	6
7	Part of skull of old specimen; antorbital region. Hind view (*oblique*)	4
8 (p. 216)	Anterior facial bones and cartilages of *Bufo ornatus*.— Adult female; 2⅛ inches long. South America. Inner view	14
9	The same object (*part*). Outer view	14
10	The same skull. Stapes and columella. Outer view . .	14

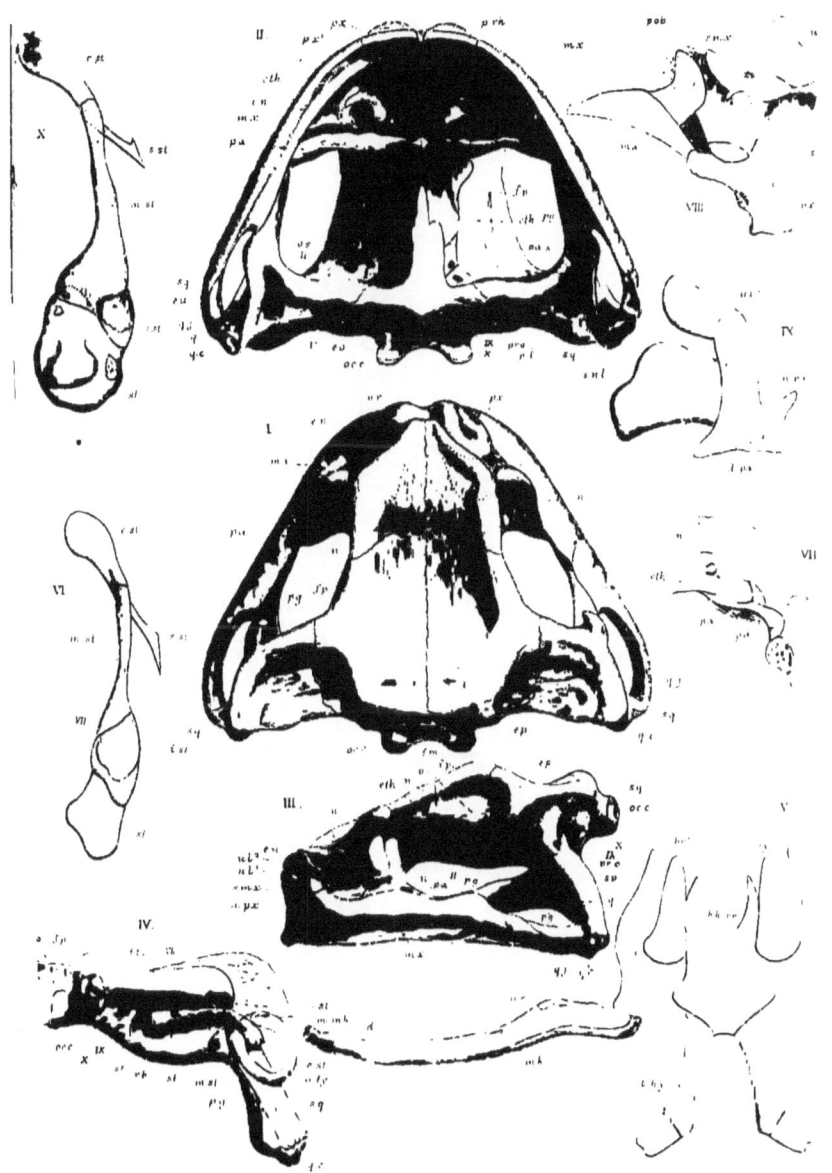

PLATE 37.

Figures.		Number of times magnified.
1 (p. 216)	Skull of *Bufo ornatus* (same as in Plate 36).—Upper view	4⅔
2	The same. Lower view	4⅔
3	The same. Side view	4⅔
4	The same. Hyo-branchials. Upper view	4⅔
5 (p. 219)	Skull of *Otilophus margaritifer* (A).—Half-grown female; 1½ inch. Venezuela. Upper view	5⅓
6	The same. Lower view	5⅓
7	The same. Side view	5⅓
8	The same. Hyo-branchials. Upper view.	5⅓
9	The same. Stapes and columella. Outer view.	10⅔
10	The same objects. Inner view	10⅔

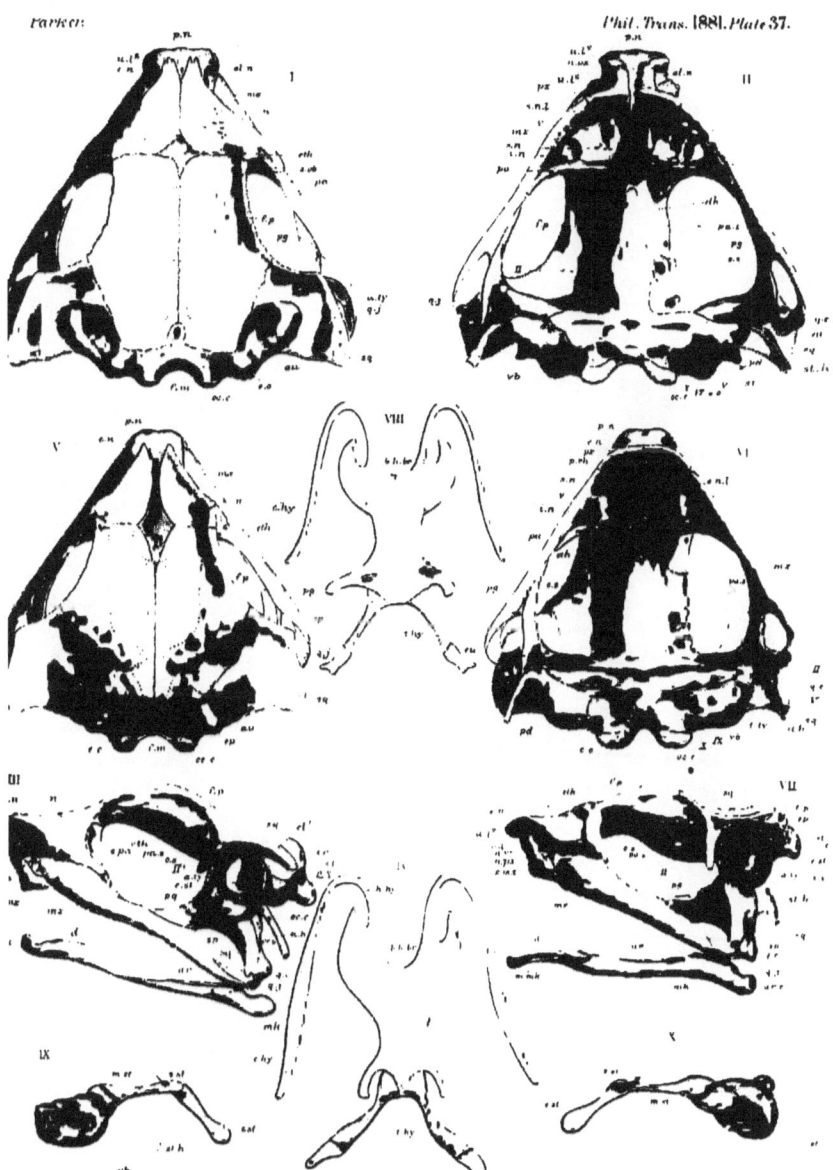

PLATE 38.

Figures.		Number of times magnified.
1 (p. 221)	Skull of *Otilophus margaritifer* (B).—Adult female; $2\frac{2}{3}$ inches long (HYRTL's prepn., Mus. Coll. Surg., England). Brazils. Side view.	4
2	The same. End view	4
3	The same species; half-grown (same as in Plate 37). End view of skull	$5\frac{1}{3}$
4 (p. 210)	Lower jaw of *Bufo lentiginosus*.—Young (same as in Plate 39). Upper view.	10
5 (p. 198)	Hyo-branchials of *Bufo pantherinus* (same as in Plate 35). Upper view.	$2\frac{2}{3}$
6 (p. 201)	Hyo-branchials of *Bufo melanostictus* (same as in Plate 35). Upper view.	4
7 (p. 206)	Skull of Tadpole of *Bufo chilensis* (B).—Total length, $\frac{5}{6}$ inch; tail, $\frac{1}{2}$ inch; hind legs, $\frac{1}{20}$ inch. Arequipa, Peru. Upper view.	15
8	The same. Lower view.	15
9 (p. 207)	Skull of Tadpole of *Bufo lentiginosus* (A).—$\frac{5}{6}$ inch long; hind legs, $\frac{1}{15}$ inch long. Penekese Island, Mass., U.S. Upper view.	10
10	The same. Lower view.	10

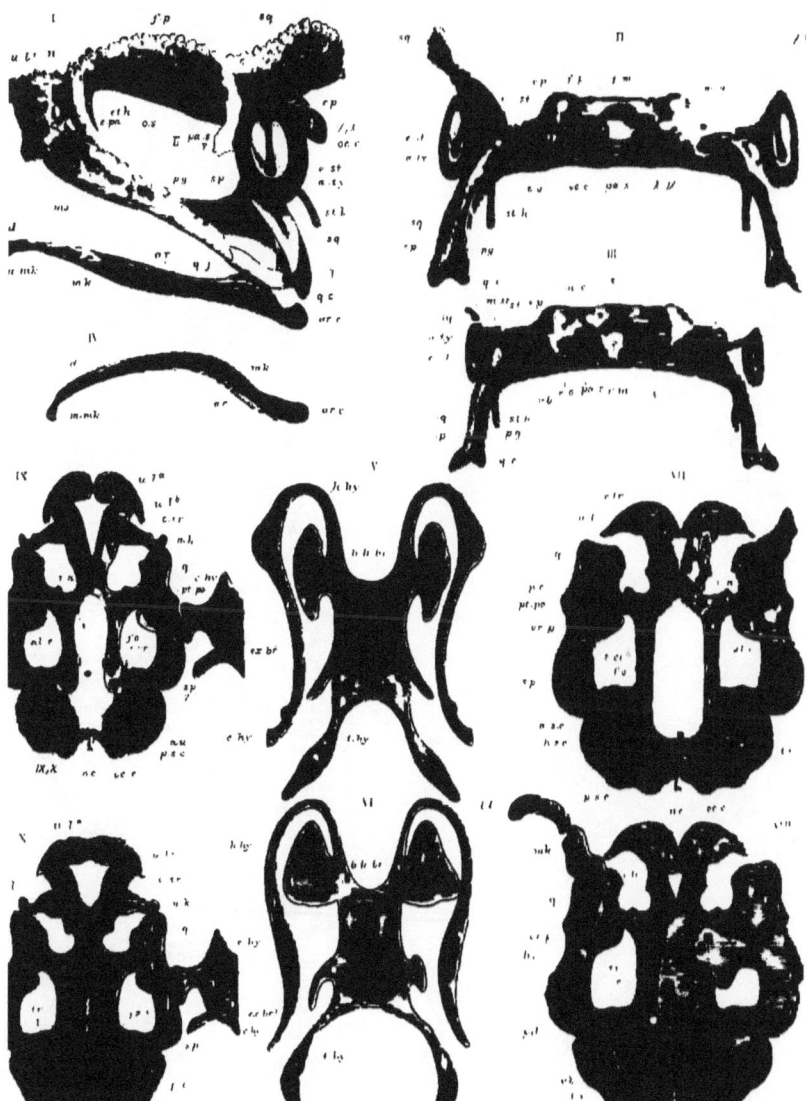

PLATE 39.

Figures.		Number of times magnified.
1 (p. 222)	Skull of *Rhinoderma Darwinii*.—Adult male; 1 inch long. Chili. Upper view . . .	8
2	The same. Lower view. . .	8
3	The same. Side view	8
4	The same. Fore part of endocranium. Lower view . .	16
5	The same. Auditory region and suspensorium. Outer view	16
6	The same. Hyo-branchials	8
7 (p. 210)	Skull of *Bufo lentiginosus* (B).—Young male; ⅜ inch long; Penekese Island, Mass., U.S. Upper view .	10
8	The same. Lower view	8
9	The same. Auditory region and suspensorium. Outer view	13⅓

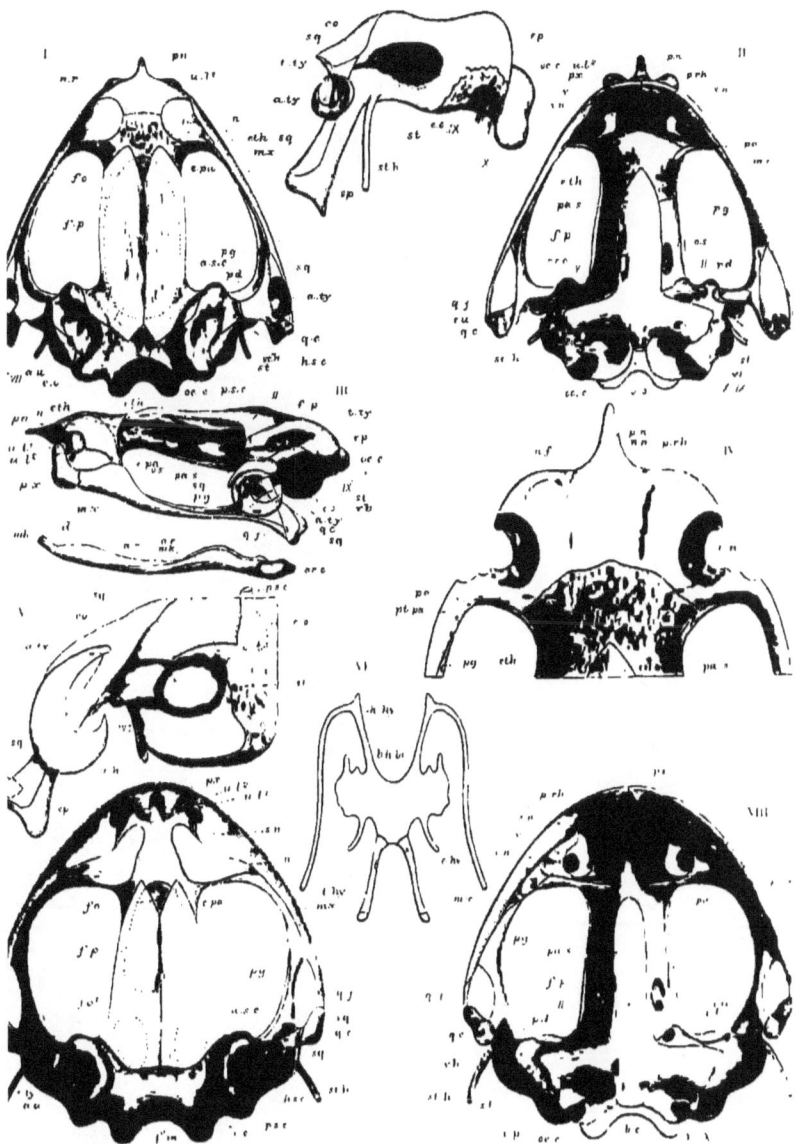

PLATE 40.

Figures.		Number of times magnified.
1 (p. 213)	Skull of *Bufo calamita.*—Adult female; 2⅜ inches long. England. Upper view .	4
2	The same. Lower view	4
3	The same. Mandible. Outer view. . .	4
4	The same. Hyo-branchials. Upper view . .	4
5	The same. Columella and stapes. Outer view	8
6 (p. 237)	Skull of *Phyrniscus lævis.*—Adult female; 1¾ inch long. Ecuador. Upper view . . .	6
7	The same. Lower view	6
8	The same. Side view	6
9	The same. Hyo-branchials. Lower view	6
10	*Phryniscus varius* (same as in Plate 41).—Part of fore face, showing pro-rhinals, &c. Inner view . .	20

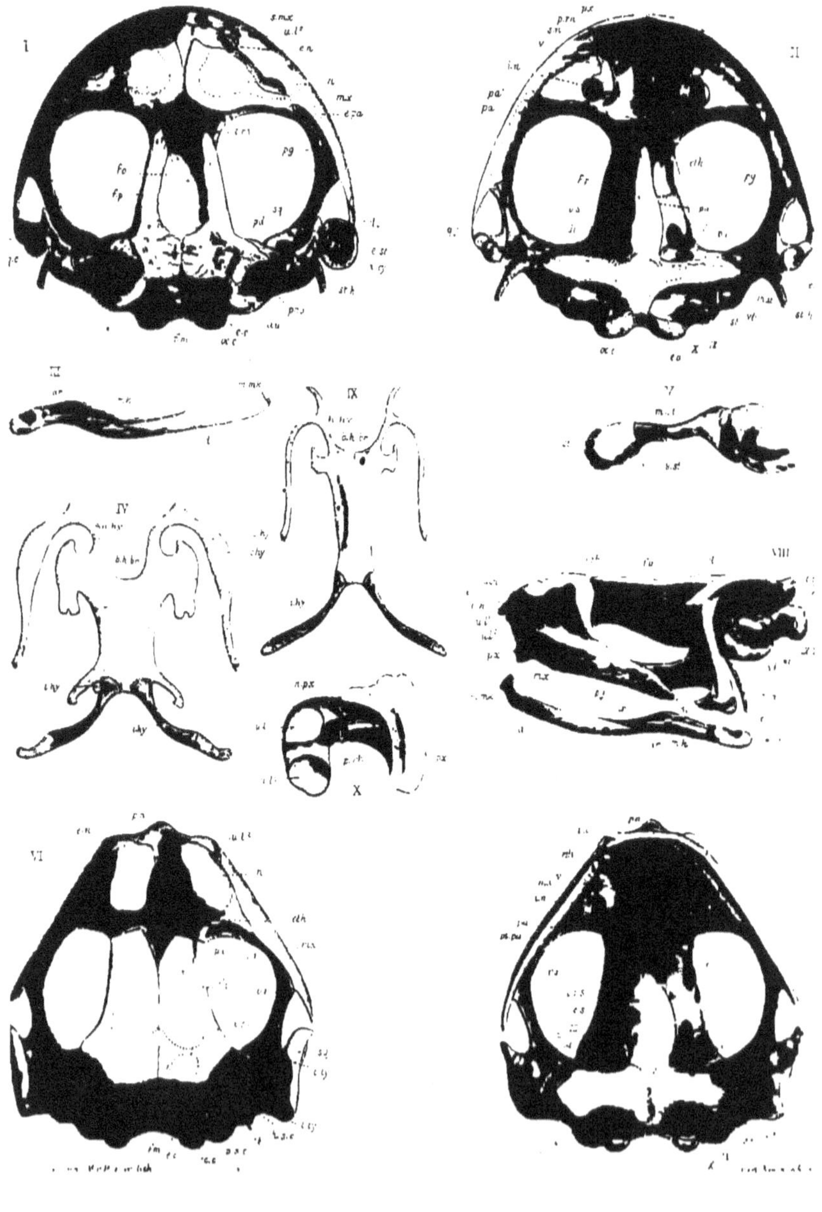

PLATE 41.

Figures.		Number of times magnified.
1 (p. 233)	Skull of *Phryniscus cruciger*.—Adult male; $1\frac{1}{6}$ inch long. Interior of Brazils. Upper view	8
2	The same. Lower view	8
3	The same. Side view	8
4	The same. Mandibles and hyo-branchials. Upper view .	8
5	The same. Stapes and columella. Outer view	16
6 (p. 236)	Skull of *Phryniscus varius*.—Adult female; $1\frac{2}{3}$ inch long. Costa Rica. Upper view	$6\frac{2}{3}$
7	The same. Lower view	$6\frac{2}{3}$
8	The same. Side view	$6\frac{2}{3}$
9	The same. Mandibles and hyo-branchials. Upper view .	$6\frac{2}{3}$

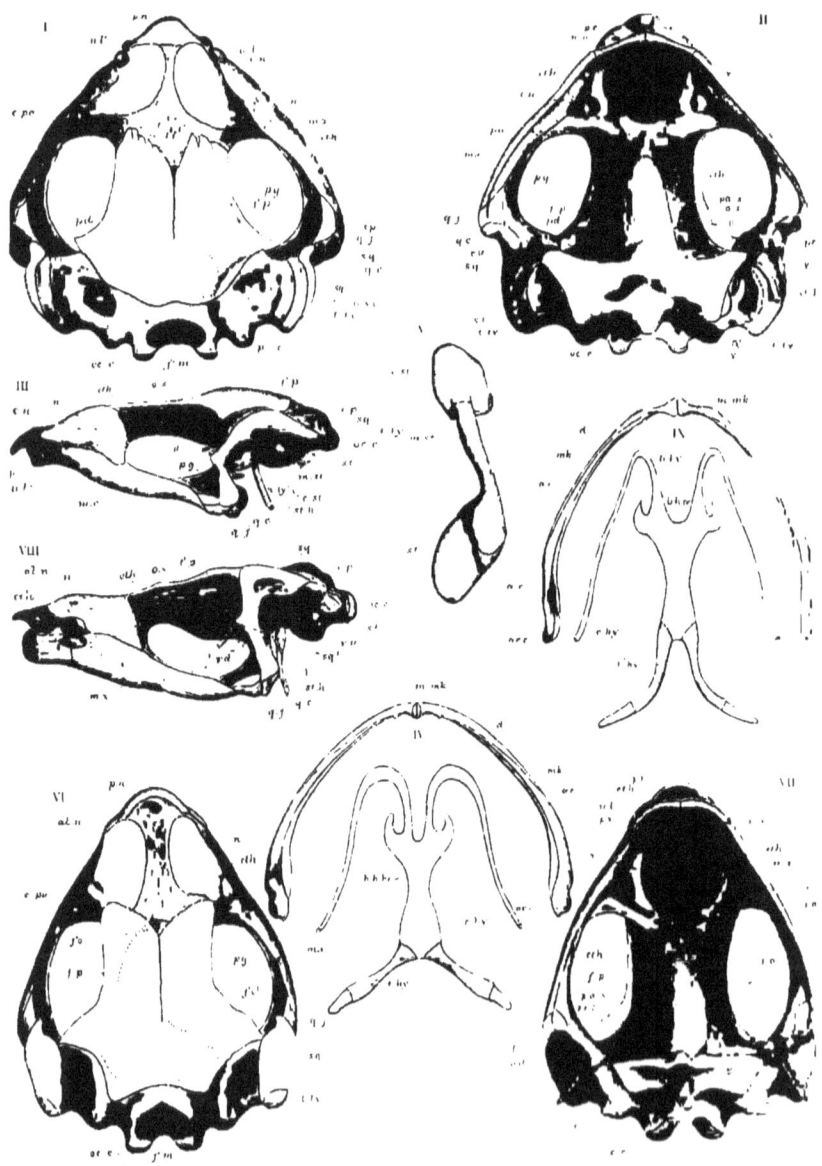

PLATE 42.

Figures.		Number of times magnified.
1 (p. 230)	Skull of *Pseudophryne Bibronii*.—Adult female; 1 inch long. New South Wales. Upper view	9
2	The same. Lower view	9
3	The same. Mandibles and hyo-branchials. Upper view.	9
4	The same. Part of palate	18
5	The same. Auditory region and suspensorium of a male; $\frac{2}{3}$ of an inch long. Outer view	18
6	The same object. Inner view	18
7	Hinder part of base of skull of female. Upper view . .	18
8 (p. 225)	Skull of *Diplopelma ornatum*, vel *rubrum*.—Adult male; 11 lines long. India. Upper view	9
9	The same. Lower view	9
10	The same. Stapes and columella. Outer view	18

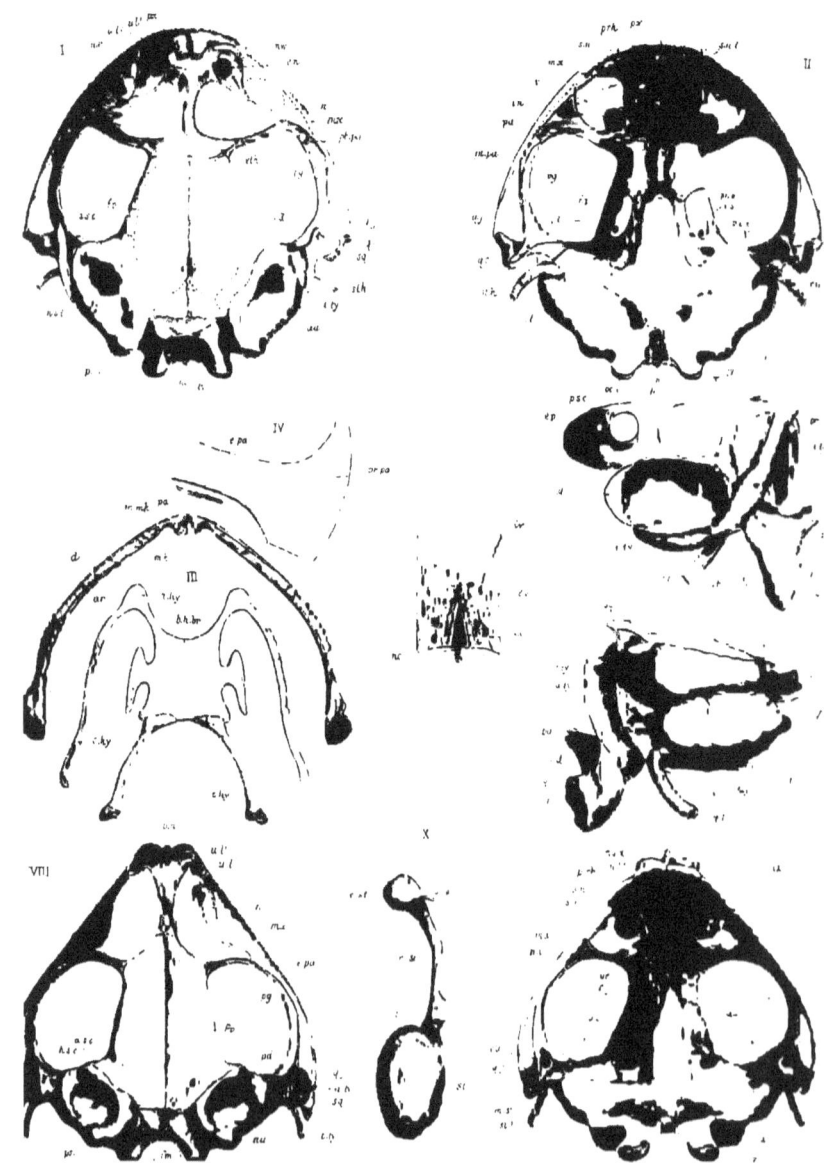

PLATE 43.

Figures.		Number of times magnified.
1 (p. 227)	Skull of *Diplopelma Berdmorei* (?).—Adult female; 1 inch 1 line long. Moulmin, Tenasserim. Upper view . .	9
2	The same. Lower view	9
3	The same. Mandible. Outer view	9
4	The same. Hyo-branchials. Lower view	9
5	The same. Stapes and columella. Outer view . . .	18
6	Mandibles and hyo-branchials of *Diplopelma ornatum*, vel *rubrum* (same as in Plate 42). Lower view . . .	9
7 (p. 240)	Skull of *Engystoma carolinense*.—Adult male; 11 lines long. Florida. Upper view	12
8	The same. Lower view	12
9	The same. Mandible. Outer view .	12
10	The same. Hyo-branchials. Lower view	12
11	The same. Suspensorium, annulus, columella, and stapes. Outer view	24
11A	The same. Sectional views of stapes . . .	24

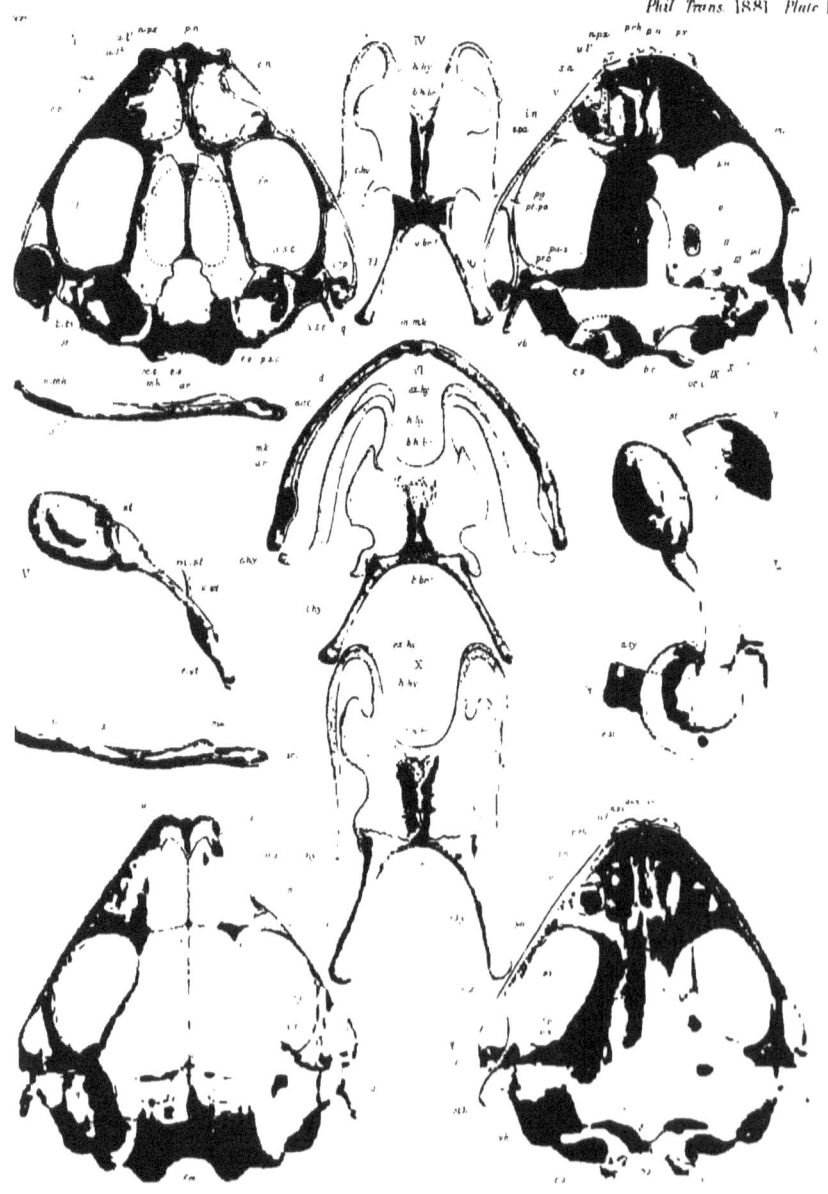

Phil. Trans. 1881. Plate 63.

PLATE 44.

Figures.		Number of times magnified.
1 (p. 247)	Skull of *Hylaplesia tinctoria.*—Adult female; 1¼ inch long. South America. Upper view	7½
2	The same. Lower view	7½
3	The same. Side view	7½
4	The same. Mandibles and hyo-branchials. Upper view	7½
5	The same. Fore part of endocranium. Lower view	7½
6	Part of nasal region of the same. Inner view	15
7	The same. Suspensorium, columella, and stapes. Outer view	15
8 (p. 243)	Skull of *Callula pulchra.*—Adult female; 2⅝ inches long. Pegu. Upper view	4
9	The same. Lower view	4
10	The same. Side view	4
11	The same. Hyo-branchials. Lower view	4
12	The same. Columella and stapes. Outer view	8

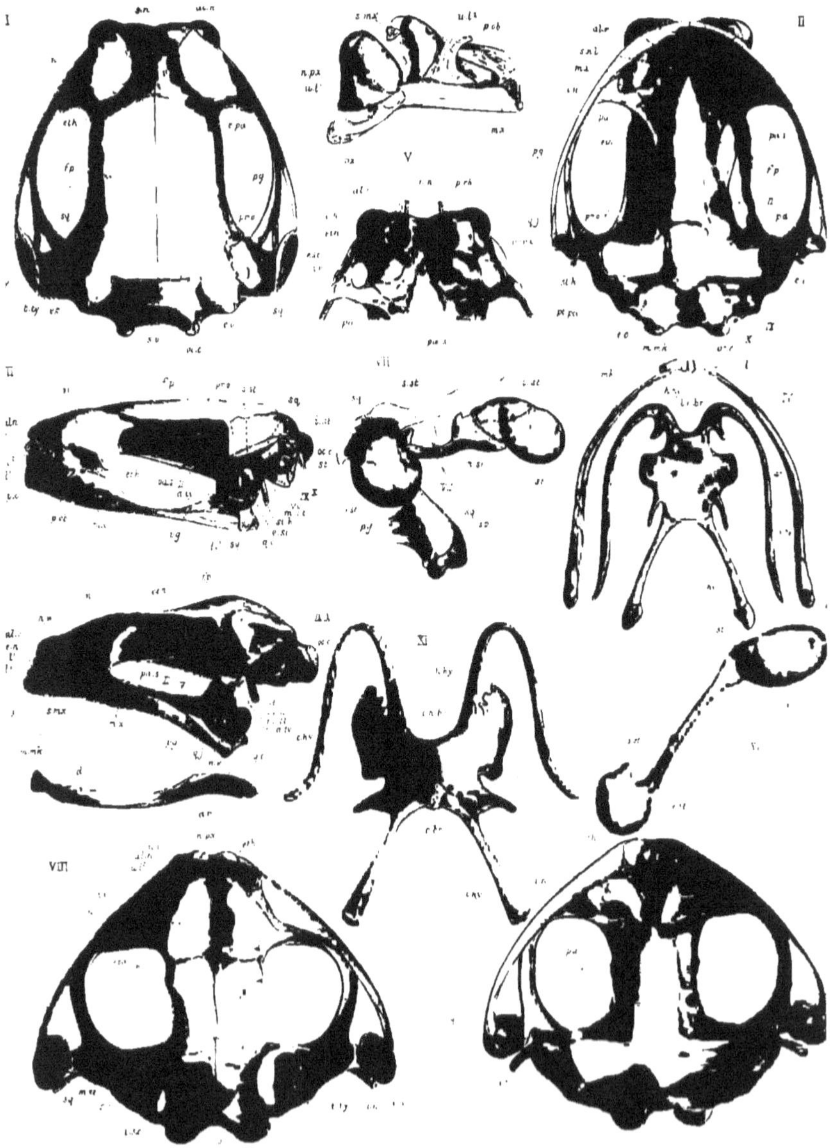

www.ingramcontent.com/pod-product-compliance
Lightning Source LLC
Chambersburg PA
CBHW051728300426
44115CB00007B/504